Undergraduate Texts in Mathematics

Editors

S. Axler
K.A. Ribet

Undergraduate Texts in Mathematics

Abbott: Understanding Analysis.

Anglin: Mathematics: A Concise History and Philosophy.
Readings in Mathematics.

Anglin/Lambek: The Heritage of Thales.
Readings in Mathematics.

Apostol: Introduction to Analytic Number Theory. Second edition.

Armstrong: Basic Topology.

Armstrong: Groups and Symmetry.

Axler: Linear Algebra Done Right. Second edition.

Beardon: Limits: A New Approach to Real Analysis.

Bak/Newman: Complex Analysis. Second edition.

Banchoff/Wermer: Linear Algebra Through Geometry. Second edition.

Berberian: A First Course in Real Analysis.

Bix: Conics and Cubics: A Concrete Introduction to Algebraic Curves.

Brémaud: An Introduction to Probabilistic Modeling.

Bressoud: Factorization and Primality Testing.

Bressoud: Second Year Calculus.
Readings in Mathematics.

Brickman: Mathematical Introduction to Linear Programming and Game Theory.

Browder: Mathematical Analysis: An Introduction.

Buchmann: Introduction to Cryptography.

Buskes/van Rooij: Topological Spaces: From Distance to Neighborhood.

Callahan: The Geometry of Spacetime: An Introduction to Special and General Relativity.

Carter/van Brunt: The Lebesgue–Stieltjes Integral: A Practical Introduction.

Cederberg: A Course in Modern Geometries. Second edition.

Chambert-Loir: A Field Guide to Algebra.

Childs: A Concrete Introduction to Higher Algebra. Second edition.

Chung/AitSahlia: Elementary Probability Theory: With Stochastic Processes and an Introduction to Mathematical Finance. Fourth edition.

Cox/Little/O'Shea: Ideals, Varieties, and Algorithms. Second edition.

Croom: Basic Concepts of Algebraic Topology.

Curtis: Linear Algebra: An Introductory Approach. Fourth edition.

Daepp/Gorkin: Reading, Writing, and Proving: A Closer Look at Mathematics.

Devlin: The Joy of Sets: Fundamentals of Contemporary Set Theory. Second edition.

Dixmier: General Topology.

Driver: Why Math?

Ebbinghaus/Flum/Thomas: Mathematical Logic. Second edition.

Edgar: Measure, Topology, and Fractal Geometry.

Elaydi: An Introduction to Difference Equations. Third edition.

Erdős/Surányi: Topics in the Theory of Numbers.

Estep: Practical Analysis in One Variable.

Exner: An Accompaniment to Higher Mathematics.

Exner: Inside Calculus.

Fine/Rosenberger: The Fundamental Theory of Algebra.

Fischer: Intermediate Real Analysis.

Flanigan/Kazdan: Calculus Two: Linear and Nonlinear Functions. Second edition.

Fleming: Functions of Several Variables. Second edition.

Foulds: Combinatorial Optimization for Undergraduates.

Foulds: Optimization Techniques: An Introduction.

Franklin: Methods of Mathematical Economics.

(continued after index)

Stephanie Frank Singer

Linearity, Symmetry, and Prediction in the Hydrogen Atom

 Springer

Stephanie Frank Singer
Philadelphia, PA 19103
U.S.A.
quantum@symmetrysinger.com

Mathematics Subject Classification (2000): Primary – 81-01, 81R05, 20-01, 20C35, 22-01, 22E70, 22C05, 81Q99; Secondary – 15A90, 20G05, 20G45

Library of Congress Cataloging-in-Publication Data
Singer, Stephanie Frank, 1964–
    Linearity, symmetry, and prediction in the hydrogen atom / Stephanie Frank
Singer.
        p. cm. — (Undergraduate texts in mathematics)
    Includes bibliographical references and index.

    1. Group theory. 2. Hydrogen. 3. Atoms. 4. Linear algebraic groups. 5. Symmetry
(Physics) 6. Representations of groups. 7. Quantum theory. I. Title. II. Series.
QC20.7.G76S56   2005
530.15′22 – dc22   2005042679

ISBN-13 978-1-4419-2035-5   e-ISBN 0-387-26369-1
e-ISBN-13 978-0-387-26369-4

springeronline.com

To my mother, Maxine Frank Singer,
who always encouraged me to follow my own instincts:
I think I may be ready to learn some chemistry now.

Contents

Preface

It just means so much more to so much more people when you're rappin' and
you know what for.

— Eminem, "Business" [Mat]

This is a textbook for a senior-level undergraduate course for math, physics
and chemistry majors. This one course can play two different but complementary roles: it can serve as a capstone course for students finishing their
education, and it can serve as motivating story for future study of mathematics.

Some textbooks are like a vigorous regular physical training program, pre-
paring people for a wide range of challenges by honing their basic skills thoroughly. Some are like a series of day hikes. This book is more like an extended trek to a particularly beautiful goal. We'll take the easiest route to the
top, and we'll stop to appreciate local flora as well as distant peaks worthy of
the vigorous training one would need to scale them.

Advice to the Student

This book was written with many different readers in mind. Some will be
mathematics students interested to see a beautiful and powerful application of
a "pure" mathematical subject. Some will be students of physics and chemistry curious about the mathematics behind some tools they use, such as

spherical harmonics. Because the readership is so varied, no single reader should be put off by occasional digressions aimed at certain other readers. For instance, in Chapter 2, we include some examples from quantum mechanics; students unfamiliar with quantum mechanics should feel free to skip these paragraphs. Similarly, readers who do not intend to continue their mathematical studies should feel free to skip the brief discussions of more advanced mathematical concepts. We have tried to label these digressions and their intended audiences clearly. In particular, readers should feel free to skip the footnotes. Some exercises require knowledge of another subject (such as topology). These exercises are clearly marked. See, e.g., Exercise 4.28. *Italicized* terms are defined close by; terms "in quotation marks" are not.

The prerequisite for this course is solid understanding of calculus and familiarity with either linear algebra or advanced quantum mechanics. We discuss prerequisites in more detail in Section 1.5.

Finally, the author wishes to offer some broader advice to students: snap out of the one course, one book mode. Talk to people in other fields. Read related material in other sources. The more you can synthesize different points of view, the more powerfully creative you will be.

Advice to the Instructor

Although this book can be used for a homogeneous audience, the author hopes that it will encourage mixed classrooms: mathematics students working with students in the physical sciences. The author has found that students in such classrooms respond well to assignments that allow them to share their particular expertise with the class. One model that has worked well in the author's experience is to replace timed tests with a final project (paper and class presentation) on a related topic of the student's choice. We have listed some paper topic suggestions in Appendix C.

The minimum plan for a semester course should be to teach Chapters 1 through 7. Chapters 8, 9, 10 and 11 (each of which depends on Chapters 1 through 7) are independent from one another and can be used to fill out the semester. Note, however, that Section 11.4 depends on the idea that the state space for the spin of the electron is $\mathbb{C}^2$. This idea (and much more) can be found in Chapter 10.

The representation theory of finite groups is not presented anywhere in this text, setting this book apart from most undergraduate books on representation theory. The author urges instructors to resist the temptation to present the theory of finite group representations before starting the text. While some students find the finite group material helpful, others find it distracting or even downright off-putting. Students interested in the finite group theory can be encouraged to study it and its beautiful physical applications (to the spectroscopy of molecules, for example) as a related topic or final project.

This is a rigorous text, except for certain parts of Chapter 3 and Chapter 4. We state Fubini's theorem and the Stone–Weierstrass theorem without proof. We do not define the Lebesgue integral or manifolds rigorously, choosing instead to write in such a way that readers familiar with the theory will find only true statements while readers unfamiliar will find intuitive, suggestive, accessible language. Finally, in the proof of Proposition 10.6, we appeal to techniques of topology that are beyond the scope of the text.

Group Theory vs. Representation Theory

The phrase "group theory" says different things to different people. To a physicist, "group theory" means what a mathematician would call "representation theory." For example, the physicists' "group theory" includes what mathematicians would call the "representation theory of algebras"; never mind that algebras are not "groups" in the technical mathematical sense. On the other hand, mathematicians use the phrase "group theory" to refer to the study of groups and groups alone. The mathematicians' "group theory" encompasses the properties and classifications of groups and subgroups, and does not often include the study of representations of Lie algebras or classifications of representations of groups. In mathematics departments, representations of groups and other objects are the subject of books, courses and lectures in "representation theory."

Acknowledgments

Many people contributed enormously to the writing of this book. Experienced editor Ann Kostant, with her regular encouragement over many years, turned me from a would-be writer into a writer. Mathematician Allen Knutson set me on the trail of this particular topic. Physicist Walter Smith bore patiently with my disruptions of his undergraduate quantum mechanics course. Mathematicians Shlomo Sternberg and Roger Howe supported my funding requests.

Thanks to the National Science Foundation for generous partial support for the project;[1] thanks to Haverford College for student assistants; thanks to the Aspen Center for Physics for the office, library and company that helped me understand the experiments behind the theory.

The colleagues and students who helped me learn the material are too numerous to list, but a few deserve special mention: Susan Tolman for many large-scale simplifications, Rebecca Goldin for suggesting excellent problems, Jared Bronski for the generating function in the proof of Proposition 4.7, Anthony Bak, Dan Heinz and Amy Ho for writing solutions to problems. Thanks to the students at George Mason University, Haverford College and the University of Illinois at Urbana Champaign for working through early drafts of the material and offering many insights and corrections.

They say that behind every successful man is a woman; I say that behind every successful woman is a housekeeper. Many thanks to Emily Lam for keeping my home clean for many years. Thanks also to Dr. Andrew D'Amico and Dr. Julia Uffner, for keeping me alive and healthy.

The deepest and most heartfelt thanks go to my readers. Keep reading, and keep in touch!

Stephanie Frank Singer
www.symmetrysinger.com
Philadelphia 2004

---

[1] Award number DUE-0125649.

1

Setting the Stage

After having been force fed in *liceo* the truths revealed by Fascist Doctrine, all revealed, unproven truths either bored me stiff or aroused my suspicion. Did chemistry theorems exist? No: therefore you had to go further, not be satisfied with the *quia*, go back to the origins, to mathematics and physics. The origins of chemistry were ignoble, or at least equivocal: the dens of the alchemists, their abominable hodgepodge of ideas and language, their confessed interest in gold, their Levantine swindles typical of charlatans or magicians; instead, at the origin of physics lay the strenuous clarity of the West — Archimedes and Euclid. I would become a physicist, *ruat coelum*: perhaps without a degree, since Hitler and Mussolini forbade it.

— Primo Levi, *The Periodic Table* [Le, pp. 52–3]

1.1   Introduction

Reading this book, you will learn about one of the great successes of 20th-century mathematics — its predictive power in quantum physics. In the process, you will see three core mathematical subjects (linear algebra, analysis and abstract algebra) combined to great effect. In particular, you will see how to make predictions about the dimensions of the basic states of a quantum system from the only two ingredients: the symmetry and the linear model of quantum mechanics. This method, known as *representation theory* to mathematicians and *group theory* to physicists and chemists, has a wide range of applications: atomic structure, crystallography, classification of manifolds with symmetry, etc.

We will find it enlightening to concentrate on one particular example of a quantum system with symmetry: the single electron in a hydrogen atom. Understanding the structure of the hydrogen atom is immensely important because the analysis generalizes easily to the structure of other atoms and determines the periodic table of the elements. We will develop just enough mathematical tools (in Chapters 2 through 6) to make predictions in Chapter 7 based solely on the physical spherical symmetry of the hydrogen atom. These predictions are equally valid for *any* quantum system with spherical symmetry. In Chapter 8 we introduce more specific information about hydrogen (specifically, the functional form of the Coulomb potential) and extend our toolset slightly to introduce some extra, hidden symmetries of the hydrogen atom; by combining these extra symmetries with the spherical symmetry, we can make much stronger predictions about the hydrogen atom (and hence the periodic table).

It is high time that this story escaped from the ivory tower in which it was born. When Pauli, Fock and Wigner did their groundbreaking work, calculus was not taken routinely by college students, let alone high schoolers. At that time, vectors and vector spaces were relatively new, and the study of groups and representations was truly esoteric, understood by very few. Now, however, many undergraduates study representation theory. At the beginning of the 21st century, many people are ready to understand the accomplishments of 20th-century scientists and mathematicians. This book is a good place to start.

1.2   Fundamental Assumptions of Quantum Mechanics

One major point of this book is to make deep predictions using only symmetry and very few assumptions about quantum mechanics. In this section we make explicit the assumptions we use and give some information about the experiments that justify these assumptions.

To appreciate this section and, more broadly, to appreciate the importance of this book's topic as a justification for mathematics, one should understand the role of theory in the physical sciences. While in mathematics the intrinsic beauty of a theory is sufficient justification for its study, the value of a theory in the physical sciences is limited to the value of the experimental predictions it makes. For example, the theory of the double-helical structure of

DNA (first proposed by Crick, Franklin and Watson in the 1950's [Ju, Part I]) suggested, and continues to suggest, experimental predictions in molecular biology. We hope, in the course of the book, to convince the reader that the mathematics we discuss (e.g., analysis, representation theory) is of scientific importance beyond its importance within mathematics proper. In order to succeed, we must use mathematics to pull testable experimental predictions from the physically-inspired assumptions of this section.

The first assumption of quantum mechanics is that each state of a mobile particle in Euclidean three-space $\mathbb{R}^3$ can be described by a complex-valued function ϕ of three real variables (called a *wave function*) satisfying

$$\int_{\mathbb{R}^3} |\phi(x, y, z)|^2 \, dx \, dy \, dz = 1. \tag{1.1}$$

To make use of this description, we must relate the function ϕ to possible experiments.

Our second quantum-mechanical assumption is that we can use the wave function ϕ to calculate the relative probabilities of all possible outcomes of any given measurement. For example, we could do an experiment to determine whether a given particle lies in the cube with unit-length sides parallel to the coordinate axes and centered at the origin (Figure 1.1); the corresponding theory says that

$$p := \int_{-1/2}^{1/2} \int_{-1/2}^{1/2} \int_{-1/2}^{1/2} |\phi(x, y, z)|^2 \, dx \, dy \, dz$$

is the probability that the particle will be found in the box, while $1 - p$ is the probability that the particle will not be found in the box. More generally, the function $|\phi|^2$ is the *probability distribution* for the position of the particle.

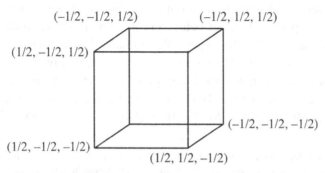

Figure 1.1. A cube with unit-length sides centered at the origin.

This means that the probability that the particle is located in a set $S \subset \mathbb{R}^3$ is given by

$$\int_S |\phi(p_x, p_y, p_z)|^2 \, dp_x \, dp_y \, dp_z. \tag{1.2}$$

(Readers familiar with Fourier transforms may be interested to know that the probability distribution of the momentum of the particle in state ϕ is given by $|\hat{\phi}|^2$, where $\hat{\phi}$ denotes the Fourier transform of ϕ.)

Of course, if we do the experiment only once, the particle will be either in or out of the box and p will be pretty much meaningless (unless $p = 1$ or $p = 0$). Quantum mechanics does not typically allow us to predict the outcome of any one experiment. The only way to find the probability p experimentally is to do the experiment many times. If we do the experiment N times and find the particle in the box i times, then the experimental value of p is i/N. Quantum mechanics provides predictions of this experimental value of p.

We usually cannot do the experiment N times on the same particle; however, we can find often a way to perform a series of identical experiments on a series of particles. We must ensure that each particle in the series starts in the particular state corresponding to the wave function ϕ. Physicists typically do this by making a machine that emits particles in large quantities, all in the same state. This is called a *beam of particles*.

Notice that the assumption that we can use the wave function ϕ to predict probabilities of various outcomes is much weaker than the corresponding assumption of classical mechanics. Classical mechanics is *deterministic*, i.e., we assume that if we know the state (position and momentum) of a classical particle such as the moon at a time t, then we can evaluate any dynamic variable (such as energy) at that same time t. Energy can be calculated from position and momentum.[1] Quantum mechanics is different, and many people find the difference disturbing. It is quite possible to know the precise quantum state of a particle without being certain of its position, momentum or energy. Not only might it be impossible to predict future behavior of a particle with certainty, it might be impossible to be certain of the outcome of a measurement done right now. Many people object to the implications of quantum mechanics, saying, "God does not play dice." These words are in a letter from Albert Einstein to Max Born [BBE']; the reader may find them

---

[1]Figuring out the position, momentum or energy at a different time t' from the state of the particle at time t is a different, harder question. Its resolution in various cases is a central motivating problem for much of classical mechanics.

in context in the epigraph to Chapter 11. But, as Einstein mentions in the very same letter, theological concerns cannot change the fact that in experiment after experiment, the assumptions of quantum mechanics yield accurate predictions about aggregate behavior.

A third assumption of quantum mechanics has to do with *observables*, such as position, momentum or energy. An observable is a numerical quantity that can be measured by an experiment. For instance, one can measure the momentum of an electron by observing the results of a collision, or the energy by observing the wavelength of an emitted photon. We will state this third assumption below, but first we must introduce some terminology. A *base state* for an observable is a state of the particle for which the measurement corresponding to the observable is certain. For example, if one measures the energy of an electron "in the lowest s-shell of the hydrogen atom," one will certainly find -13.6 electron-volts.[2] Even though many things about this electron are uncertain (its position and momentum, for example), its energy is certain, and hence the lowest s-shell is a base state for energy. There are many base states for the energy observable. On the other hand, not every wave function is a base state for the energy. For example, a wave function that is zero outside a unit cube and equal to one on the unit cube (describing a particle that must be in the unit cube but is equally likely to be anywhere inside the cube) is not a base state for the energy.

The third fundamental assumption of quantum mechanics states that any wave function can be expressed as a superposition of base states of any observable. Consider, for example, the energy observable. Any function ϕ of three real variables satisfying Equation 1.1 can be decomposed as a weighted sum.[3] of wave functions describing states with energy values that are certain. In other words, suppose ϕ_1 and ϕ_2 are base states for the energy of a certain system, and consider a state in which the particle has probability 3/4 of being found in state ϕ_1 and probability 1/4 of being found in state ϕ_1; such a state

---

[2] An electron-volt (abbreviated "eV") is a unit of energy equal to 1.6×10^{-19} joules. It is the amount of energy required to move one electron through a one-volt potential difference.

[3] For this statement to be precisely true, we must let integrals count as sums. We must also be willing to use base states that do not satisfy Equation 1.1 For example, in studying the behavior of a slightly bound electron in a lattice of atoms (such as a semiconductor) one introduces base states such as $e^{i(k_x x + k_y y + k_z z)}$ ([FLS, II-13-4]). To study these ideas rigorously from a mathematical perspective, one studies "continuous spectrum" and "spectral measures," as in [RS, Section VII.2].

has the form

$$\frac{\sqrt{3}}{2}\phi_1 + \frac{e^{i\theta}}{2}\phi_2,$$

where θ is a real number. More generally, one sees expressions such as $\sum c_n \phi_n$ or $\sum \langle \psi | \phi_n \rangle | \phi_n \rangle$.

In its full generality, our third fundamental quantum-mechanical assumption says that the same kind of decomposition is possible with base states of the position observable, the momentum observable or indeed any observable. In other words, every observable has a *complete set of base states*. Typically the information about the base states and the value of the observable on each base state is collected into a mathematical object called a *self-adjoint linear operator*. The base states are the eigenvectors and the corresponding values of the observable are the eigenvalues. For more information about this point of view, see [RS, Section VIII.2].

Our next assumption is that we can use the superposition of base states to predict the probabilities of experimental outcomes. For example, consider the energy observable. Suppose we have a finite linear combination

$$\phi = \sum_{k=1}^{n} c_k \phi_k,$$

where ϕ satisfies Equation 1.1, each c_k is a complex number, each ϕ_k satisfies Equation 1.1 and there are distinct real numbers $\lambda_1, \ldots, \lambda_n$ such that for each k the wave function ϕ_k is a base state for the energy observable corresponding to the value λ_k. In other words, measuring the energy of a particle in the state corresponding to the wave function k is certain to yield the value λ_k. Here is our quantum mechanical assumption: if we measure the energy of a particle in the state described by the wave function ϕ, we will find one of the values $\lambda_1, \ldots, \lambda_n$; what is more, the probability of measuring the energy to be λ_k is $|c_k|^2$. In full generality, the assumption applies to any observable (not just the energy observable in our example) and to more general linear combinations, such as infinite linear combinations and integrals. But the essential idea is the same: the squares of the absolute values of the coefficients of a superposition of base states give the probabilities of measurements corresponding to the base states.

There is a practical shortcut for calculating probabilities from base states. For example, suppose that the observable A has exactly one base state ψ corresponding to a certain real number λ. Suppose we would like to predict the probability p that a particle in a certain state ϕ will yield the result λ when we measure A. Rather than expand the state ϕ into base states for the observable A, we can simply calculate the coefficient of the base state ψ and take the square of the absolute value. The formula is

$$p = \left| \int_{\mathbb{R}^3} \psi^*(x, y, z)\phi(x, y, z)dx\,dy\,dz \right|^2 . \tag{1.3}$$

Finally, we will assume the *Pauli exclusion principle*. The simplest form of the exclusion principle is that no two electrons can occupy the same quantum state. This is a watered-down version, designed for people who may not understand linear algebra. A stronger statement of the Pauli exclusion principle is: no more than n particles can occupy an n-dimensional subspace of the quantum mechanical state space. In other words, if $\phi_1, \ldots, \phi_n$ are wave functions of n particles, then the set $\{\phi_1, \ldots, \phi_n\}$ must be a linearly independent set. We will review these linear algebraic concepts in Chapter 2.

Let us summarize the quantum mechanical assumptions.

1. Each state of a particle moving in $\mathbb{R}^3$ is described by a complex-valued function ϕ of three real variables satisfying

$$\int_{\mathbb{R}^3} |\phi(x, y, z)|^2 \, dx\,dy\,dz = 1.$$

2. The aggregate outcomes of one position measurement repeated on many particles in the state corresponding to a wave function ϕ can be predicted from ϕ.

3. Fix any observable. Then any wave function ϕ satisfying Equation 1.1 can be written as a superposition of base states of that observable.

4. Fix any observable and any wave function ϕ. The probabilities governing repeated measurements of the observable on particles in the state corresponding to ϕ can be calculated from the coefficients in the expression of ϕ as a superposition of base states for the given observable. To calculate these probabilities it suffices to calculate quantities of the form

$$\left| \int_{\mathbb{R}^3} \psi^*(x, y, z)\phi(x, y, z)dx\,dy\,dz \right|^2 .$$

5. Pauli exclusion principle: no two electrons can occupy the same state simultaneously.

We remark that all these assumptions are stated for the dynamics of the particle. To model other aspects of the particle (such as spin), complex-valued functions on $\mathbb{R}^3$ will not suffice. In Chapter 11 we incorporate other aspects into the model. So, while the fundamental assumptions above are not the only assumptions used in analyses of quantum systems, they suffice for the analysis up through Chapter 9.

1.3   The Hydrogen Atom

Hydrogen (H) is the simplest and lightest atom in the periodic table. We drink it every day: it is an essential component of water; in fact, "hydro-gen" means "water-generating." It has played a crucial role in many developments of modern physics. In this book we will model the hydrogen atom by a single quantum particle (the electron) moving in a spherically symmetric force field (created by the proton in the nucleus). There are certainly more sophisticated models available — for example, it is more precise to model the hydrogen atom as the mutual interaction of two particles, a proton and an electron[4] — but our model is simple and quite accurate.

To demonstrate the accuracy of our mathematical model, we must consider the experimental evidence. Scientifically speaking, it is a bit of a cheat to make "predictions" about a phenomenon whose experimental behavior is already understood; pedagogically, however, it is beyond reproach. When excited (for example, by heat), hydrogen gas will emit light. (This is true of other gases as well: the distinctive colors of neon signs and sodium streetlights depend on the same basic phenomenon.) Some important early experiments on the structure of the hydrogen atom consisted of exciting hydrogen gas and splitting the emitted light with a prism before collecting it on a photographic plate. The prism sends differently colored light in different directions, so that each color corresponds to a particular position on the plate. Most positions on the plate collected no light, but a few positions on the plate collected a lot of light — these are the black stripes in Figure 1.2. The data collected indicated that only a few specific colors were emitted by the gas. These colors make up the *spectrum* of hydrogen. The study of quantum systems by experiments that measure light or, more generally, electromagnetic radiation is called *spectroscopy*.

---

[4]See for example [FLS, III-12].

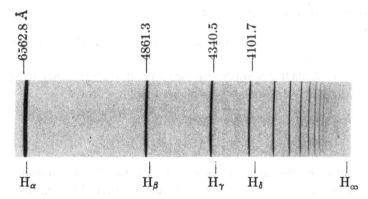

Figure 1.2. An image produced by exciting hydrogen gas and separating the outgoing light with a prism, reprinted from [Her, Fig. 1, p. 5]. Specifically, this is the emission spectrum of the hydrogen atom in the visible and near ultraviolet region. The label H_∞ marks the position of the limit of the series of wavelengths.

The strongest, most easily discerned set of lines were called the *principal spectrum*. After the principal spectrum, there are two series of lines, the *sharp spectrum* and the *diffuse spectrum*. In addition, there was a fourth series of lines, the *Bergmann* or *fundamental spectrum*.

In the spectroscopy literature, a color is usually labeled by the corresponding wavelength of light (in angstroms Å) or by the reciprocal of the wavelength (in cm^{-1}), called the *wave number*. One angstrom equals 10^{-10} meters, while one centimeter equals 10^{-2} meters, so to convert from wavelength to wave number one must multiply by a factor of 10^8:

$$\text{wave number in cm}^{-1} = \frac{10^8}{\text{wave length in Å}}.$$

As a concrete example, consider the strongest spectral line of hydrogen, corresponding to a wavelength of about 1200Å. The corresponding wave number is

$$\frac{10^8}{1200} = 8.3 \times 10^4 \quad (\text{in cm}^{-1}).$$

The wave number is natural because it is proportional to the energy of a photon of the given frequency. More specifically, we have

$$\text{energy} = hc \text{ (wave number)},$$

where $h = 6.6 \times 10^{-27}$ erg-seconds is *Planck's constant*, and $c = 3.0 \times 10^{10}$ cm/sec is the speed of light. Thus the strongest spectral line of hydrogen corresponds to the energy difference

$$(6.6 \times 10^{-27}) \times (3.0 \times 10^{10}) \times (8.3 \times 10^4) = 1.6 \times 10^{-11} \quad \text{(in ergs)}.$$

There is a formula that describes all the wave numbers obtained for spectral lines of hydrogen: every such wave number is of the form

$$R_H \left(\frac{1}{j^2} - \frac{1}{k^2} \right), \tag{1.4}$$

where j and k are natural numbers with $j < k$ and R_H is a constant. Conversely, as far as experiments can tell, there is a spectral line at most wave numbers of the given form. Formula 1.4 was first established from experimental data, not from any theoretical calculation. The value of R_H has been determined experimentally with great precision; the known value is approximately

$$R_H = 1.1 \times 10^5 \text{cm}^{-1}.$$

For example, when $j = 1$ and $k = 2$ the formula predicts a spectral line of wave number

$$\left(1.1 \times 10^5 \text{cm}^{-1} \right) (0.75) = 8.3 \times 10^4 \quad \text{(in cm}^{-1}),$$

that is, the strongest spectral line of hydrogen. Furthermore, taking $j = 1$ in Equation 1.4 and letting k vary, we obtain all the wave numbers corresponding to the principal spectrum; taking $j = 2$ yields the sharp spectrum; taking $j = 3$ yields the diffuse spectrum; and taking $j = 4$ yields the fundamental spectrum. Niels Bohr proposed that the electron hydrogen atom had a discrete set of possible orbits and possible energies, and that each spectral line corresponded to the energy difference between two states (see [Her, p. 13]). The energy values can be taken to be

$$-\frac{\hbar c R_H}{(n + 1)^2} \tag{1.5}$$

as n varies over the nonnegative integers. The number n is called the *principal quantum number*.

Other experiments showed the finer structure of the hydrogen spectrum. Some of these experiments were spectroscopic; some measured the angle of deflection of atoms as they pass through a magnetic field; some experiments were done on *alkali* atoms (i.e., the atoms in the first column of the periodic table, whose behavior is similar to hydrogen's) and the results extrapolated back to hydrogen. These experiments are described in detail in the books of Herzberg [Her] and Hochstrasser [Ho]. Experiments involving a magnetic field used *Stern–Gerlach machines*, described in the Feynman Lectures [FLS, III-5] and pictured in Figure 10.3.

To describe the results of these experiments, it is useful to introduce the *azimuthal quantum number* ℓ. States corresponding to the "sharp" spectral lines on the photographic plates (often labeled s) have $\ell = 0$; those corresponding to "principal" lines (labeled p) have $\ell = 1$; those corresponding to "diffuse" lines (labeled d) have $\ell = 2$ and those corresponding to "fundamental" lines (labeled f) have $\ell = 3$. The experiments showed that each spectral line of hydrogen with at least one state of azimuthal quantum number ℓ contains $2(2\ell + 1)$ different states with azimuthal quantum number ℓ. Because these spectral lines split in the presence of a magnetic field, the new split lines were labeled by the *magnetic quantum number* m. The magnetic quantum number could take any of the $2\ell + 1$ values $-\ell, 1 - \ell, \ldots, \ell - 1, \ell$. Similarly, the *spin quantum number* s takes either of the values $\pm 1/2$.

Up to and including Chapter 7, we make very few assumptions; in particular, we do not need to know the functional form of the force on the electron. We assume only that this force is spherically symmetric. Yet, armed with some powerful undergraduate-level mathematics (plus Fubini's Theorem and the Stone–Weierstrass Theorem), we can make meaningful predictions from the meager assumptions of the basic model of quantum mechanics and spherical symmetry.

We will see in Chapter 7 that our model predicts the existence of states indexed by the quantum numbers ℓ and m but fails to predict the factor of two introduced by the spin quantum number s. The beauty of this prediction is that it is close to the experimental data — off only by a measly factor of two! — even though the assumptions are quite meager. We discuss spin in Chapter 10. Readers who have seen these predictions come out of the analysis of the Schrödinger equation should note that the predictions of Chapter 7 use neither the concept of energy nor the theory of observables. In other words, we will make these powerful predictions from symmetry considerations alone.

When we include in our model an explicit formula for the energy of the system, we can make stronger predictions. The energy observable for the hydrogen atom is completely described by the *Schrödinger operator*,

$$\mathbf{H} := -\frac{\hbar^2}{2\mathbf{m}} \left(\partial_x^2 + \partial_y^2 + \partial_z^2 \right) - \frac{e^2}{\sqrt{x^2 + y^2 + z^2}},$$

where **m** is the mass of the electron, $\hbar$ is Planck's constant divided by 2π and **e** is the charge of the electron.[5] One may write the defining equation more succinctly as

$$\mathbf{H} := -\frac{\hbar^2}{2\mathbf{m}}\nabla^2 - \frac{\mathbf{e}^2}{r}.$$

The differential operator **H** describes the energy observable in the sense that the *eigenfunctions* of this differential operator, i.e., wave functions ϕ_E satisfying $\mathbf{H}\phi_E = E\phi_E$, with $E \in \mathbb{R}$, are the base states of the energy observable (see Assumption 3 of Section 1.2) and the probability of getting the result E' from an energy measurement of an electron in the state ϕ_E is

$$1, \quad \text{if } E = E'$$
$$0, \quad \text{if } E \neq E'$$

(see Assumption 4 of Section 1.2).

The function $-\mathbf{e}^2/\sqrt{x^2 + y^2 + z^2}$ is called the *Coulomb potential*. It has the same functional form as the gravitational potential energy function in the classical two-body problem of the motion of a planet around the sun. For this reason the hydrogen atom is called the quantum version of the classical celestial mechanics problem. In the classical case, energy is a function on the state space, while in the quantum case energy is an operator. Hence the Coulomb potential term is an operator: it operates on ϕ by multiplication. Just as the classical problem has extra symmetries associated to the *Runge–Lenz vector* (whose direction determines the direction of the major axis of the orbit and whose length determines the eccentricity), the quantum two-body system has extra symmetries corresponding to "Runge–Lenz operators." We introduce these operators in Section 8.6.

This model makes definite predictions about energy observations. For example, from the experimentally observed spectrum of hydrogen one can calculate the energy levels up to the addition of an arbitrary constant. One can choose this constant so that the *ionization energy* of the hydrogen electron is 0, i.e., so that any electron with energy $E > 0$ has enough energy to escape the attracting force of the hydrogen nucleus. With this choice of constant, one can deduce from the experimental data that the only possible observable energy values for an electron bound in a hydrogen atom are

$$E_n := \frac{-\mathbf{m}\mathbf{e}^4}{2\hbar^2(n+1)^2},$$

---

[5]Numerically $\mathbf{m} = 9.1 \times 10^{-28}$ in grams, $\hbar = 1.1 \times 10^{-27}$ in units of erg-seconds and $\mathbf{e} = 1.6 \times 10^{-19}$ in units of coulombs [To, pp. 277, 463].

n (principal)	ℓ (azimuthal)	total number of states
0	0	2
1	0, 1	8
2	0, 1, 2	18
$\vdots$	$\vdots$	$\vdots$
n	$0, \ldots, n$	$2(n+1)^2$

Figure 1.3. Table of the number of states for a given energy, i.e., for a given value of the principal quantum number n.

where n is a nonnegative integer called the *principal quantum number*. Moreover, there is an experimentally verifiable relationship between the principal quantum number n and the possible azimuthal quantum numbers ℓ of the states at the nth energy level.

The total number of different states with principal quantum number n is obtained from the sum

$$\sum_{\ell=0}^{n} 2(2\ell+1) = 2(n+1)^2,$$

since the number of states of azimuthal quantum number ℓ and principal quantum number n is

$$2(2\ell+1), \quad \text{if } \ell \leq n;$$
$$0, \qquad\quad \text{if } \ell > n.$$

In Section 8.6 we will see that symmetry considerations alone, without any appeals to special functions or series solutions, will allow us to predict these results from the model, up to a factor of two.

1.4   The Periodic Table

The *periodic table of the elements* is a list of all known types of atoms, arranged in a way to highlight similarities and differences in chemical properties of the atoms. See Figure 1.4. One can view the periodic table as a mnemonic for the known experimental properties of the various elements. For example, the elements of the last column, helium, neon, argon, krypton, xenon, radon and ununoctium, are called *noble gases* because they are particularly unreactive. On the other end, spectral data for the *alkali* atoms lithium, sodium and potassium, all elements of the first column, strongly resemble the data for hydrogen. There are other ways to arrange the table — see Figure 1.5.

1	2	3	4	5	6	7	8	9	10	11	12	13	14	15	16	17	18
H 1																	He 2
Li 3	Be 4											B 5	C 6	N 7	O 8	F 9	Ne 10
Na 11	Mg 12											Al 13	Si 14	P 15	S 16	Cl 17	Ar 18
K 19	Ca 20	Sc 21	Ti 22	V 23	Cr 24	Mn 25	Fe 26	Co 27	Ni 28	Cu 29	Zn 30	Ga 31	Ge 32	As 33	Se 34	Br 35	Kr 36
Rb 37	Sr 38	Y 39	Zr 40	Nb 41	Mo 42	Tc 43	Ru 44	Rh 45	Pd 46	Ag 47	Cd 48	In 49	Sn 50	Sb 51	Te 52	I 53	Xe 54
Cs 55	Ba 56	57–70 * / Lu 71	Hf 72	Ta 73	W 74	Re 75	Os 76	Ir 77	Pt 78	Au 79	Hg 80	Tl 81	Pb 82	Bi 83	Po 84	At 85	Rn 86
Fr 87	Ra 88	89–102 ** / Lr 103	Rf 104	Db 105	Sg 106	Bh 107	Hs 108	Mt 109	Ds 110	Uuu 111	Uub 112	113	Uuq 114	115	Uuh 116	117	Uuo 118

*	La 57	Ce 58	Pr 59	Nd 60	Pm 61	Sm 62	Eu 63	Gd 64	Tb 65	Dy 66	Ho 67	Er 68	Tm 69	Yb 70
**	Ac 89	Th 90	Pa 91	U 92	Np 93	Pu 94	Am 95	Cm 96	Bk 97	Cf 98	Es 99	Fm 100	Md 101	No 102

Figure 1.4. The most common form of the periodic table of the elements.

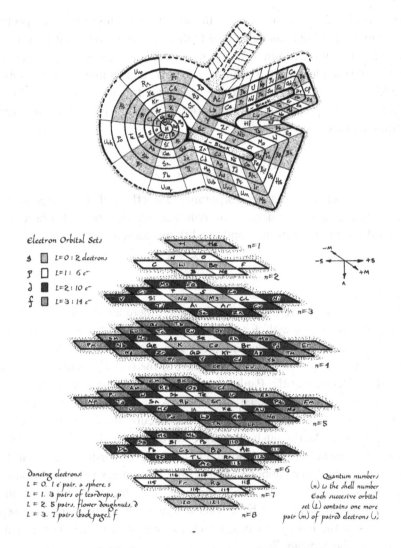

Figure 1.5. Three uncommon versions of the periodic table [Tw, pp. 8–9]. For more variations, see [Hei].

Why should the spectral data for the alkali atoms resemble the spectral data for hydrogen? Our model of the hydrogen atom, along with the Pauli exclusion principle (Section 1.2) and some other assumptions, provides an answer. For example, consider lithium, the third element in the periodic table. Its nucleus has a positive charge of three and it tends to attract three electrons. The Schrödinger operator for the behavior of a single electron in the presence of a lithium nucleus is

$$\mathbf{H}_L := -\frac{\hbar^2}{2\mathbf{m}}\nabla^2 - \frac{\mathbf{Z}e^2}{r},$$

where $\mathbf{Z}$ is a constant factor incorporating the effect of the charge of the nucleus. By the same argument as for hydrogen, the only possible observable energy values for an electron bound to a lithium nucleus are

$$E_n^L := \frac{-\mathbf{Z}^2\mathbf{m}e^4}{2\hbar^2(n+1)^2},$$

where n is a nonnegative integer. Furthermore, there are two states with energy E_0^L and six states with E_1^L. If we assume that the three electrons in a lithium atom do not affect one another, then the lowest–energy state of a lithium atom will have one electron in each of the two E_0^L states and one in an E_1^L state. Recall that the Pauli exclusion principle says that no two electrons can occupy the same state simultaneously. The two E_0^L electrons are called *inner electrons* and we say that they occupy the *innermost shell* of the lithium atom. Analogously, the E_1^L electron is called an *outer electron*. Because the outer electron is more likely to change its energy state than the inner ones, spectral lines obtained by exciting lithium gas will correspond to one electron changing states, and so will resemble the hydrogen spectrum. The model would make even better predictions if one incorporated the negative charge of the inner electrons, which cancels some of the charge of the nucleus, into the constant $\mathbf{Z}$.

The same argument can be made for each alkali atom: because there is only one outer electron, one can model an alkali atom as a hydrogen-like atom with one electron and a "nucleus" made up of the true nucleus and the inner electrons. As above, this argument hinges on the fact that the inner electrons tend to be in the lowest possible states, while the Pauli exclusion principle forbids any two electrons from occupying the same state. And indeed, spectral data for alkali atoms resembles spectral data for hydrogen. Moreover, the chemical properties of the alkali atom is similar. For example, each combines easily with chlorine to form a salt such as potassium chloride, lithium chloride or sodium chloride (better known as table salt). These chemical combinations are natural because there is only a single electron in the outer shell of each alkali atom.

More generally, one can model a many-electron atom (such as carbon with six electrons) simply and fairly accurately by assuming that the forces of the inner electrons on the outer electrons can be approximated by a repellent force at the origin, and that the outer electrons exert no force on one another. The repellent force is often called the *shielding force*, since the inner electrons shield the outer electrons from the full force of attraction of the nucleus. The chemical properties of an element will depend heavily on the number of electrons in (or missing from) the outer shell. In fact, each row of the periodic table corresponds to a particular energy level, i.e., to a particular outer shell. Because our model (including Chapter 8) predicts the numbers of electron states in each shell, it predicts the lengths of the rows of the periodic table. From Section 8.6 we can read off the predictions of our model: the rows of the periodic table cannot have any length other than the double of a square; i.e., the rows must be of length 2, 8, 18, etc., i.e., each row must have length $2(n + 1)^2$ for some nonnegative integer n. We invite the reader to count the number of elements in each row of the periodic table. For example, notice that there are two rows with $2 \times (3 + 1)^2 = 32$ elements. As before, the theory of spin (see Chapter 10) contributes an important factor of two.

The prediction of the structure of the periodic table from symmetries is one of the great successes of representation theory. It is more than just an application of mathematical techniques to calculations that arise in physics (such as the use of complex analysis to calculate contour integrals). It is an example of the foundational importance of mathematics in physics.

1.5 Preliminary Mathematics

In this section we list the mathematical background material assumed by the text.

Readers should have linear algebra at their fingertips, either metaphorically or literally. We will use linear algebraic concepts freely. For example, we will need to use determinants and traces of matrices, as well as diagonal matrices. Some readers may wish to keep a linear algebra reference handy as they work through this book. Any college-level linear algebra text will do. The author particularly likes the elementary text by Shifrin and Adams [SA] and the more advanced (and very interesting) text by Lax [La]. Readers should also know calculus well.

Otherwise the exposition in this book is self-contained. However, we will mention many related topics, and we strongly urge the reader to make connections with what she already knows about or is curious about. In particular, a reader who knows some quantum mechanics, abstract algebra, analysis or topology might want to keep the relevant books available for reference. We encourage instructors to put related books on reserve. The books referred to most in these pages are Rudin's undergraduate analysis text [Ru76], Artin's abstract algebra text [Ar] and the *Feynman Lectures on Physics* [FLS].

Another book well worth exploring is *Lie Groups and Physics* [St], by Sternberg. There are so many wonderful ideas and stories about mathematics and physics in this book that it can be a bit bewildering at first, but the persevering reader will be well rewarded. In particular, Sternberg discusses the structure of the hydrogen atom and the periodic table; almost every idea in the book you are reading now is contained (in more abbreviated form) in Sternberg's book.

We use common (but not universal) mathematical notation and terminology for functions. When we define a function, we indicate its domain (the objects it can accept as arguments), the target space (the kind of objects it puts out as values) and a rule for calculating the value from the argument. For example, if we wish to introduce a function f that takes a complex number to its absolute value squared, we write

$$f : \mathbb{C} \to \mathbb{R}$$
$$z \mapsto |z|^2 .$$

Note that z is a *dummy variable*: the definition would have the same meaning if we replaced it by x, m, ξ or any other letter. The general form is:

$$\text{function} : \text{domain} \to \text{target space}$$

$$\begin{matrix} \text{dummy} \\ \text{variable} \end{matrix} \mapsto \begin{matrix} \text{function evaluated} \\ \text{at dummy variable.} \end{matrix}$$

One common function is the *identity function*. On any space S we define the identity function $I : S \to S$ by

$$I(s) := s$$

for each $s \in S$.

Next we introduce some useful terminology. A function f is *injective* if it is one-to-one, i.e., if $f(x) = f(y)$ implies that $x = y$. The *image* of a function $f : S \to T$ is its range, i.e., the set

$$\{t \in T : f(s) = t \text{ for some } s \text{ in the domain of } f\}.$$

Note that the target space need not equal the image. For example, the image space of the squaring function defined above is $\mathbb{R}^{\geq 0}$, which is a proper subset of the range $\mathbb{R}^{\geq 0}$. A function f is *surjective (onto its target space T)* if the image is equal to the target space. The *preimage (under f)* of a subset U of the target space T, denoted $f^{-1}[U]$, is the set of all s in the domain of f such that $f(s) \in U$; in other words,

$$f^{-1}[U] := \{s \in S : f(s) \in U\}.$$

Similarly, the *image (under f)* of a subset U of the domain is the set $f[U] := \{f(u) : u \in U\}$.

We will often define functions in terms of other functions. For example, The *composition* of two functions f and g is the function

$$f \circ g : \quad f^{-1}[\text{domain of } g] \to \text{target space of } f$$
$$x \mapsto f(g(x)).$$

Another common way of defining a new function is by *restriction*. Suppose f is a function with domain S. Suppose $\tilde{S}$ is a subset of S. Then the restriction $f|_{\tilde{S}}$ of f to $\tilde{S}$ is the function with domain $\tilde{S}$ defined by

$$f\Big|_{\tilde{S}}(x) := f(x)$$

for any $x \in \tilde{S}$. Note that if $\tilde{S}$ is a proper subset of S (i.e., if $S \neq \tilde{S}$), then $f|_{\tilde{S}}$ is not the same as f: if $x \in S$ but $x \notin \tilde{S}$ then $f(x)$ is well defined but $f|_{\tilde{S}}(x)$ is meaningless. For an example, see Figure 1.6. If a function $f : S \to T$ is injective and surjective, then one can define the *inverse function* $f^{-1} : T \to S$ by

$$f^{-1}(t) := s,$$

where s is the unique element of S such that $f(s) = t$. Note that $g = f^{-1}$ if and only if $f \circ g$ is the identity function on T and $g \circ f$ is the identity function on S.

Homogeneous polynomials play an important role in our story.

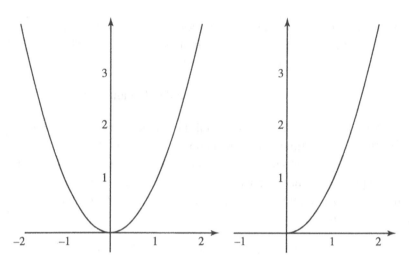

Figure 1.6. The graph of the squaring function defined on all of $\mathbb{R}$, and the graph of its restriction to $\mathbb{R}^{\geq 0}$.

Definition 1.1 *A function f is* homogeneous of degree n *on a Euclidean space $\mathbb{R}^d$ (or, more simply,* homogeneous*) if, for every $r \in \mathbb{R}$ and every $x \in \mathbb{R}^d$ we have $f(rx) = r^n f(x)$.*

For example, the polynomial $xy + z^2$ is homogeneous of degree 2, since

$$(rx)(ry) + (rz)^2 = r^2(xy + z^2).$$

On the other hand, the polynomial $x^2 + 1$ is of *mixed degree*, that is, not homogeneous of any degree. See Exercise 1.9.

We use a perhaps unfamiliar but elegant notation for partial derivatives. In many standard textbooks the partial derivative of a function f with respect to a variable y is denoted

$$\frac{\partial f}{\partial y}.$$

We prefer (and encourage the reader to use) the more modern notation

$$\partial_y f.$$

Not only is this notation more succinct, but it also suggests the sophisticated (and correct!) point of view that ∂_y is itself a mathematical object worthy of study[6]. In fact, ∂_y is a linear operator; for more details on this topic, see [Si,

---

[6]Even more elegant, but almost never used, is the notation ∂_2 to indicate differentiation with respect to the second slot, obviating the need to assign a name (such as y) to the variable in the second slot.

Section 2.2]. Many such *partial differential operators* will play a significant role in Chapter 8.

One partial differential operator plays an important role in the first several chapters: the *Laplacian*,

$$\nabla^2 := \partial_x^2 + \partial_y^2 + \partial_z^2,$$

defined on twice-differentiable functions of three variables x, y and z. For example, because

$$\partial_x^2 e^{-x^2-y^2-z^2} = (4x^2 - 2)e^{-x^2-y^2-z^2}$$

(and similarly for y and z), we have

$$\nabla^2 e^{-x^2-y^2-z^2} = (4x^2 + 4y^2 + 4z^2 - 6)e^{-x^2-y^2-z^2}.$$

The equation $\nabla^2 f = 0$ is called *Laplace's equation*, and a function f satisfying $\nabla^2 f = 0$ is called a *harmonic* function.

We denote the set of complex numbers by $\mathbb{C}$. Readers should be familiar with complex numbers and how to add and multiply them, as described in many standard calculus texts. We use i to denote the square root of -1 and an asterisk to denote complex conjugation: if x and y are real numbers, then $(x + iy)^* := x - iy$. Later in the text, we will use the asterisk to denote the conjugate transpose of a matrix with complex entries. This is perfectly consistent if one thinks of a complex number $x + iy$ as a one-by-one complex matrix $(\ x + iy\ )$. See also Exercise 1.6. The *absolute value* of a complex number, also known as the *modulus*, is denoted

$$|x + iy| := \sqrt{x^2 + y^2}.$$

We also define the *real part*

$$\Re(x + iy) := x$$

and the *imaginary part*

$$\Im(x + iy) := y.$$

It is often convenient to write a complex number in the *polar form* $re^{i\theta}$, where θ is a real number and r is a nonnegative real number. For a beautiful, idiosyncratic exposition exploiting the full power of a geometric interpretation of the complex numbers, see Needham [N].

Not quite so standard, but not difficult, is the idea of complex-valued functions of real variables and derivatives of such functions. If we have a complex-valued function f of three real variables, x, y and z, we can define its partial derivatives by the same formulas used to define partial derivatives of real-valued functions. More generally, any algebraic calculations that are possible with real-valued functions are also possible with complex-valued functions. For the readers' convenience, we state a few properties formally.[7]

Proposition 1.1 *Suppose $f: \mathbb{R}^n \to \mathbb{C}$ is a complex-valued function. Define its real part $f_R: \mathbb{R}^n \to \mathbb{R}$ and its imaginary part $f_I: \mathbb{R}^n \to \mathbb{R}$ as the real-valued functions satisfying $f_R + if_I = f$. Then f is differentiable if and only if both f_R and f_I are differentiable. Furthermore, any derivative of f is equal to the sum of the corresponding derivative of f_R plus the complex number i times the corresponding derivative of f_I.*

For example, if f is a function of x, y and z, then

$$\partial_x \partial_y f = \partial_x \partial_y f_R + i \partial_x \partial_y f_I.$$

The familiar rules for combining derivatives with sums, products and quotients apply to complex-valued functions.

Proposition 1.2 *If f and g are differentiable, complex-valued functions of one real variable, then $(f + g)' = f' + g'$, $(fg)' = f(g') + (f')g$ and, wherever g is nonzero, $\left(\frac{1}{g}\right)' = \frac{-g'}{g^2}$. (The superscript $'$ denotes the derivative.)*

One can also define integration easily.

Definition 1.2 *Suppose $f = f_R + if_I$ is a complex-valued function and S is a set on which an integral $\int_S$ is defined for real-valued functions. Then we define*

$$\int_S f := \int_S f_R + i \int_S f_I.$$

This integral satisfies all the algebraic rules of integration. Also, integration respects conjugation.

Proposition 1.3 *Suppose S is a set on which an integral $\int_S$ is defined and f is a complex-valued, integrable function on S. Then*

$$\left(\int_S f\right)^* = \int_S (f^*).$$

---

[7] Proofs can be found in Chapters 5 and 6 of Rudin's undergraduate text [Ru76].

Proof.

$$\left(\int_S f\right)^* = \left(\int_S f_R + i \int_S f_I\right)^* = \int_S f_R - i \int_S f_I = \int_S (f^*).$$

$\square$

In Chapter 8 matrix exponentiation will play a crucial role. If n is a non-negative integer and M is an $n \times n$ matrix, we define

$$\exp M := \sum_{k=0}^{\infty} \frac{1}{k!} M^k.$$

For example,

$$\exp \begin{pmatrix} 0 & \pi & 0 \\ -\pi & 0 & 0 \\ 0 & 0 & 0 \end{pmatrix} = \begin{pmatrix} -1 & 0 & 0 \\ 0 & -1 & 0 \\ 0 & 0 & 1 \end{pmatrix}.$$

See Exercise 1.8. We will need several properties of matrix exponentiation.

Proposition 1.4 *Suppose M_1 and M_2 are $n \times n$ matrices with complex entries. Suppose T is an invertible $n \times n$ matrix with complex entries. Then:*

1. *the sum $\sum_{k=0}^{\infty} \frac{1}{k!} M_1^k$ converges to an $n \times n$ matrix with complex entries;*

2. *$\exp M_1$ is an invertible matrix, with inverse $\exp(-M_1)$;*

3. *$\exp(T M_1 T^{-1}) = T (\exp M_1) T^{-1}$;*

4. *if M_1 and M_2 commute, i.e., if $M_1 M_2 = M_2 M_1$, then*

$$\exp M_1 \exp M_2 = \exp(M_1 + M_2);$$

5. *$\partial_t \exp(t M_1) = M_1 \exp(t M_1) = \exp(t M_1) M_1$.*

The proof of this proposition follows fairly easily from the definition of matrix exponentiation and standard techniques of vector calculus. See any linear algebra textbook, such as [La, Chapter 9].

We will use spherical coordinates on the two-sphere

$$S^2 := \left\{ \begin{pmatrix} x \\ y \\ z \end{pmatrix} : x^2 + y^2 + z^2 = 1 \right\}.$$

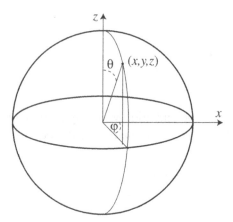

Figure 1.7. Spherical coordinates on S^2.

Following the physicists' convention, we use ϕ for longitude and θ for *colatitude*, i.e., the angle of formed by a point, the center of the sphere and the north pole. We can express Cartesian coordinates in terms of spherical coordinates on the two-sphere S^2 as follows:

$$\begin{pmatrix} x \\ y \\ z \end{pmatrix} = \begin{pmatrix} \sin\theta\cos\phi \\ \sin\theta\sin\phi \\ \cos\theta \end{pmatrix}.$$

See Figure 1.7. To integrate a function $f(\theta, \phi)$ on the two-sphere S^2, recall the formula for surface integration:

$$\int_{S^2} f = \int_0^{2\pi} \int_0^{\pi} f(\theta, \phi)\sin\theta\, d\theta d\phi.$$

Note that $\sin\theta\, d\theta d\phi$ is the natural surface area coming from the Euclidean geometry of the space $\mathbb{R}^3$ in which the two-sphere S^2 sits.

In our discussion of spherical harmonics we will use an expression of the three-dimensional Laplacian in spherical coordinates. For this we need spherical coordinates not just on S^2 but on all of three-space. The third coordinate is r, the distance of a point from the origin. We have, for arbitrary $(x, y, z)^T \in \mathbb{R}^3$,

$$\begin{pmatrix} x \\ y \\ z \end{pmatrix} = \begin{pmatrix} r\sin\theta\cos\phi \\ r\sin\theta\sin\phi \\ r\cos\theta \end{pmatrix}.$$

The derivation of the formula for the Laplacian in spherical coordinates is a healthy exercise in proper application of the chain rule for functions of several variables (Exercise 1.12). The answer is

$$\nabla^2 = \partial_r^2 + \frac{2}{r}\partial_r + \frac{1}{r^2}\partial_\theta^2 + \frac{\cos\theta}{r^2\sin\theta}\partial_\theta + \frac{1}{r^2\sin^2\theta}\partial_\phi^2.$$

We will also use spherical coordinates on the three-sphere

$$S^3 := \left\{ \begin{pmatrix} u \\ x \\ y \\ z \end{pmatrix} : u^2 + x^2 + y^2 + z^2 = 1 \right\}.$$

One way to visualize the three-sphere S^3 is to think of a movie, with u playing the role of time. For times before -1 or after 1 there is nothing on the three-dimensional "screen"; at time $u = -1$ exactly there is one point visible at the spatial point $(0, 0, 0)^T$; more generally, for $u \in [-1, 1]$ there is a two-sphere of radius $\sqrt{1 - u^2}$ visible on the three-dimensional screen. (One can also interpret the fourth dimension as color. See Exercise 1.10.) We can write

$$\begin{pmatrix} u \\ x \\ y \\ z \end{pmatrix} = \begin{pmatrix} \cos\psi \\ \sin\psi \sin\theta \cos\phi \\ \sin\psi \sin\theta \sin\phi \\ \sin\psi \cos\theta \end{pmatrix}.$$

In the movie analogy we can think of θ and ϕ as the colatitude and the longitude on the visible two-sphere. The new variable ψ varies from 0 to π, and the radius of the visible sphere is $\sin\psi$. The natural volume element on the three-sphere S^3 coming from the four-dimensional "volume" in $\mathbb{R}^4$ is $\sin^2\psi \sin\theta\, d\psi d\theta d\phi$. In other words, to integrate a function $f(\psi, \theta, \phi)$ over the three-sphere S^3 we calculate

$$\int_{S^3} f = \int_0^{2\pi} \int_0^\pi \int_0^\pi f(\phi, \theta, \psi) \sin^2\psi \sin\theta\, d\psi d\theta d\phi.$$

We invite the reader to check this formula in Exercise 1.11.

We will find it convenient to use the *algebra* $\mathbf{Q}$ *of quaternions*. This is a real four-dimensional vector space. We pick a basis and name it $\{1, \mathbf{i}, \mathbf{j}, \mathbf{k}\}$; then we define a multiplication on the vector space by the rules

$$\text{for any } \mathbf{q} \in \mathbf{Q}, \ 1\mathbf{q} = \mathbf{q}1 = \mathbf{q}$$
$$\mathbf{ij} = -\mathbf{ji} = \mathbf{k}$$
$$\mathbf{jk} = -\mathbf{kj} = \mathbf{i}$$
$$\mathbf{ki} = -\mathbf{ik} = \mathbf{j},$$

along with the usual distributive law for multiplication. More explicitly, we have, for any real numbers u, x, y, z,

$$(u + x\mathbf{i} + y\mathbf{j} + z\mathbf{k})(\tilde{u} + \tilde{x}\mathbf{i} + \tilde{y}\mathbf{j} + \tilde{z}\mathbf{k})$$

$$\left.\begin{aligned} := \ & (u\tilde{u} - x\tilde{x} - y\tilde{y} - z\tilde{z}) \\ & + (u\tilde{x} + x\tilde{u} + y\tilde{z} - z\tilde{y})\mathbf{i} \\ & + (u\tilde{y} + y\tilde{u} + z\tilde{x} - x\tilde{z})\mathbf{j} \\ & + (u\tilde{z} + z\tilde{u} + x\tilde{y} - y\tilde{x})\mathbf{k}. \end{aligned}\right\} \tag{1.6}$$

There is a conjugation defined on the quaternions:

$$(u + x\mathbf{i} + y\mathbf{j} + z\mathbf{k})^* := u - (x\mathbf{i} + y\mathbf{j} + z\mathbf{k}).$$

A *unit quaternion* is a quaternion $u + x\mathbf{i} + y\mathbf{j} + z\mathbf{k}$ such that

$$u^2 + x^2 + y^2 + z^2 = 1.$$

Other concepts we will use freely include: partial derivatives, trigonometric identities, the natural numbers $\mathbb{N} := \{1, 2, 3, \ldots\}$, basic properties of integration and proof by induction. Interested readers will find a nice introduction to proof by induction in [Sp, Chapter 2].

The reader may notice that we choose to distinguish between an equals sign that defines a term (":=", with the colon facing the term being defined) and an equals sign that states the equality of two terms that are already well defined ("=").

An understanding of Fourier theory is not required for this text. However, Fourier series are an essential part of any mathematician's or physicist's education, and we encourage readers to remedy any ignorance. The Feynman Lectures has an introduction that musicians will particularly enjoy [FLS, I-50]; more mathematical introductions can be found in Davis [Da, Chapter 3], Rudin [Ru76, Chapter 8] and Dym and McKean [DyM, Chapter 1] (in order of increasing sophistication). Fourier transforms are ubiquitous in physics; their mathematical theory is analogous to, but more subtle than, the theory of Fourier series. See Rudin's more advanced book [Ru74, Chapter 9] or Dym and McKean [DyM, Chapter 2]. We use $\hat{f}$ to denote the Fourier transform of f. Because many readers will have encountered Fourier series and transforms, we will use them in some examples and remarks. Less experienced readers should feel free to skim or skip these digressions.

1.6 Spherical Harmonics

Physicists are familiar with many special functions that arise over and over again in solutions to various problems. The analysis of problems with spherical symmetry in $\mathbb{R}^3$ often appeal to the *spherical harmonic functions*, often called simply *spherical harmonics*. Spherical harmonics are the restrictions of homogeneous harmonic polynomials of three variables to the sphere S^2. In this section we will give a typical physics-style introduction to spherical harmonics. Here we state, but do not prove, their relationship to homogeneous harmonic polynomials; a formal statement and proof are given Proposition A.2 of Appendix A.

Physics texts often introduce spherical harmonics by applying the technique of *separation of variables* to a differential equation with spherical symmetry. This technique, which we will apply to Laplace's equation, is a method physicists use to find solutions to many differential equations. The technique is often successful, so physicists tend to keep it in the top drawer of their toolbox. In fact, for many equations, separation of variables is guaranteed to find all nice solutions, as we prove in Proposition A.3.

Faced with a partial differential equation (i.e., an equation involving derivatives with respect to more than one independent variable), one can often construct some solutions by looking for solutions that are the product of functions of one variable. We will apply this technique, called *separation of variables*, to find harmonic functions of three variables. Recall from Section 1.5 that a function is harmonic if and only if it satisfies Laplace's equation, which we write in spherical coordinates (see Exercise 1.12):

$$0 = \left(\partial_r^2 + \frac{2}{r}\partial_r + \frac{1}{r^2}\partial_\theta^2 + \frac{\cos\theta}{r^2\sin\theta}\partial_\theta + \frac{1}{r^2\sin^2\theta}\partial_\phi^2 \right)\psi, \qquad (1.7)$$

where ψ is an unknown function of (r, θ, ϕ). To apply the technique of separation of variables to this equation, suppose that there is a solution of the form

$$\psi(r, \theta, \phi) = R(r)\Theta(\theta)\Phi(\phi), \qquad (1.8)$$

where R, Θ and Φ are differentiable functions of one variable. On the face of it, this is quite a bold supposition: in general such a solution might not exist. But when such solutions do exist, our supposition will help us find them. Such a supposition is called an *ansatz*.[8] For example, the equation $\nabla^2\psi = 0$ gives

---

[8]From the German word *Ansatz*, which means something close to "hypothesis" or "setup" but does not have an exact English equivalent.

enough information about the functions R, Θ and Ψ that we will be able to find them. To this end we multiply Equation 1.7 by r^2/ψ (why? because it ends up working), plug in Equation 1.8 and calculate:

$$0 = \frac{r^2}{\psi}\left(\partial_r^2 + \frac{2}{r}\partial_r + \frac{1}{r^2}\partial_\theta^2 + \frac{\cos\theta}{r^2\sin\theta}\partial_\theta + \frac{1}{r^2\sin^2\theta}\partial_\phi^2\right)R(r)\Theta(\theta)\Phi(\phi)$$

$$= \left(\frac{r^2 R''(r)}{R(r)} + \frac{2r R'(r)}{R(r)}\right) + \left(\frac{\Theta''(\theta)}{\Theta(\theta)} + \frac{\cos\theta}{\sin\theta}\frac{\Theta'(\theta)}{\Theta(\theta)} + \frac{1}{\sin^2\theta}\frac{\Phi''(\phi)}{\Phi(\phi)}\right).$$

The crucial observation is that the first parenthesis in the last expression depends only on r, while the second parenthesis depends only on θ and ϕ. Because the sum of the two parentheses is 0, each one must be constant in r, θ and ϕ. Let us repeat the argument in a slightly different form. Rearranging the equation above we find

$$\left(-\frac{r^2 R''(r)}{R(r)} - \frac{2r R'(r)}{R(r)}\right) = \left(\frac{\Theta''(\theta)}{\Theta(\theta)} + \frac{\cos\theta}{\sin\theta}\frac{\Theta'(\theta)}{\Theta(\theta)} + \frac{1}{\sin^2\theta}\frac{\Phi''(\phi)}{\Phi(\phi)}\right).$$
$$(1.9)$$

The right-hand side is constant in r, so the left-hand side must also be constant in r. Contrariwise, both sides are constant in θ and ϕ. In other words, the variables are separated into different terms, a happy accident that we can exploit. We started with one differential equation involving three variables, and ended up with two separate equations, one involving one variable, and one involving two variables. Thus we have reduced the problem (finding solutions to the original equation) to two simpler problems. Of course, this simplification works only if our supposition (that there are solutions of the given form) turns out to be true.

Let us first find some solutions to the equation for R. We will make another ansatz, i.e., another supposition: we will look for solutions of the form $R(r) = r^\ell$ for some nonnegative integer ℓ. In other words, we will look for homogeneous solutions to Laplace's equation. Then $R'(r) = \ell r^{\ell-1}$ and $R''(r) = \ell(\ell-1)r^{\ell-2}$. Such an ℓ must satisfy

$$\text{constant} = -\frac{r^2\ell(\ell-1)r^{\ell-2}}{r^\ell} - \frac{2r\ell r^{\ell-1}}{r^\ell} = -\ell(\ell+1),$$

which is true for any nonnegative integer ℓ.

Next, we must find corresponding solutions for Θ and Φ. According to Equation 1.9, if $R(r) = r^\ell$, then we must have

$$-\ell(\ell+1) = \left(\frac{\Theta''(\theta)}{\Theta(\theta)} + \frac{\cos\theta}{\sin\theta}\frac{\Theta'(\theta)}{\Theta(\theta)} + \frac{1}{\sin^2\theta}\frac{\Phi''(\phi)}{\Phi(\phi)}\right).$$
$$(1.10)$$

Functions $\Theta(\theta)\Phi(\phi)$ such that Θ and Φ solve this equation are called *spherical harmonic functions of degree ℓ*. We can find solutions by separating variables again. Multiplying both sides by $\sin^2\theta$ and rearranging we have

$$-\frac{\Phi''(\phi)}{\Phi(\phi)} = \ell(\ell+1)\sin^2\theta + \frac{\Theta''(\theta)}{\Theta(\theta)}\sin^2\theta + \frac{\Theta'(\theta)}{\Theta(\theta)}\sin\theta\cos\theta.$$

Because the left-hand side is constant in θ and the right-hand side is constant in ϕ, both must be constant.

Next we find solutions for Φ. It is known from the theory of ordinary differential equations that the only solutions of $\Phi''/\Phi = \text{constant}$ are of the form $\Phi(\phi) = e^{im\phi}$. In our situation, ϕ is an angular variable, so Φ must satisfy $\Phi(\phi + 2\pi) = \Phi(\phi)$ for all $\phi \in \mathbb{R}$. So a legitimate solution requires $m \in \mathbb{Z}$, and in this case we have

$$-\frac{\Phi''(\phi)}{\Phi(\phi)} = m^2.$$

Finally we must solve the equation

$$\ell(\ell+1)\sin^2\theta + \frac{\Theta''(\theta)}{\Theta(\theta)}\sin^2\theta + \frac{\Theta'(\theta)}{\Theta(\theta)}\sin\theta\cos\theta = m^2$$

for Θ. While the solutions we found before (r^ℓ and $e^{im\phi}$) are probably familiar to most readers, the functions that solve this equation are more obscure. A change of variables will let us rewrite this equation. Define $P\colon [-1,1] \to \mathbb{R}$ by $P(\cos\theta) = \Theta(\theta)$, where $\theta \in [0, \pi]$. Then $\Theta'(\theta) = -P'(\cos\theta)\sin\theta$ and $\Theta''(\theta) = P''(\cos\theta)\sin^2\theta - P'(\cos\theta)\cos\theta$, and so we can rewrite the differential equation as

$$\ell(\ell+1)\sin^2\theta + \frac{P''(\cos\theta)}{P(\cos\theta)}\sin^4\theta - \frac{P'(\cos\theta)}{P(\cos\theta)}(2\cos\theta\sin^2\theta) = m^2.$$

Setting $t := \cos\theta$ and recalling that $\sin^2 + \cos^2 = 1$ we find

$$\ell(\ell+1)(1-t^2) + (1-t^2)^2\frac{P''(t)}{P(t)} + 2t(t^2-1)\frac{P'(t)}{P(t)} = m^2. \qquad (1.11)$$

Equation 1.11 is known as the *Legendre equation* and it has solutions for integers m with $m^2 \le \ell^2$, as the reader may check in Appendix A. The solutions $\mathbf{P}_{\ell,m}$ to the Legendre equation are called *Legendre functions*. Putting it all together we have a harmonic function

$$R(r)\Theta(\theta)\Phi(\phi) = r^\ell P_{\ell,m}(\cos\theta)e^{im\phi} \qquad (1.12)$$

for some nonnegative integer ℓ, some integer m and a function $P_{\ell,m}$ satisfying the Legendre equation (Equation 1.11).

The angular part $Y_{\ell,m} := P_{\ell,m}(\cos\theta)e^{im\phi}$ of the solution (1.12) is a *spherical harmonic function*. It turns out that there is a nonzero $P_{\ell,m}$ whenever ℓ is a nonnegative integer and m is an integer with $|m| \le \ell$. In Appendix A we will prove this and other facts about spherical harmonic functions. The number ℓ is called the *degree* of the spherical harmonic. From Equation 1.10 we see that each spherical harmonic of degree ℓ satisfies the equation

$$\left(\partial_\theta^2 + \frac{\cos\theta}{\sin\theta}\partial_\theta + \frac{1}{\sin^2\theta}\partial_\phi^2\right)Y_{\ell,m} = -\ell(\ell+1). \tag{1.13}$$

There is one spherical harmonic functions of degree $\ell = 0$:

$$Y_{0,0}(\theta,\phi) := \frac{1}{2\sqrt{\pi}};$$

three of degree $\ell = 1$:

$$Y_{1,1}(\theta,\phi) := -\frac{\sqrt{3}}{2\sqrt{2\pi}}\sin\theta\, e^{i\phi}$$

$$Y_{1,0}(\theta,\phi) := \frac{\sqrt{3}}{2\sqrt{\pi}}\cos\theta$$

$$Y_{1,-1}(\theta,\phi) := \frac{\sqrt{3}}{2\sqrt{2\pi}}\sin\theta\, e^{-i\phi};$$

and five of degree $\ell = 2$:

$$Y_{2,2}(\theta,\phi) := \sqrt{\frac{15}{32\pi}}\sin^2\theta\, e^{2i\phi}$$

$$Y_{2,1}(\theta,\phi) := -\sqrt{\frac{15}{8\pi}}\sin\theta\cos\theta\, e^{i\phi}$$

$$Y_{2,0}(\theta,\phi) := \sqrt{\frac{5}{16\pi}}(3\cos^2\theta - 1)$$

$$Y_{2,-1}(\theta,\phi) := \sqrt{\frac{15}{8\pi}}\sin\theta\cos\theta\, e^{-i\phi}$$

$$Y_{2,-2}(\theta,\phi) := \sqrt{\frac{15}{32\pi}}\sin^2\theta\, e^{-2i\phi}.$$

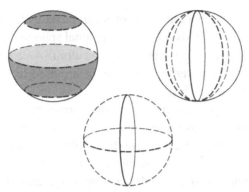

Figure 1.8. The top left sphere shows the positive (shaded) and negative (unshaded) regions for the real-valued function $Y_{2,0}$. The top right sphere shows the pure real (solid) and pure imaginary (dashed) meridian for the function $Y_{2,2}$. The bottom picture shows the zero points (double-dashed) as well as the pure real (solid) and pure imaginary (dashed) meridians of $Y_{2,1}$. There are colored versions of these pictures available on the internet. See, for instance, [Re].

Since spherical harmonics are functions from the sphere to the complex numbers, it is not immediately obvious how to visualize them. One method is to draw the domain, marking the sphere with information about the value of the function at various points. See Figure 1.8. Another way to visualize spherical harmonics is to draw polar graphs of the Legendre functions. See Figure 1.9. Note that for any ℓ, m we have $|Y_{\ell,m}| = |\mathbf{P}_{\ell,m}|$. So the Legendre function carries all the information about the magnitude of the spherical harmonic.

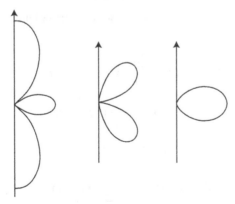

Figure 1.9. Polar graphs of, left to right, $|\mathbf{P}_{2,2}|$, $|\mathbf{P}_{2,1}|$ and $|\mathbf{P}_{2,0}|$. Rotate each graph around the vertical axis to obtain the spherical graph of the absolute value of the spherical harmonics. Three-dimensional versions of these pictures, with color added to indicate the phase $e^{im\phi}$, are available on the internet. See for instance [Sw].

The construction of spherical harmonics can be extended to other dimensions. For example, V. Fock uses four-dimensional spherical harmonics in his article on the $SO(4)$ symmetry of the hydrogen atom — see Chapter 9. Spherical harmonic functions of various dimensions are used in many spherically symmetric problems in physics.

It turns out that the spherical harmonic functions correspond exactly to the restrictions of homogeneous harmonic polynomials in three variables. This is not too surprising given that we found the spherical harmonics by taking the angular part ($r = 1$) of solutions to the equation $\nabla^2 \psi = 0$ (the defining property of harmonic functions) with the supposition that the radial part has the form r^ℓ. A proof of the exact correspondence is in Appendix A. For the moment, we will simply verify that the spherical harmonics above are indeed restrictions of harmonic polynomials. Recall that on the unit sphere we have $(x, y, z) = (\sin\theta \cos\phi, \sin\theta \sin\phi, \cos\theta)$. For $\ell = 0$ we have a constant function, which is a polynomial of degree 0. For $\ell = 1$ it is easy to compute from the definitions that

$$Y_{1,1}(\theta, \phi) = -\frac{\sqrt{3}}{2\sqrt{2\pi}}(x + iy)$$

$$Y_{1,0}(\theta, \phi) = \frac{\sqrt{3}}{2\sqrt{\pi}}z$$

$$Y_{1,-1}(\theta, \phi) = \frac{\sqrt{3}}{2\sqrt{2\pi}}(x - iy).$$

For $\ell = 2$ we must make use of some trigonometric identities to see that

$$Y_{2,2}(\theta, \phi) = \frac{\sqrt{6}}{4}(x^2 - y^2 + 2ixy)$$

$$Y_{2,1}(\theta, \phi) = \frac{\sqrt{6}}{2}(xz + iyz)$$

$$Y_{2,0}(\theta, \phi) = z^2 - \frac{1}{2}(x^2 - y^2)$$

$$Y_{2,-1}(\theta, \phi) = -\frac{\sqrt{6}}{2}(xz - iyz)$$

$$Y_{2,-2}(\theta, \phi) = \frac{\sqrt{6}}{4}(x^2 - 2ixy - y^2).$$

The right-hand side of each equation is a homogeneous polynomial of degree two in x, y and z. Each is harmonic, as the reader may check by direct calculation. For example,

$$\left(\partial_x^2 + \partial_y^2 + \partial_z^2\right) \frac{\sqrt{6}}{4}(x^2 - 2ixy - y^2) = \frac{\sqrt{6}}{4}(2 + 0 - 2) = 0.$$

Relating the spherical harmonic functions introduced here to the homogeneous harmonic polynomials is not logically necessary in this book. Morally, however, the calculation is well worth doing, in the name of better communication between mathematics and physics. Because this calculation is a bit tricky, we have postponed it to Appendix A.

1.7 Equivalence Classes

We will encounter *equivalence relations* and *equivalence classes* several times in our story. Equivalence is ubiquitous in mathematics. Because mathematicians insist on defining every object rigorously, and rigor often requires technical details, we need a mechanism to suppress any details that are irrelevant to our main point. The reader may have encountered this technique before in studying vectors and indefinite integration. In many courses, vectors are introduced as *directed line segments*, or arrows from one point to another point. See, for example, Marsden, Tromba and Weinstein [MTW]. This geometric image is very useful for developing intuition about vectors. For instance, one can interpret vector addition as a picture of a parallelogram made up of these arrows. See Figure 1.10.

In indefinite integration (also knows as antidifferentiation), there is extra information in the constant of integration. It is, strictly speaking, incorrect to say that *"the* antiderivative of x is $\frac{1}{2}x^2$" because $\frac{1}{2}x^2$ is only one of many antiderivatives, including $\frac{1}{2}x^2 + 1.7 \times 10^3$. But the statement is correct in spirit: the difference between $\frac{1}{2}x^2$ and any antiderivative of x is irrelevant for most purposes. Equivalence classes are the mathematician's way to make precise the notion of irrelevant ambiguity.

Definition 1.3 *A relation* $\sim$ *on a set S is called an* equivalence relation *if and only if* $\sim$ *is reflexive (for all $a \in S$, $a \sim a$), symmetric (for all $a, b \in S$, $a \sim b$ if and only if $b \sim a$) and transitive ($a \sim b$ and $b \sim c$ implies $a \sim c$). Given a set S, an equivalence relation* $\sim$ *and an element $a \in S$, the* equivalence class *of a is the set*

$$[a] := \{b \in S : b \sim a\}$$

and the set of all equivalence classes of elements of S is denoted $S/\!\sim$.

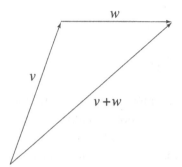

Figure 1.10. Addition of vectors.

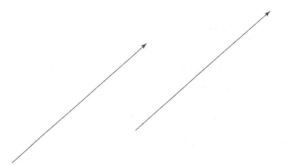

Figure 1.11. These two arrows represent the same vector.

The expression $S/\sim$ is often pronounced "S modulo equivalence" or "S mod equivalence." If possible and convenient, we refer to the equivalence by name: for example, vectors are "directed line segments modulo translation" and antiderivatives are "functions modulo constants". We leave the details of applying Definition 1.3 to vectors and antiderivatives to the interested reader in Exercises 1.19 and 1.20.

However, arrows with the same length and direction represent the same vector, regardless of the placement of the arrows. See Figure 1.11.

As an example, consider the set S of functions from $\mathbb{R} \to \mathbb{R}$. Define an equivalence relation on S by $f \sim g$ if and only if f and g agree on all but a finite number of points in their domain. More formally, the condition for equivalence is that the set $\{x \in \mathbb{R}: f(x) \neq g(x)\}$ should be finite.[9] It is not hard to check that $\sim$ is indeed an equivalence relation: since $\{x \in \mathbb{R}: f(x) \neq f(x)\}$

---

[9]Physicists should note that here, as in much of the rest of the mathematical literature, "finite" means "not infinite," and thus 0 is a finite number. Physicists often use "finite" to mean "nonzero." In this book, when we want to specify that a certain number is not zero and not negative, we will write that it is *strictly positive*.

is empty, the relation is reflexive; since the set $\{x \in \mathbb{R}: f(x) \neq g(x)\}$ equals the set $\{x \in \mathbb{R}: g(x) \neq f(x)\}$ the relation is symmetric; and since

$$\{x \in \mathbb{R}: f(x) \neq h(x)\} \subset \{x \in \mathbb{R}: f(x) \neq g(x)\} \cup \{x \in \mathbb{R}: g(x) \neq h(x)\}$$

and the union of two finite sets is finite, the relation is transitive. Each element of $S/\sim$ is an equivalence class of functions.

Equivalence classes become interesting and powerful when we consider which operations (such as evaluation, addition, or integration) *survive the equivalence*. Continuing with our example, consider the equivalence class $[\sin] \in S/\sim$. Each element f of $[\sin]$ is a function from $\mathbb{R}$ to $\mathbb{R}$. But while it makes sense to evaluate each f at 0, the value obtained will depend on the choice of $f \in c$. For instance, if we define $f: \mathbb{R} \to \mathbb{R}$ by

$$f(x) := \begin{cases} \sin x & x \neq 0 \\ -17 & x = 0 \end{cases}$$

then both f and sin belong to the equivalence class $[\sin]$ but $f(0) = -17 \neq 0 = sin(0)$. In other words, there is no natural way to evaluate $[\sin]$ at 0. So we say that evaluation *does not survive the equivalence*. On the other hand, we can make sense of addition in $S/\sim$. The sum of two equivalence classes will be an equivalence class. To justify addition in $S/\sim$ we must show that the equivalence class of the sum of two functions depends only on the equivalence classes from which the two functions are chosen. More explicitly, consider two equivalence classes b and c in $S/\sim$. Consider arbitrary $f_b, g_b \in b$ and f_c, g_c in c. If we can show that $(f_b + f_c) \sim (g_b + g_c)$, then we can legitimately define $b + c := [f_b + f_c]$. To this end, note that

$$\{x \in \mathbb{R}: f_b(x) + f_c(x) \neq g_b(x) + g_c(x)\}$$
$$\subset \{x \in \mathbb{R}: f_b(x) \neq g_b(x)\} \cup \{x \in \mathbb{R}: f_c(x) \neq g_c(x)\},$$

and since $f_b \sim g_b$ and $f_c \sim g_c$ the union is finite. So we can add equivalence classes. In other words, *addition survives the equivalence*. We leave it to the reader to show that we can multiply and integrate equivalence classes, and that addition and multiplication satisfy the usual algebraic rules. See Exercises 1.21 and 1.22.

A note on terminology: operations that survive the equivalence are sometimes called *well defined on equivalence classes*. A function on the original set S taking the same value on every element of an equivalence class is called an *invariant of the equivalence relation*. We will see an example of an invariant of an equivalence class in our introduction to tensor products in Section 2.6.

1.8   Exercises

Exercise 1.1 *Check that the expression*

$$\frac{\hbar c R_H}{(n+1)^2}$$

has units of energy.

Exercise 1.2 (Induction) *Show that for any n in the natural numbers*

$$\sum_{\ell=0}^{n-1}(2\ell + 1) = n^2.$$

Notice that the preceding exercise relates the dimensions $(2\ell+1)$ of the orbital types of the hydrogen atom to the lengths $(2n^2)$ of the rows of the periodic table.

Exercise 1.3 (Induction) *Show that for any nonnegative integer n and for any complex number λ such that $\lambda \neq 1$ we have*

$$\sum_{k=0}^{n}\lambda^{2k-n} = \frac{\lambda^{n+1} - \lambda^{-n-1}}{\lambda - \lambda^{-1}}.$$

Exercise 1.4 (Used in Proposition 4.8)   *For each nonnegative integer n, consider the function $f_n : [-1, 1] \to \mathbb{R}$ defined by*

$$f_n(x) := \Re\left(\left(x + i\sqrt{1 - x^2}\right)^n\right),$$

Show that for each n, the function f_n is a polynomial of degree n. Also show that for any n we have

$$f_n(x) = \Re\left(\left(x - i\sqrt{1 - x^2}\right)^n\right).$$

Exercise 1.5 (Geometry of multiplication in $\mathbb{C}$) *The complex plane can be considered as a two-dimensional real vector space, with basis $\{1, i\}$. Show that multiplication by any complex number c is a linear transformation. Find the matrix for multiplication by i in the given basis. Find the matrix for multiplication by $e^{i\theta}$, where θ is a real number. Find the matrix for multiplication by $a + ib$, where a and b are real numbers.*

Exercise 1.6 *Consider the function f from the complex plane $\mathbb{C}$ to the set of two-by-two real matrices defined by*

$$f(x+iy) := \begin{pmatrix} x & y \\ -y & x \end{pmatrix}.$$

Show that this function respects the asterisk notation, i.e., that for any $z \in \mathbb{C}$ we have $f(z^*) = f(z)^*$. Does this function respect complex addition and multiplication? I.e., is it true that $f(z_1+z_2) = f(z_1) + f(z_2)$ and $f(z_1 z_2) = f(z_1) f(z_2)$ for any $z_1, z_2 \in \mathbb{C}$? Find the determinant of $f(z)$.

Exercise 1.7 *Consider the function $f : \mathbb{R} \to \mathbb{C}$ defined by*

$$f(t) := \cos t + i \sin t.$$

Show that $f'(t) = if(t)$. (We remark that this makes Euler's formula $e^{it} = \cos t + i \sin t$ plausible.)

Exercise 1.8 *In this exercise you will calculate $\exp tM$, where t is any real number and*

$$M := \begin{pmatrix} 0 & \pi & 0 \\ -\pi & 0 & 0 \\ 0 & 0 & 0 \end{pmatrix},$$

in two different ways.

1. *Diagonalize the matrix M, i.e., find a diagonal matrix D and an invertible matrix N such that $M = NDN^{-1}$. Show that*

$$tM = N(tD)N^{-1}.$$

   Calculate $\exp(tD)$. Finally, use Proposition 1.4 to derive $\exp(tM)$ from $\exp(tD)$.

2. *Recall from calculus the Taylor series expansions for $\sin t$ and $\cos t$ around $t = 0$. Now calculate M^n for each nonnegative integer n. Using the definition of $\exp$ as an infinite sum, find an expression for $\exp tM$ in terms of $\sin t$ and $\cos t$.*

Finally, find $\exp M$.

Exercise 1.9 *Find a homogeneous function of degree $1/2$ on $\mathbb{R}^2$. Find a homogeneous function on $\mathbb{R}^3$ that is not continuous. Show that if a degree n polynomial is homogeneous of degree m, then $n = m$. Is every homogeneous function a polynomial?*

Exercise 1.10 *Consider* $\mathbb{R}^4 = \{(u, x, y, z)^T : u, x, y, z \in \mathbb{R}\}$. *Interpreting u as a color variable, with u = −1 corresponding to red and u = 1 corresponding to purple, with the interval* $[-1, 1]$ *corresponding to the spectrum of the rainbow[10], what is the three-sphere* S^3? *What is the hypercube?*

Exercise 1.11 *In this exercise you will derive the volume element for the three-sphere* S^3 *in* $\mathbb{R}^4$. *Define a function*

$$F : [0, \pi] \times [0, \pi] \times [0, 2\pi] \to \mathbb{R}^4$$

$$\begin{pmatrix} \psi \\ \theta \\ \phi \end{pmatrix} \mapsto \begin{pmatrix} \cos \psi \\ \sin \psi \sin \theta \cos \phi \\ \sin \psi \sin \theta \sin \phi \\ \sin \psi \cos \theta \end{pmatrix}.$$

Calculate the partial derivatives $\partial_\psi F$, $\partial_\theta F$ *and* $\partial_\phi F$. *Each of these is a vector in* $\mathbb{R}^4$. *Find the volume of the parallelepiped they span. Then show that the volume element on the three-sphere is* $\sin^2 \psi \sin \theta d\psi d\theta d\phi$.

Exercise 1.12 (Used in Section 1.6 and Proposition A.3)   *Show that in spherical coordinates we have*

$$\nabla^2 = \partial_r^2 + \frac{2}{r}\partial_r + \frac{1}{r^2}\partial_\theta^2 + \frac{\cos\theta}{r^2 \sin\theta}\partial_\theta + \frac{1}{r^2 \sin^2\theta}\partial_\phi^2.$$

(Hint: this is an exercise in careful, correct application of the chain rule for functions of several variables.)

Exercise 1.13 *Show that the total surface area of the two-sphere* S^2 *is* 4π. *Show that the total* surface volume *(i.e., the three-dimensional volume, not the four-dimensional volume) of the three-sphere* S^3 *is* $2\pi^2$.

Exercise 1.14 *Show that the multiplication of quaternions is* associative, *i.e., that for any* $q_1, q_2, q_3 \in \mathbf{Q}$ *we have*

$$(q_1 q_2)q_3 = q_1(q_2 q_3).$$

---

[10]This interpretation leads to a cute proof that any loop in $\mathbb{R}^4$ can be unknotted. Suppose someone hands you a loop in $\mathbb{R}^4$, even a very knotted-up one. Interpreting the fourth dimension as color, you have a string in three dimensions whose color varies continuously. It is legitimate to pass one part of the string through another, as long as the two pieces are different colors. But you can change the color of any segment continuously, so you can undo the three-dimensional knot by passing any troublesome strands through each other!

Exercise 1.15 (Used in Section 4.1) *Show that the product of two unit qua-ternions is a unit quaternion. (Hint: Brute calculation will suffice, but the geometry of $\mathbb{R}^4$ may provide more insight: think of the right-hand quaternion in the multiplication as a unit vector in $\mathbb{R}^4$, think of the left-hand quaternion as a linear transformation of $\mathbb{R}^4$.)*

Exercise 1.16 *Find a list[11] of the algebraic axioms for $\mathbb{R}$. For each axiom, either prove the corresponding statement for the quaternions $\mathbf{Q}$ or find a counterexample in $\mathbf{Q}$.*

Exercise 1.17 *Suppose $\lambda_1, \lambda_2, \ldots$ are distinct eigenvalues of an energy operator. Suppose that $\phi_1, \phi_2, \ldots$ are the associated eigenvectors. Consider the state corresponding to the wave function*

$$c_1\phi_1 + c_3\phi_3,$$

where c_1 and c_3 are complex numbers satisfying $|c_1|^2 + |c_3|^2 = 1$. Now imagine measuring the energy of such a state. Is it possible to obtain the value λ_2?

Exercise 1.18 *Draw the zero points (on the sphere) of the real and imaginary parts of the spherical harmonics of degrees 0, 1 and 2.*

Exercise 1.19 *Define an* arrow *in $\mathbb{R}^3$ to be an ordered pair (p_1, p_2), where p_1 and p_2 are each a triple of real numbers. (Think of p_1 as the initial point and p_2 as the endpoint.) Define a relation on the set of arrows by $(p_1, p_2) \sim (q_1, q_2)$ if and only if $p_2 - p_1 = q_2 - q_1$. Show that this is an equivalence relation. Now think of each arrow as a point in $\mathbb{R}^6$. Does the usual addition in $\mathbb{R}^6$ survive the equivalence relation? If so, is the resulting addition on equivalence classes of arrows the same as the addition of 3-vectors you learned in linear algebra? What about scalar multiplication in $\mathbb{R}^6$? Find an injective and surjective linear function from $\mathbb{R}^6/\sim$ to $\mathbb{R}^3$. (Hint: it will help to introduce some notation for "$(r_1, r_2, r_3, r_4, r_5, r_6) \in \mathbb{R}^6$ lies in the equivalence class corresponding to $(s_1, s_2, s_3) \in \mathbb{R}^3$.")*

Exercise 1.20 *Fix an interval $[a, b]$ in $\mathbb{R}$. Consider the set S of differentiable functions on $[a, b]$. Define a relation $\sim$ on S by*

$$f \sim g \text{ if and only if } f' = g'.$$

---

[11] One is in Rudin [Ru76, Def. 1.12]).

Show that ~ satisfies the criteria of Definition 1.3. Show that f ~ g if and only if f − g is a constant function. Show that addition, scalar multiplication and differentiation are well defined on equivalence classes. Show that evaluation is not *well defined: given a point c ∈ [a, b], find two functions in S that are equivalent but take different values at c. On the other hand, differences of evaluations are well defined: show that f(b) − f(a) is well defined on equivalence classes.*

Exercise 1.21 *Show that multiplication of equivalence classes of functions (as defined in Section 1.7) is well defined. Show that addition and multiplication of equivalence classes of functions satisfy some but not all the standard field axioms (such as the distributive law, existence of 0, etc.). The list of field axioms is available in many texts, including [Ru76, Definition 1.12]. Which axioms hold, and which fail?*

Exercise 1.22 *Consider an equivalence class c of functions as defined in Section 1.7. Show that if any one element of c is Riemann integrable on an interval [a, b] ⊂ ℝ, then every element of c is Riemann integrable on [a, b]. Show that the value of the definite integral does not depend on the choice of function in the equivalence class. Hence the real number $\int_a^b c$ is well defined.*

Exercise 1.23 (Used in Section 10.1) *Let C[−1, 1] denote the set of continuous, complex-valued functions on the interval [−1, 1]. Let 0 denote the zero function on [−1, 1]. Define a relation ~ on C[−1, 1] \ {0} by*

$$f \sim g \text{ if and only if } \exists c \in \mathbb{C} \text{ such that } f = cg.$$

Show that ~ is an equivalence relation.

Does addition of functions survive the equivalence? Does scalar multiplication (by complex numbers) survive the equivalence? Does multiplication of two functions survive the equivalence?

Exercise 1.24 *Find another example of a meaningful equivalence relation from your own experience. Define the relation rigorously and prove that it is an equivalence relation. Which relevant operations survive for equivalence classes?*

Exercise 1.25 (Useful in Chapter 4.2) *Suppose R is a 3 × 3 matrix with real entries. Show that the following three conditions are equivalent:*

1. $R^T R = I$;

2. $(Rx) \cdot (Ry) = x \cdot y$ *for all x, y in* $\mathbb{R}^3$;

3. $\|Rx\| = \|x\|$ *for all* $x \in \mathbb{R}^3$.

2
Linear Algebra over the Complex Numbers

Charles Wallace accepted the explanation serenely. Even Calvin did not seem perturbed. "Oh, *dear*," Meg sighed. "I guess I *am* a moron. I just don't get it."

"That is because you think of space only in three dimensions," Mrs. Whatsit told her. "We travel in the fifth dimension. This is something you can understand, Meg. Don't be afraid to try."

—M. L'Engle, *A Wrinkle in Time* [L'E, p. 76]

In this chapter we introduce complex linear algebra, that is, linear algebra where complex numbers are the scalars for scalar multiplication. This may feel like review, even to readers whose experience is limited to real linear algebra. Indeed, most of the theorems of linear algebra remain true if we replace $\mathbb{R}$ by $\mathbb{C}$: because the axioms for a real vector space involve only addition and multiplication of real numbers, the definition and basic theorems can be easily adapted to any set of scalars where addition and multiplication are defined and reasonably well behaved,[1] and the complex numbers certainly fit the bill. However, the examples are different. Furthermore, there are theorems (such as Proposition 2.11) in complex linear algebra whose analogues over the reals are false. We will recount but not belabor old theorems, concentrating on new ideas and examples. The reader may find proofs in any number of

---

[1]More generally, any field can be used as the scalars for vector spaces. A vector space is an example of an even more general concept, namely, a module over a ring. Details can be found in many abstract algebra textbooks, e.g., Artin [Ar].

linear algebra texts. For detailed proofs we recommend the book by Shifrin and Adams [SA]; for a sophisticated perspective we recommend the one by Lax [La].

2.1   Complex Vector Spaces

In this section we define and discuss complex vector spaces. We give many examples, especially of vector spaces of functions. Such vector spaces do not usually figure prominently in introductory courses on linear algebra, but the vector nature of functions is crucial in many areas of math and physics.

Definition 2.1 *Consider a set V, together with an addition operation*

$$V \times V \to V$$

(denoted by $+$) and a scalar multiplication

$$\mathbb{C} \times V \to V$$

(denoted by juxtaposition). Such a set V is a complex vector space *if and only if the addition and scalar multiplication satisfy the usual algebraic properties of (e.g., real) vector spaces, such as associativity and distribution. Specifically, for all $u, v, w \in V$ and all $b, c \in \mathbb{C}$, we have*

1. *Commutativity: $u + v = v + u$.*

2. *Associativity: $(u + v) + w = u + (v + w)$.*

3. *Zero vector: there is a vector $0 \in V$ such that $0 + v = v$ for every $v \in V$.*

4. *Distributivity: $c(u + v) = cu + cv$.*

5. *Respect of field operations: $b(cu) = (bc)u$, $(b+c)u = bu+cu$, $1u = u$ and $0u = 0$, where the zero on the left-hand side is $0 \in \mathbb{C}$ and the zero on the right-hand side is $0 \in V$.*

Note in particular that because the definition specifies that the range of each of the operations is V, the space V must be *closed under addition and scalar multiplication*. That is, the sum of two elements of V must be itself an element of V; likewise, the multiple of an element of V by a complex number must be an element of V. In many examples of vector spaces the addition and

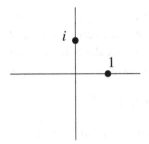

Figure 2.1. Is $\mathbb{C}$ a line or plane?

scalar multiplication naturally satisfy the usual algebraic properties. Often verification that a given set is a vector space boils down to checking that the set is closed under addition and scalar multiplication.

For example, the real line $\mathbb{R}$ is not a complex vector space under the usual multiplication of real numbers by complex numbers. It is possible for the product of a complex number and a real number to be outside the set of real numbers: for instance, $(i)(3) = 3i \notin \mathbb{R}$. So the real line $\mathbb{R}$ is not closed under complex scalar multiplication.

The *trivial complex vector space* has one element, the zero vector 0. Addition is defined by $0 + 0 := 0$; for any complex number c, define the scalar multiple of 0 by c to be 0. Then all the criteria of Definition 2.1 are trivially true. For example, to check distributivity, note that for any $c \in \mathbb{C}$ we have

$$c(0 + 0) = c(0) = 0 = 0 + 0 = c(0) + c(0).$$

The simplest nontrivial example of a complex vector space is $\mathbb{C}$ itself. Adding two complex numbers yields a complex number; multiplication of a vector by a scalar in this case is just complex multiplication, which yields a complex number (i.e., a vector in $\mathbb{C}$). Mathematicians sometimes call this complex vector space the *complex line*. One may also consider $\mathbb{C}$ as a real vector space and call it the complex plane. See Figure 2.1.

For every natural number n we can define a complex vector space

$$\mathbb{C}^n := \{(c_1, \ldots, c_n) \colon c_1, c_2, \ldots, c_n \in \mathbb{C}\}.$$

Addition and scalar multiplication are defined component by component: $(b_1, \ldots, b_n) + (c_1, \ldots, c_n) := (b_1 + c_1, \ldots, b_n + c_n)$ and, for any complex scalar s, we have $s(c_1, \ldots, c_n) := (sc_1, \ldots, sc_n)$. This space can be called "$\mathbb{C}$-to-the-n" or "complex n-space."

Readers familiar with quantum mechanics may recognize that complex vector spaces are often important. For example, in the study of spin-1/2 particles the complex vector space $\mathbb{C}^2 := \{(c_1, c_2) : c_1, c_2 \in \mathbb{C}\}$ plays an important role. For instance, if physicists are considering a Stern–Gerlach machine oriented along the z-axis they would describe an arbitrary spin state of an electron by an expression of the form

$$c_+ |{+}\mathbf{z}\rangle + c_- |{-}\mathbf{z}\rangle \,,$$

where c_+ and c_- are complex numbers satisfying $|c_+|^2 + |c_-|^2 = 1$ and $|{+}\mathbf{z}\rangle$ and $|{-}\mathbf{z}\rangle$ are two convenient states. Any object of the form $|\cdot\rangle$ is called a *ket*; the information between the vertical line and the angle bracket usually helps the reader identify which state the ket is meant to denote. In the theory of quantum computing, one often finds the vector space $\mathbb{C}^2$ used to describe the state of a *qubit*. A typical expression is

$$c_0 |0\rangle + c_1 |1\rangle \,,$$

where c_0 and c_1 are complex numbers satisfying $|c_0|^2 + |c_1|^2 = 1$. It is no accident that this expression matches the previous one: a qubit is just a spin-1/2 particle. See Chapter 10.

In another physics application, the Dirac equation for states of an electron in relativistic space-time requires wave functions taking values in the complex vector space $\mathbb{C}^4 := \{(c_1, c_2, c_e, c_4) : c_1, c_2, c_3, c_4 \in \mathbb{C}\}$. These wave functions are called *Dirac spinors*.

Various vector spaces of complex-valued polynomials will arise in our analysis of the hydrogen atom and the periodic table. Consider first the set of polynomials functions from $\mathbb{C}$ to $\mathbb{C}$. One element of this set is $x \mapsto e^{i\pi/5}x^2 + 1$, and every element of the set is of the form

$$\begin{array}{c} \mathbb{C} \to \mathbb{C} \\ x \mapsto c_n x^n + c_{n-1}x^{n-1} + \cdots + c_1 x + c_0 \end{array} , \tag{2.1}$$

where the nonnegative number n is called the *degree* and the complex numbers $c_0, \ldots, c_n$, with $c_n \neq 0$ are called the *coefficients*. This set is closed under addition and complex scalar multiplication: the sum of two polynomials with complex coefficients is itself a polynomial with complex coefficients; likewise the product of a complex number and a polynomial with complex coefficients is again a polynomial with complex coefficients.

If we consider polynomials with real coefficients, we get a real vector space that is not a complex vector space: it is not closed under multiplication by complex scalars. For instance, the polynomial $x \mapsto x$ is a polynomial with real coefficients: we have $c_1 = 1$ and $c_0 = 0$ in Formula 2.1. But if we multiply by i, we get the polynomial $x \mapsto ix$, with $c_1 = i \notin \mathbb{R}$. So the set of polynomials with real coefficients and the natural scalar multiplication is not a complex vector space. We leave it to the reader to show that it is a real vector space.

Notice that we did not use the domain of the polynomial functions in our arguments in the previous paragraphs. A mathematician's natural reaction to such an observation is to think about a way to define the object in question without mentioning the unused information. These vector spaces (along with the natural multiplication of polynomials by polynomials) are studied in abstract algebra under the name of *polynomial rings*. Interested readers might consult Artin's book [Ar, Chapters 10-11] for more details and related ideas.

If a subset W of a vector space V satisfies the definition of a vector space, with addition and scalar multiplication defined by the same operation as in V, then W is called a *vector subspace* or, more succinctly, a *subspace* of V. For example, the *trivial subspace* $\{0\}$ is a subspace of any vector space.

A more interesting example involves the vector space $\mathcal{P}_3$ of complex-coefficient polynomials in three variables. Let $\mathbb{H}$ denote the subset of $\mathcal{P}_3$ containing only *harmonic* polynomials, i.e., only polynomials p in three variables satisfying $\nabla^2 p = 0$. Then $\mathbb{H}$ is a subspace of the vector space $\mathcal{P}_3$. To justify this claim, it suffices to check that $\mathbb{H}$ is closed under addition and scalar multiplication. But if $\nabla^2 p_1 = 0$ and $\nabla^2 p_2 = 0$, then $\nabla^2(p_1 + p_2) = 0$; and if $c \in \mathbb{C}$, then $\nabla^2(cp_1) = c\nabla^2 p_1 = 0$. So $\mathbb{H}$ is a subspace.

Another example we will find useful is the complex vector space

$$C[-1, 1] := \{\text{continuous complex-valued functions on } [-1, 1]\}.$$

Because the sum of two continuous functions is continuous, and any scalar multiple of a continuous function is continuous, $C[-1, 1]$ is indeed a vector space.

Readers who are still uncomfortable with thinking of functions as vectors should take the time to review this section carefully and do some exercises. These vector spaces are fundamental to our analysis of the hydrogen atom. In particular, we will look at the function space containing the wave functions for the hydrogen atom, and we will work with various subspaces of that function space.

2.2 Dimension

Now that we have defined complex vector spaces, we can introduce *dimension*.

First we recall the notion of a *finite basis* of a vector space.

Definition 2.2 *Let V be a complex vector space. Let B be a finite subset of V. Then B is a* finite basis *of V if and only if every vector v in V can be written as a linear combination of vectors in B (i.e., B* spans *V) and for each v the linear combination is unique (i.e., B is a* linearly independent *set).*

Definition 2.3 *A complex vector space is* finite-dimensional *if it has a finite basis. Any complex vector space that is not finite-dimensional is* infinite-dimensional.

For the remainder of this section we concentrate on finite-dimensional spaces. In this section, and whenever we are clearly discussing a finite-dimensional space, we may use the word "basis" to refer to a finite basis.

Suppose that V is a finite-dimensional complex vector space. By the definition this means that V has a finite basis. It turns out that all the different bases of V must be the same size. This is geometrically plausible for real Euclidean vector spaces, where one can visualize a basis of size one determining a line, a basis of size two determining a plane, and so on. The same is true for complex vector spaces. A key part of the proof, useful in its own right, is the following fact.

Proposition 2.1 *Suppose V is a finite-dimensional vector space with basis $\{v_1, \ldots, v_n\}$. Suppose $\{u_1, \ldots, u_m\}$ is a linearly independent subset of V. Then $m \leq n$.*

An easy corollary is:

Proposition 2.2 *Let V be a finite-dimensional complex vector space. Suppose $\{v_1, \ldots, v_n\}$ and $\{u_1, \ldots, u_m\}$ are both bases of V. Then $n = m$.*

This proposition makes the following definition possible and powerful.

Definition 2.4 *Let V be a finite-dimensional complex vector space. Suppose that $\{v_1, \ldots, v_n\}$ is a finite basis of V. Then the* dimension *of V is n.*

Proposition 2.2 ensures that the dimension of V is the same no matter which basis we use to calculate it.

Readers familiar with spin systems may recall that the study of spin yields a physical example of different bases for the same complex vector space. For instance, to study an electron, or any other particle of spin-1/2, one uses a basis of two kets. Which kets one chooses depends on the orientation of the Stern–Gerlach machine (real or imagined). One might use $|+\mathbf{z}\rangle$ and $|-\mathbf{z}\rangle$ as a basis for one calculation and $|+\mathbf{x}\rangle$ and $|-\mathbf{x}\rangle$ for another. No matter what

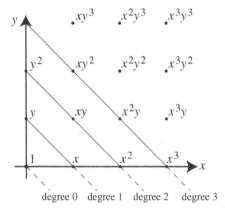

Figure 2.2. A picture of the bases of homogeneous polynomials of degree ℓ in two variables for $\ell = 0, 1, 2$ and 3.

orientation is chosen, there are always two basis kets for a spin-1/2 particle. Similarly, a spin-1 particle requires three kets in each basis. In general, the study of a spin-s particle requires a complex vector space of dimension $2s+1$.

Let us calculate, for future reference, the dimension of the complex vector space of homogeneous polynomials (with complex coefficients) of degree n on various Euclidean spaces. Homogeneous polynomials of degree n on the real line $\mathbb{R}$ are particularly simple. This complex vector space is one-dimensional for each n. In fact, every element has the form cx^n for some $c \in \mathbb{C}$. In other words, the one-element set $\{x^n\}$ is a finite basis for the homogeneous polynomials of degree n on the real line.

Homogeneous polynomials (with complex coefficients) of degree n on $\mathbb{R}^2$ (or on $\mathbb{C}^2$) form a complex vector space of dimension $n+1$. We call this space $\mathcal{P}^n$. If we call our variables x and y, then there is a finite basis of $\mathcal{P}^n$ of the form $\{x^n, x^{n-1}y, x^{n-2}y^2, \dots, xy^{n-1}, y^n\}$. Because this basis has $n + 1$ elements, the dimension of the complex vector space is $n + 1$. We can represent this basis geometrically by noting that each basis element corresponds to a way of writing n as the sum of two nonnegative integers; this implies that the size of the basis is the number of integer lattice points on the line $x + y = n$ in the first quadrant of $\mathbb{R}^2$. (An integer lattice point is a point whose coefficients are all integers.) See Figure 2.2.

Likewise, one can obtain the dimension of the vector space $\mathcal{P}^\ell_3$ of homogeneous polynomials of degree ℓ in three variables by counting the number of lattice points on the plane $x+y+z = \ell$ in the first octant of $\mathbb{R}^3$. See Figure 2.3. This geometric picture makes it clear that the answer is a triangle number;

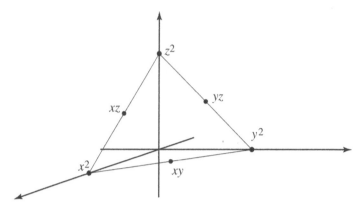

Figure 2.3. A picture of the basis of homogeneous polynomials of degree two in three variables.

careful accounting shows that the number is precisely $(\ell + 1)(\ell + 2)/2$. We will use this triangular picture in Section 7.1. See especially Figure 7.1.

In the end, dimension is important physically because we can associate a certain complex vector space to each orbital type, and the dimension of the complex vector space tells us how many different states can fit in each orbital of that type. Roughly speaking, this insight, along with the Pauli exclusion principle, determines the number of electrons that fit simultaneously into each shell. These numbers determine the structure of the periodic table.

2.3   Linear Transformations

The notion of a linear transformation is crucial. A function from a (complex) vector space to a (complex) vector space is a *(complex) linear transformation* if it preserves addition and (complex) scalar multiplication. Here is a more explicit definition.

Definition 2.5 *Let V and W be complex vector spaces, and let T be a function from V into W. Then T is a* complex linear transformation *if and only if, for every v_1 and v_2 in V and every $c \in \mathbb{C}$ we have*

$$T(v_1 + v_2) = T(v_1) + T(v_2)$$
$$\text{and } T(cv_1) = cT(v_1).$$

The vector space V is called the domain *of T. The vector space W is called the* target space *of T.*

Note that complex conjugation is *not* a complex linear transformation. While it satisfies the additive condition, it does not preserve complex scalar multiplication: if we let $T : \mathbb{C} \to \mathbb{C}$ denote complex conjugation and take $c = i$ and $v_1 = 1$ we find $T(cv_1) = T(i) = -i \neq i = iT(1) = cT(v_1)$.

Any $m \times n$ matrix (m rows and n columns) with complex entries determines a complex linear transformation from $V := \mathbb{C}^n$ to $W := \mathbb{C}^m$ (with the standard choice of basis). For example, the 1×2 matrix $\begin{pmatrix} i & i \end{pmatrix}$ is the linear transformation taking an element of $\mathbb{C}^2$ to i times the sum of the two entries: for any $(c_1, c_2)^T \in \mathbb{C}^2$ we have

$$\begin{pmatrix} i & i \end{pmatrix} \begin{pmatrix} c_1 \\ c_2 \end{pmatrix} = i(c_1 + c_2).$$

We leave it to the reader to check that this example satisfies the definition of a complex linear transformation.

What about the converse: does any linear transformation determine a matrix? This question raises two issues. First, if the domain is infinite-dimensional, the question is more complicated. Mathematicians usually reserve the word "matrix" for a finite-dimensional matrix (i.e., an array with a finite number of rows and columns). Physicists often use "matrix" to denote a linear transformation between infinite-dimensional spaces, where mathematicians would usually prefer to say "linear transformation." Second, even in finite-dimensional spaces, one must specify bases in domain and target space to determine the entries in a matrix. We discuss this issue in more detail in Section 2.5 for the special case of linear operators.

Readers already familiar with quantum mechanics may have seen many examples of complex linear transformations. For example, in the study of spin-1/2 systems, it is convenient to define *projection operators* by

$$\Pi_+ \left(c_+ \left| +\mathbf{z} \right\rangle + c_- \left| -\mathbf{z} \right\rangle \right) := c_+ \left| +\mathbf{z} \right\rangle$$
$$\Pi_- \left(c_+ \left| +\mathbf{z} \right\rangle + c_- \left| -\mathbf{z} \right\rangle \right) := c_- \left| -\mathbf{z} \right\rangle .$$

Both functions Π_+ and Π_- are linear transformations from $\mathbb{C}^2$ to $\mathbb{C}^2$. Notice that $\Pi_+ + \Pi_-$ is the identity transformation from $\mathbb{C}^2$ to $\mathbb{C}^2$. The typical physics notation for Π_+ is $\left| +\mathbf{z} \right\rangle \left\langle +\mathbf{z} \right|$, and a typical calculation looks like this:

$$\left| +\mathbf{z} \right\rangle \left\langle +\mathbf{z} \right| \left(c_+ \left| +\mathbf{z} \right\rangle + c_- \left| -\mathbf{z} \right\rangle \right)$$
$$= c_+ \left| +\mathbf{z} \right\rangle \left\langle +\mathbf{z} \right| + \mathbf{z} \rangle + c_- \left| +\mathbf{z} \right\rangle \left\langle +\mathbf{z} \right| - \mathbf{z} \rangle = c_+ \left| +\mathbf{z} \right\rangle ,$$

since $\left\langle +\mathbf{z} \right| + \mathbf{z} \rangle = 1$ and $\left\langle +\mathbf{z} \right| - \mathbf{z} \rangle = 0$.

To define a linear transformation, it suffices to define it on a basis.

Proposition 2.3 *Suppose V is a finite-dimensional complex vector space and* $\{v_1, \ldots, v_n\}$ *is a basis of V. Suppose W is a complex vector space. Suppose* $f: \{v_1, \ldots, v_n\} \to W$ *is a function. Then there is a unique linear transformation* $T: V \to W$ *such that for any* $k = 1, \ldots, n$ *we have*

$$T(v_k) = f(v_k).$$

There is an example of this type of definition at the end of the section.

Proof. First we will define T. Let v be an arbitrary element of V. Then there is a unique n-tuple of complex numbers $c_1, \ldots, c_n$ such that $v = c_1 v_1 + \cdots + c_n v_n$. Set

$$T(v) := c_1 f(v_1) + \cdots + c_n f(v_n).$$

Next we must check that T is linear. Suppose $b \in \mathbb{C}$ and $v \in V$, and suppose that $v = c_1 v_1 + \cdots + c_n v_n$ as above. Then $bv = (bc_1)v_1 + \cdots + (bc_n)v_n$, and this expansion is unique. Hence we have

$$T(bv) = (bc_1)f(v_1) + \cdots + (bc_n)f(v_n) = bT(v).$$

The proof of the additive property of linear transformations is similar.

It is easy to see that for each $k = 1, \ldots, n$ we have $T(v_k) = f(v_k)$; the only remaining task is to show that T is unique. Suppose $\tilde{T}$ is another linear transformation satisfying all of the criteria. Then, for any $v \in V$ we have $v = c_1 v_1 + \cdots + c_n v_n$ and

$$\tilde{T}(v) = c_1 \tilde{T}(v_1) + \cdots + c_n \tilde{T}(v_n) = c_1 T(v_1) + \cdots + c_n T(v_n) = T(v).$$

So $\tilde{T} = T$. Hence T is unique. $\qquad\square$

Similarly, one can define a linear transformation by defining it on a spanning set, but in this case one must check consistency conditions.

Proposition 2.4 *Suppose V is a finite-dimensional complex vector space and S is a subset of V that spans V. Suppose W is a complex vector space. Suppose* $f: S \to W$ *is a function. Then there is a unique linear transformation* $T: V \to W$ *such that for any* $s \in S$ *we have*

$$T(s) = f(s)$$

if and only if, for any complex numbers $c_1, \ldots, c_k$ *and any* $s_1, \ldots, s_k \in S$ *such that* $c_1 s_1 + \cdots c_k s_k = 0$, *we have*

$$c_1 f(s_1) + \cdots + c_k f(s_k) = 0.$$

This equation is called a consistency condition.

Proof. First we will define T. Let v be an arbitrary element of V. Because S spans V and V is finite-dimensional, v can be written as a linear combination of elements of S: there are complex numbers $c_1, \ldots, c_n$ and elements $s_1, \ldots, s_n$ of S such that $v = c_1 s_1 + \cdots + c_n s_n$. Then we set

$$T(v) := c_1 f(s_1) + \cdots + c_n f(s_n).$$

Note that the expression of v as a linear combination of elements of S is not unique. To show that T is well defined, we must check that any other expression for v would yield the same value for $T(v)$. Suppose that there are complex numbers $\tilde{c}_1, \ldots, \tilde{c}_m$ and elements $\tilde{s}_1, \ldots, \tilde{s}_m$ such that $v = \tilde{c}_1 \tilde{s}_1 + \cdots + \tilde{c}_m \tilde{s}_m$. Then

$$0 = (c_1 s_1 + \cdots + c_n s_n) - (\tilde{c}_1 \tilde{s}_1 + \cdots + \tilde{c}_m \tilde{s}_m).$$

So by the consistency condition we have

$$0 = (c_1 f(s_1) + \cdots + c_n f(s_n)) - \left(\tilde{c}_1 \tilde{f}(s_1) + \cdots + \tilde{c}_m \tilde{f}(s_m) \right).$$

So

$$c_1 f(s_1) + \cdots + c_n f(s_n) = \tilde{c}_1 \tilde{f}(s_1) + \cdots + \tilde{c}_m \tilde{f}(s_m),$$

and hence $T(v)$ is well defined.

Now because V is finite-dimensional there must be a subset of S that is a basis for V. We can now apply Proposition 2.3 to conclude that T is linear and uniquely defined. Finally, it is easy to see that $T(s) = f(s)$ for any $s \in S$. $\square$

Not only are linear transformations necessary for the very definition of a representation in Chapter 6, but they are useful in calculating dimensions of vector spaces — see Proposition 2.5. Linear transformations are at the heart of homomorphisms of representations and many other constructions. We will often appeal to the propositions in this section as we construct linear transformations. For example, we will use Proposition 2.4 in Section 5.3 to define the tensor product of representations.

2.4   Kernels and Images of Linear Transformations

A linear transformation T from a space V to a space W determines a certain subspace (the *kernel*) of V and a certain subspace (the *image*) of W. In this section we define kernel and image and use them to introduce several important concepts. The first is the Fundamental Theorem of Linear Algebra, an important tool for counting dimensions. We apply it to the Laplacian to calculate dimensions of spaces of homogeneous harmonic polynomials. Finally, we introduce isomorphisms of vector spaces.

The reader may recall that the *kernel* of a linear transformation $T : V \to W$ is the set of vectors *annihilated* by T, that is, the set of vectors $v \in V$ such that $Tv = 0$. The *image* is the set of vectors $w \in W$ such that there exists $v \in V$ with $Tv = w$. It is easy to check that the kernel of T is a subspace of V and the image of T is a subspace of W.

We will need the Fundamental Theorem of Linear Algebra in Section 7.1.

Proposition 2.5 (Fundamental Theorem of Linear Algebra) *For any linear transformation T with finite-dimensional domain V we have*

$$\dim V = \dim(\ker T) + \dim(\text{Image } T). \tag{2.2}$$

The dimension of the image of T is often called the *rank of T*. This theorem is also known as the *rank-nullity theorem*.

Several examples of linear transformations can be mined from the *Laplacian* ∇^2, the sum of the second partial derivatives with respect to each coordinate. For example, the three-dimensional Laplacian (in Cartesian coordinates) is the partial differential operator $\nabla^2 = \partial_x^2 + \partial_y^2 + \partial_z^2$. Note that the kernel of the Laplacian in the space of polynomials in three variables is precisely the vector space $\mathbb{H}$ of harmonic polynomials. Although we speak informally of "the" three-dimensional Laplacian, as if there is only one, we can construct many different linear transformations from one formula by considering different domains. According to Definition 2.5, in order to specify a linear transformation, one must specify its domain V and its target space W. Changing the target space W makes no more than a cosmetic change to the linear transformation, but restricting or enlarging the domain can affect the dimensions of the kernel and image. For example, consider

$$\mathcal{P}_3^0 := \{\text{constant complex-valued functions on } \mathbb{R}^3\}.$$

(We will reserve the plainer symbol $\mathcal{P}^n$ for homogeneous polynomials of degree n in only two variables, a star player in our drama.) The set $\mathcal{P}_3^0$ is a complex vector space of dimension 1: the set containing only the function $f : \mathbb{R}^3 \to \mathbb{C}, (x, y, z) \mapsto 1$ is a basis. Let T denote the restriction of the Laplacian to $\mathcal{P}_3^0$. Because all derivatives of constant functions are zero, the kernel of this T is all of $\mathcal{P}_3^0$, while its image is the zero-dimensional complex vector space $\{0\}$. We can verify the Fundamental Theorem of Linear Algebra (Proposition 2.5) in this example:

$$\dim \mathcal{P}_3^0 = 1 = 1 + 0 = \dim(\text{kernel} T) + \dim(\text{image} T).$$

Restricting the Laplacian to the set $\mathcal{P}_3^2$ of homogeneous quadratic polynomials of degree two on $\mathbb{R}^3$ yields another example. Here the domain $\mathcal{P}_3^2$ is a complex vector space of dimension 6 with basis $\{x^2, xy, xz, y^2, yz, z^2\}$. Every homogeneous quadratic polynomial can be written in the form $c_1 x^2 + c_2 xy + c_3 xz + c_4 y^2 + c_5 yz + c_6 z^2$, where $c_1, \ldots, c_6$ are complex numbers. Applying the Laplacian to this expression we find

$$\nabla^2(c_1 x^2 + c_2 xy + c_3 xz + c_4 y^2 + c_5 yz + c_6 z^2) = 2c_1 + 2c_4 + 2c_6.$$

So we can take $\mathcal{P}_3^0$ as the target space. The image of the linear transformation is all of $\mathcal{P}_3^0$, since we can get any constant function by setting $c_4 = c_6 = 0$ and setting c_1 to half the desired value. We can use the calculation above to find the kernel as well: it is the set $\{c_1 x^2 + c_2 xy + c_3 xz + c_4 y^2 + c_5 yz + c_6 z^2 : c_1 + c_4 + c_6 = 0\}$. One can check that a basis of the kernel is $\{xy, xz, yz, x^2 - y^2, 2z^2 - x^2 - y^2\}$. So the kernel is five-dimensional and we can check Proposition 2.5 in this case:

$$\dim \mathcal{P}_3^2 = 6 = 5 + 1 = \dim(\text{kernel}) + \dim(\text{image}).$$

Recall from Section 1.5 that any function in the kernel of the Laplacian (on any space of functions) is called a *harmonic* function. In other words, a function f is harmonic if $\nabla^2 f = 0$. The harmonic functions in the example just above are the *harmonic homogeneous polynomials of degree two*. We call this vector space $\mathbb{H}^2$. In Exercise 2.23 we invite the reader to check that the following set is a basis of $\mathbb{H}^2$:

$$\{(x + iy)^2, (x + iy)z, (x + iy)(x - iy), (x - iy)z, (x - iy)^2\}.$$

Restrictions of homogeneous harmonic polynomials play an important role in our analysis.

Definition 2.6 *Suppose ℓ is a nonnegative integer. Define the vector space of* homogeneous harmonic polynomials of degree ℓ in three variables

$$\mathbb{H}^\ell := \{p \in \mathcal{P}_3^\ell : \nabla^2 p = 0\}$$

and the vector space of restrictions of these to the sphere by

$$\mathcal{Y}^\ell := \{p|_{S^2} : p \in \mathbb{H}^\ell\}.$$

Finally, we define

$$\mathcal{Y} := \bigcup_{\ell=0}^{\infty} \mathcal{Y}^{\ell}.$$

In other words, a function $p : S^2 \to \mathbb{C}$ is an element of the vector space $\mathcal{Y}^{\ell}$ if and only if there is a polynomial $q \in \mathbb{H}^{\ell}$ such that $p(s) = q(s)$ for every point $s \in S^2 \subset \mathbb{R}^3$. The elements of $\mathcal{Y}^{\ell}$ are precisely the *spherical harmonics of order* ℓ, as we show in Appendix A.

In Section 7.1 we will use this characterization of homogeneous harmonic polynomials as a kernel of a linear transformation (along with the Fundamental Theorem of Linear Algebra, Proposition 2.5) to calculate the dimensions of the spaces of the spherical harmonics.

Isomorphisms are particularly important linear transformations because they tell us that domain and range are the same as far as vector space operations are concerned.

Definition 2.7 *Suppose V and W are vector spaces and $T : V \to W$ is a linear transformation. If T is invertible and $T^{-1} : W \to V$ is a linear transformation, then we say that T is an* isomorphism of vector spaces *(or isomorphism for short) and that V and W are* isomorphic *vector spaces.*

In practice, there is an easier criterion to check in situations where we do not need to calculate the inverse explicitly.

Proposition 2.6 *Suppose V and W are vector spaces. A linear transformation $T : V \to W$ is an isomorphism of vector spaces if and only if it is injective and surjective (i.e., if the kernel of T is the trivial vector space $\{0\}$ and the range of T is all of W).*

Proof. Suppose T is an isomorphism of vector spaces. Then the inverse function T^{-1} exists, so T must be injective. Moreover, the function T^{-1} has domain W, so the image of T must be W as well, i.e., T is surjective. On the other hand, suppose that T is injective and surjective. Then T^{-1} has domain W and image V. We must show that T^{-1} is a linear transformation. Let w_1 and w_2 be arbitrary elements of W. Since T is surjective, there are elements v_1 and v_2 of V such that $w_1 = T(v_1)$ and $w_2 = T(v_2)$. Then

$$T^{-1}(w_1 + w_2) = T^{-1}(T(v_1) + T(v_2)) = T^{-1} \circ T(v_1 + v_2)$$
$$= v_1 + v_2 = T^{-1}(w_1) + T^{-1}(w_2).$$

So T^{-1} satisfies the additive criterion of Definition 2.5. If $c \in \mathbb{C}$, we have

$$T^{-1}(cw_1) = T^{-1} \circ T(cv_1) = cv_1 = cT^{-1}(w_1),$$

so T^{-1} satisfies the scalar multiplication criterion of Definition 2.5.   □

For example, there is an isomorphism from $\mathbb{C}^3$ to the vector space $\mathcal{P}_3^1$ of homogeneous polynomials of degree one in three variables (x, y, z), given by

$$\begin{pmatrix} 1 \\ 0 \\ 0 \end{pmatrix} \mapsto x$$

$$\begin{pmatrix} 0 \\ 1 \\ 0 \end{pmatrix} \mapsto y$$

$$\begin{pmatrix} 0 \\ 0 \\ 1 \end{pmatrix} \mapsto z.$$

By Proposition 2.3, these formulas define a linear transformation. By Exercise 2.17 this transformation is an isomorphism.

On the other hand, many linear transformations are not isomorphisms. For one example, define a linear transformation from $\mathbb{C}^2$ to the vector space $\mathcal{P}_3^1$ by

$$\begin{pmatrix} 1 \\ 0 \end{pmatrix} \mapsto y$$

$$\begin{pmatrix} 0 \\ 1 \end{pmatrix} \mapsto z.$$

This linear transformation is not an isomorphism because it is not surjective: there is no element of $\mathbb{C}^2$ that maps to the polynomial x.

We warn the reader that the word "isomorphism" is used in many different contexts in mathematics. It generally refers to an injective and surjective function that respects some mathematical operations, and whose inverse also respects those operations. Often it is up to the reader to infer from context exactly what the "isomorphism" in question respects.

All of the concepts of this section — kernel, image, Fundamental Theorem, homogeneous harmonic polynomials and isomorphisms — come up repeatedly in the rest of the text.

2.5 Linear Operators

We will often want to consider linear transformations from a vector space V to itself. Such a transformation is called a *linear operator on V*. In this

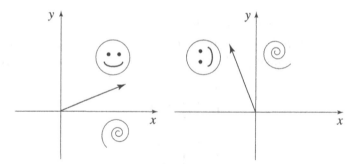

Figure 2.4. Counterclockwise rotation by $\pi/2$ around the origin. The two vectors really are the same length.

case, we can take advantage of powerful technology unavailable in the more general case ($T : V \to W$ with $V \neq W$).

First we consider the relation between a linear operator and its matrices. Consider the real linear operator T on $\mathbb{R}^2$ (with the usual basis) defined by the matrix

$$\begin{pmatrix} 0 & -1 \\ 1 & 0 \end{pmatrix}. \tag{2.3}$$

This linear operator corresponds to rotation through an angle of $\pi/2$ around the origin, counterclockwise. See Figure 2.4. The standard basis is so standard that we think of this matrix as *the* matrix of the linear operator. But in another basis, this operator has another matrix. For example, if we take the new basis $((2, 0)^T, (0, 1)^T)$, the matrix of the same linear operator is

$$\begin{pmatrix} 0 & -\frac{1}{2} \\ 2 & 0 \end{pmatrix}. \tag{2.4}$$

Why? This kind of computation, using two different bases at once, can be confusing. We will use two typefaces to distinguish expressions in the standard basis from expressions in the new basis. Thus the new basis, written in the new basis, is $((\mathbf{1}, \mathbf{0})^T, (\mathbf{0}, \mathbf{1})^T)$. Notice that our favorite rotation takes $(\mathbf{1}, \mathbf{0})^T$, otherwise known as $(2, 0)^T$, to the vector $(0, 2)$, otherwise known as $(\mathbf{0}, \mathbf{2})^T$. Similarly, the rotation takes $(\mathbf{0}, \mathbf{1})^T$ to $(-\frac{1}{2}, \mathbf{0})^T$. Since the columns of any matrix are the images of the basis vectors under the linear transformation represented by the matrix, these calculations show that the matrix given above is correct. See Figure 2.5.

The general recipe relating the matrix A of a linear operator in the standard basis to the matrix $\tilde{A}$ in a new basis involves the matrix B whose columns are

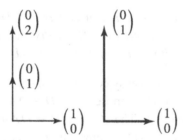

Figure 2.5. Change of basis.

the basis vectors of the new basis (expressed in the standard basis). We have

$$A = B\tilde{A}B^{-1}. \tag{2.5}$$

See Figure 2.6. The expression on the right-hand side of the equation above

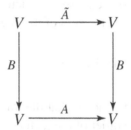

Figure 2.6. A commutative diagram for $A = B\tilde{A}B^{-1}$.

has many names (as befits an operation of great practical and theoretical importance): *conjugation by B*, *similarity transformation*, and others. We exhort each reader to justify this formula carefully (why isn't it $B^{-1}\tilde{A}B$?) and promise that a fluent understanding of the relationship between changing bases and multiplying on left or right by B or B^{-1} in various situations is well worth the effort.

As another example of the relationship between geometry (changing bases) and algebra (working with matrices), consider diagonal matrices. Recall that a matrix A is said to be *diagonal* if and only if $A_{jk} = 0$ whenever $j \neq k$. In other words, the matrix A is diagonal if and only if each standard basis vector is an eigenvector of A. This implies that the matrix A of a linear operator T in a particular basis will be diagonal if and only if every vector in the given basis is an eigenvector of T. The following proposition about diagonal matrices will be useful in Section 6.5.

Proposition 2.7 *Suppose A is an $n \times n$ matrix and D is a diagonal $n \times n$ matrix. Suppose the diagonal entries of D are distinct. Then A is diagonal if and only if A commutes with D, i.e., if and only if $AD - DA = 0$.*

Proof. If A is diagonal, then an easy computation shows that $AD - DA = 0$. To prove the other implication, suppose that $AD - DA = 0$. Let e_i denote the ith standard basis vector of $\mathbb{C}^n$. Then $0 = (AD - DA)e_i = D_{ii}Ae_i - DAe_i$. So Ae_i is an eigenvector of D with eigenvalue D_{ii}. Because $D_{ii} \neq D_{jj}$ unless $i = j$, it follows that Ae_i must be a scalar multiple of e_i for each i. Hence A must be diagonal. $\qquad\qquad\qquad\qquad\qquad\qquad\qquad\qquad\qquad\qquad\qquad\qquad$ □

Two important complex numbers associated to any particular complex linear operator T (on a finite-dimensional complex vector space) are the trace and the determinant. These have algebraic definitions in terms of the entries of the matrix of T in any basis; however, the values calculated will be the same no matter which basis one chooses to calculate them in. We define the *trace* of a square matrix A to be the sum of its diagonal entries:

$$\operatorname{Tr} A := \sum_{j=1}^{n} A_{jj}.$$

One important property of the trace is that the trace of a product of two matrices does not depend on the order of the factors.

Proposition 2.8 *Suppose A and B are two $n \times n$ matrices. Then $\operatorname{Tr}(AB) = \operatorname{Tr}(BA)$.*

We will use this Proposition in Section 8.1.

Proof. Note that

$$\operatorname{Tr}(AB) = \sum_{j=1}^{n}(AB)_{jj} = \sum_{j=1}^{n}\sum_{k=1}^{n} A_{ji}B_{ij} = \sum_{k=1}^{n}(BA)_{ii} = \operatorname{Tr}(BA).$$

$\qquad\qquad\qquad\qquad\qquad\qquad\qquad\qquad\qquad\qquad\qquad\qquad\qquad\qquad$ □

It follows that if A and $\tilde{A}$ are related by conjugation (as in Equation 2.5), then the traces of A and $\tilde{A}$ are equal:

$$\operatorname{Tr} A = \operatorname{Tr}(B\tilde{A}B^{-1}) = \operatorname{Tr}(\tilde{A}B^{-1}B) = \operatorname{Tr}\tilde{A}.$$

Because all different matrices of one linear operator are related by conjugation, this observation allows us to define the trace of a linear operator.

Definition 2.8 *Suppose T is a linear operator on a finite-dimensional vector space. Then the trace of T is the trace of the matrix of T in any basis.*

So the trace of the counterclockwise rotation through the angle $\pi/2$ (see Figure 2.4) is $0 + 0 = 0$.

We will make extensive use of the trace in Chapters 4 through 6, when we define and exploit the notion of "characters." In particular, in the proof of Proposition 6.8 we will use the following proposition.

Proposition 2.9 *Suppose V is a finite-dimensional vector space and Π is a linear operator on V such that $\Pi^2 = \Pi$. (Such a linear operator is called a* projection.*) Let W denote the image of Π. Then* $\operatorname{Tr}\Pi = \dim W$.

Proof. The trick is to choose a nice basis in which to calculate the trace of Π. First choose a basis $\{w_1, \ldots, w_k\}$ of W. Note that $k = \dim W$. Next, choose $\{v_1, \ldots, v_k\} \subset V \setminus W$ such that $\{w_1, \ldots, w_j, v_1, \ldots, v_m\}$ is a basis of V.

Now consider Πw for $w \in W$. For any $w \in W$, there is a $v \in V$ such that $\Pi v = w$. So
$$\Pi w = \Pi^2 v = \Pi v = w.$$
In particular, if w is one of our basis vectors, say $w = w_j$, then we know that $A_{jj} = 1$.

Next consider Πv for $v \in V \setminus W$. By the definition of W, we have $\Pi v \in W$. In particular, in the expression of Πv_j in terms of basis vectors, the coefficient of v_j must be zero. Hence $A_{(k+j)(k+j)} = 0$.

Finally, we compute that
$$\operatorname{Tr}\Pi = \sum_{j=1}^{k+m} A_{jj} = \sum_{j=1}^{k} 1 = k = \dim W.$$
$\square$

In Section 11.3 we will use the following generalization of Proposition 2.8.

Proposition 2.10 *Suppose V and W are finite-dimensional vector spaces and $A\colon V \to W$ and $B\colon W \to V$ are linear transformations. Then*
$$\operatorname{Tr}(AB) = \operatorname{Tr}(BA).$$

Proof. Fix any two bases of V and W. Let $\hat{A}$ and $\hat{B}$ denote the matrices of the linear transformations with respect to the bases. Then
$$\operatorname{Tr}(AB) = \sum_{i=1}^{\dim W}\sum_{j=1}^{\dim V} (\hat{A})_{ij}(\hat{B})_{ji} = \sum_{j=1}^{\dim V}\sum_{i=1}^{\dim W} (\hat{B})_{ji}(\hat{A})_{ij} = \operatorname{Tr}(BA).$$
$\square$

The definition of the determinant of a linear operator is analogous to the definition of the trace. We start with the determinant of a matrix, which should be familiar from a linear or abstract algebra textbook such as Artin [Ar, Section 1.3]. It is a fact of linear algebra that $\det(AB) = (\det A)(\det B)$ for any two square matrices A and B of the same size. Hence for any matrices A and $\tilde{A}$ related by Equation 2.5, we have

$$\det(\tilde{A}) = \det(BAB^{-1}) = \det(B)\det(A)\det(B^{-1})$$
$$= \det(A)\det(BB^{-1}) = \det(A)\det(I) = \det(A).$$

As before, this calculation allows us to use the determinant of a matrix to define the determinant of a linear operator.

Definition 2.9 *Suppose T is a linear operator. Then the* determinant of T *is the determinant of the matrix of T in any basis.*

So, for example, we can use the matrix of Formula 2.3 to see that the determinant of our favorite rotation is $(0)(0) - (1)(-1) = 1$. Note that we could just as well have used the matrix of Formula 2.4 to calculate the same answer: $(0)(0) - (2)(-\frac{1}{2}) = 1$. No one familiar with the geometric interpretation of the determinant will be surprised by this result: the determinant of a matrix with real entries is always the signed volume of the image of the unit square (or cube, or higher-dimensional cube), with a negative sign if the linear transformation changes the orientation. For more on this topic, see Lax [La, Chapter 5].

Next we define *eigenvalues* and *eigenvectors*.

Definition 2.10 *Suppose V is a vector space and T is a linear operator on V. We say that λ is an eigenvalue of T with eigenvector v if and only if $v \neq 0$ and*

$$Tv = \lambda v.$$

In this case we also say that v is an eigenvector associated to the eigenvector λ.

Consider, for example, our old friend, rotation by $\pi/2$ around the origin. See Figure 2.4. For every nonzero vector v, the vector Tv points in a direction different from v. Hence it is impossible to have $Tv = \lambda v$ for any real λ and nonzero real two-vector v. So this real linear operator has no real eigenvalues.

However, we can use the same matrix to define a complex linear operator S on the complex vector space $\mathbb{C}^2$ (with the usual basis). Unlike T, the linear operator S has two eigenvalues, $\pm i$, with associated eigenvectors $(1, \pm i)^T$,

since
$$\begin{pmatrix} 0 & -1 \\ 1 & 0 \end{pmatrix} \begin{pmatrix} 1 \\ \pm i \end{pmatrix} = \begin{pmatrix} \mp i \\ 1 \end{pmatrix} = (\mp i) \begin{pmatrix} 1 \\ \pm i \end{pmatrix}.$$

Proposition 2.11 *Suppose V is a complex vector space of dimension $n \in \mathbb{N}$. Suppose $T : V \to V$ is a complex linear operator. Then T has at least one eigenvalue (and at least one corresponding eigenvector).*

Proof. Consider the *characteristic polynomial* of T, that is, $\det(\lambda I - T)$. Because of the λ^n term, this complex-coefficient polynomial has degree $n > 0$. Hence, by the Fundamental Theorem of Algebra,[2] this polynomial has at least one complex root. In other words, there exists a $\lambda \in \mathbb{C}$ such that $\det(\lambda I - T) = 0$. This implies that there is a nonzero $v \in V$ such that $(\lambda I - T)v = 0$. Hence $\lambda v = T v$ and λ is an eigenvalue of T.   □

This proof does not give a method for finding real eigenvalues of real linear operators, because the Fundamental Theorem of Algebra does not guarantee real roots for polynomials with real coefficients. Proposition 2.11 does not hold for infinite-dimensional complex vector spaces either. See Exercise 2.28.

Eigenvalues and eigenvectors play a large role in the analysis of quantum mechanical systems. We will not use this technology until Chapter 8, where we will find the hidden symmetries of the hydrogen atom. We introduce it here as an important example of linear algebra in quantum mechanics. The reader may have encountered the term "Hamiltonian operator"; e.g., the Schrödinger operator is the Hamiltonian operator for the electron in the hydrogen atom. The eigenvalues of the Schrödinger operator (on the appropriate vector space of functions) turn out to be the possible observable energies of the electron in the hydrogen atom. As we discussed in Section 1.3, these are numbers of the form
$$E_n := \frac{-\mathbf{m}\mathbf{e}^4}{2\hbar^2(n+1)^2},$$
for some nonnegative integer n, where $\mathbf{m}$ is the mass of the electron, $\mathbf{e}$ is charge on the electron, and $\hbar$ is Planck's constant divided by 2π. It turns out that an eigenvector (also called an *eigenfunction* or an *eigenstate*) for a particular eigenvalue E_n corresponds to a state of the electron whose energy is sure to be measured to be E_n, as we discussed in Section 1.2.

The existence of eigenvalues for linear transformations is what makes representation theory so much more powerful than abstract group theory. Rep-

---

[2]For a proof of the Fundamental Theorem of Algebra, see any abstract algebra textbook, such as Artin [Ar, Section 13.9].

resentation theory is all about interpreting abstract group elements (which do not necessarily have eigenvalues) as linear operators (which do have eigenvalues). The added power of the eigenvalues may be the reason that when physicists and chemists speak of "group theory" they really mean representation theory; why should they bother with abstract (i.e., eigenvalueless) group theory at all? Still, this does not explain the use of the term "group theory" to describe representation theory of objects (such as Lie algebras) that are not groups at all. The reader may wish to keep in mind this discrepancy in nomenclature, especially in Chapter 4.

2.6   Cartesian Sums and Tensor Products

In this section we introduce two different ways of building a new vector space by combining old ones.

One way to combine vector spaces is to take a *Cartesian sum*. (Mathematicians sometimes call this a *Cartesian product*. Another common term is *direct sum*.)

Definition 2.11 *Suppose* $V_1, \ldots, V_n$ *are vector spaces over the same scalar field. The* Cartesian sum *of these vector spaces, denoted* $V_1 \oplus \cdots \oplus V_n$ *or*

$$\bigoplus_{k=1}^{n} V_k,$$

is the set

$$\{(v_1, \ldots, v_n) : \text{for } k = 1, \ldots, n \text{ we have } v_k \in V_k\},$$

with addition defined by

$$(v_1, \ldots, v_n) + (w_1, \ldots, w_n) := (v_1 + w_1, \ldots, v_n + w_n)$$

and scalar multiplication defined by

$$c(v_1, \ldots, v_n) := (cv_1, \ldots, cv_n).$$

When the summands $V_1, \ldots, V_n$ are *linearly independent subspaces* of one space W, then the Cartesian sum $\bigoplus_{k=1}^{n} V_k$ is isomorphic to the subspace of W spanned by $\bigcup_{k=1}^{n} V_k$. Let us be a bit more explicit. A set $\{V_1, \ldots, V_n\}$ of subspaces of one vector space W is said to be linearly independent whenever the following condition holds: If $v_1 \in V_1, \ldots, v_n \in V_n$, then $\sum_{k=1}^{n} v_k = 0$

if and only if $v_1 = \cdots = v_n = 0$. So for example, in $\mathbb{C}^3$ the pair $\{\mathbb{C} \times \mathbb{C} \times \{0\}, \{0\} \times \{0\} \times \mathbb{C}\}$ is linearly independent, while the pair $\{\mathbb{C} \times \mathbb{C} \times \{0\}, \mathbb{C} \times \{0\} \times \{0\}\}$ is not. If $V_1, \ldots, V_n$ is a set of linearly independent subspaces, then there is an isomorphism between $\bigoplus_{k=1}^{n} V_k$ and the span of $\bigcup_{k=1}^{n} V_k$ given by

$$(v_1, \ldots, v_n) \mapsto v_1 + \cdots + v_n.$$

We will often use this isomorphism implicitly, letting $\bigoplus_{k=1}^{n} V_k$ denote the subspace of W spanned by $\bigcup_{k=1}^{n} V_k$ and writing $v_1 + \cdots + v_n$ instead of $(v_1, \ldots, v_n)$.

Thus, for example, $\mathbb{C}^n$ is equal (as a complex vector space) to the Cartesian sum of n copies of $\mathbb{C}$:

$$\mathbb{C}^n = \bigoplus_{k=1}^{n} \mathbb{C}.$$

Here we think of the first copy of $\mathbb{C}$ as the set of vectors of the form $(c, 0, \ldots, 0)$, the second copy of $\mathbb{C}$ as the set of vectors of the form $(0, c, 0, \ldots, 0)$, and so on.

Note that vector space operations are required. Thus, while we can use spherical coordinates to write any element of $\mathbb{R}^3 \setminus \{0\}$ uniquely as a triple (ρ, θ, ϕ), where $\rho \in (0, \infty)$, $\theta \in [0, \pi]$ and $\phi \in [-\pi, \pi)$, the expression "$(0, \infty) \oplus [-\pi, \pi) \oplus [0, \pi]$" is nonsense, because none of the three intervals is a vector space.[3]

There are natural projections defined on any Cartesian sum.

Definition 2.12 *Suppose $V_1 \oplus \cdots \oplus V_n$ is a Cartesian sum of vector spaces. For any summand V_k we can define a linear transformation:*

$$\Pi_k \colon \bigoplus_{j=1}^{n} V_j \to V_k$$

$$(v_1, \ldots, v_n) \mapsto v_k.$$

This linear transformation is called the projection onto the kth summand, *or* projection onto V_k.

We will use these projections in Section 5.2 and in the proof of Proposition 6.5.

---

[3] One can, however, speak of the *Cartesian product* of sets, without vector space operations. So, merely as sets, $\mathbb{R}^3 \setminus \{0\}$ and the Cartesian product $(0, \infty) \times [-\pi, \pi) \times [0, \pi]$ are equal.

Another useful way to construct a vector space from other vector spaces is to take what mathematicians call a *tensor product* and physicists call a *direct product*. We will need to consider tensor products of representations in Section 5.3. In this section we will define and discuss tensor products of vector spaces.

Warning: physicists use the word "tensor" to describe objects that arise in the theory of general relativity (such as the metric tensor or the curvature tensor), among other places. Although these objects are indeed tensors in the sense we will define below, they are also more complicated: they involve multiple coordinate systems. We warn the reader that this section will not address the issues raised by multiple coordinate systems. Thus a reader who has been confused by such physicists' tensors may not be fully satisfied by our discussion here.[4]

Since many people find the definition difficult, we start with two examples. First, consider the space $\mathbb{C}^2$ of column 2-vectors and $(\mathbb{C}^3)^*$ of row 3-vectors with complex entries. Matrix multiplication gives us a way of multiplying elements of $\mathbb{C}^2$ and $(\mathbb{C}^3)^*$; for instance,

$$
e_2 \otimes e_2^* := \begin{pmatrix} 0 \\ 1 \end{pmatrix} \begin{pmatrix} 0 & 1 & 0 \end{pmatrix} = \begin{pmatrix} 0 & 0 & 0 \\ 0 & 1 & 0 \end{pmatrix}.
$$

The rules of matrix multiplication ensure that the result is always a 2×3 matrix. The *tensor product of $\mathbb{C}^2$ and $(\mathbb{C}^3)^*$* (denoted $\mathbb{C}^2 \otimes (\mathbb{C}^3)^*$) is the set of 2×3 matrices spanned by these multiples. The span consists of all 2×3 matrices with complex entries, since we can construct any matrix with the ij-th entry equal to one and all other entries zero by taking the product $e_i \otimes e_j^*$. These matrices form a basis of the set of 2×3 matrices. Thus $\mathbb{C}^2 \otimes (\mathbb{C}^3)^*$ is the set of 2×3 matrices with complex entries. Notice that taking the span of the products nets us more than just the products. For instance, while we can write the matrix

$$
\begin{pmatrix} 1 & 0 & 0 \\ 0 & 1 & 0 \end{pmatrix} \tag{2.6}
$$

---

[4]Such a reader might find relief in *differential geometry*, the mathematical study of multiple coordinate systems. There are many excellent standard texts, such as Isham's book [I]; for a gentle introduction to some basic concepts of differential geometry, try [Si]. A text that discusses "covariant" and "contravariant" tensors is Spivak's introduction to differential geometry [Sp, Volume I, Chapter 4]. For a quick introduction aimed at physical calculations, try Joshi's book [Jos].

as a sum of two products $(e_1 \otimes e_1^* + e_2 \otimes e_2^*)$, we cannot write it as a single product $v \otimes w := vw$ for any column 2-vector v and row 3-vector w. See Exercise 2.27. A second example to keep in mind is products of polynomials. Consider the six-dimensional complex vector space V of homogeneous polynomials in four variables, u, v, x and y that are of degree one in u and v together, and are of degree two in x and y together. One basis for this vector space is

$$\{ux^2, uxy, uy^2, vx^2, vxy, vy^2\}. \tag{2.7}$$

Recall the vector space $\mathcal{P}^\ell$ of homogeneous polynomials in two variables defined in Section 2.2. The vector space V is the *tensor product of* $\mathcal{P}^1$ *and* $\mathcal{P}^2$, denoted $\mathcal{P}^1 \otimes \mathcal{P}^2$. In other words, the elements of V are precisely the linear combinations of terms of the form $p(u, v)q(x, y)$, where p is a homogeneous polynomial of degree one and q is a homogeneous polynomial of degree two. Note that, given an element $r(u, v, x, y)$ of $\mathcal{P}^1 \otimes \mathcal{P}^2$, there are many different ways to write it as a linear combination of products. For example,

$$ux^2 + ivx^2 = (u)(x^2) + (v)(ix^2) = (u + iv)(x^2) = (v - iu)(ix^2).$$

The same phenomenon occurs in $\mathbb{C}^2 \otimes (\mathbb{C}^3)^*$. We have

$$
\begin{pmatrix} 1 & 0 & 0 \\ 0 & 7e^{i\pi/9} & 0 \end{pmatrix} = \begin{pmatrix} 1 \\ 0 \end{pmatrix} \begin{pmatrix} 1 & 0 & 0 \end{pmatrix} + \begin{pmatrix} 0 \\ 1 \end{pmatrix} \begin{pmatrix} 0 & 7e^{i\pi/9} & 0 \end{pmatrix}
$$

$$
= \begin{pmatrix} i \\ 0 \end{pmatrix} \begin{pmatrix} -i & 0 & 0 \end{pmatrix} + \begin{pmatrix} 0 \\ 1 \end{pmatrix} \begin{pmatrix} 0 & 7e^{i\pi/9} & 0 \end{pmatrix}
$$

$$
= \begin{pmatrix} 1 \\ 1 \end{pmatrix} \begin{pmatrix} 1 & 0 & 0 \end{pmatrix} + \begin{pmatrix} 0 \\ 1 \end{pmatrix} \begin{pmatrix} -1 & 7e^{i\pi/9} & 0 \end{pmatrix}.
$$

Recall from Section 1.7 that the standard mathematical way to deal with irrelevant ambiguity is to define an equivalence relation and work with equivalence classes. In this case of tensors, the irrelevant ambiguity arises from the different ways of writing the same object as a linear combination of products. We will use this insight to define tensor products. Suppose V and W are complex vector spaces. Consider the complex vector space VW generated by the set

$$S := \{(v, w): v \in V, w \in W\}.$$

In other words, VW is equal to the set

$$\left\{ \sum_{j=1}^n c_j(v_j, w_j): n \in \mathbb{N}, \forall i \ c_i \in \mathbb{C}, v_i \in V \text{ and } w_i \in W \right\},$$

where the only allowable manipulation of sums is to replace $c_1(v, w) + c_2(v, w)$ by $(c_1 + c_2)(v, w)$. The vector space VW is huge! In this vector space, $(v_1 + v_2, w)$ is *not* the same as $(v_1, w) + (v_2, w)$. The set S is a basis for VW; its elements are linearly independent. Our definition of the equivalence relation reflects our intuition about what we would like the tensor product to be. Think about the rules we use naturally to calculate in the two examples above. For example, we want $(v_1 + v_2, w)$ to be equivalent to $(v_1, w) + (v_2, w)$, and similarly for sums in the second slot. Also, for any complex number c, we want $c(v, w)$, (cv, w), and (v, cw) to be equivalent to one another, as they are for matrix multiplication and multiplication of polynomials. We call these the *computation rules*.

Definition 2.13 *Suppose $c_1(v_1, w_1) + \cdots + c_n(v_n, w_n)$ and $\tilde{c}_1(\tilde{v}_1, \tilde{w}_1) + \cdots + \tilde{c}_{\tilde{n}}(\tilde{v}_{\tilde{n}}, \tilde{w}_{\tilde{n}})$ are elements of VW. Then we define*

$$c_1(v_1, w_1) + \cdots + c_n(v_n, w_n) \sim \tilde{c}_1(\tilde{v}_1, \tilde{w}_1) + \cdots + \tilde{c}_{\tilde{n}}(\tilde{v}_{\tilde{n}}, \tilde{w}_{\tilde{n}})$$

if and only if we can get from

$$c_1(v_1, w_1) + \cdots + c_n(v_n, w_n)$$

to

$$\tilde{c}_1(\tilde{v}_1, \tilde{w}_1) + \cdots + \tilde{c}_{\tilde{n}}(\tilde{v}_{\tilde{n}}, \tilde{w}_{\tilde{n}})$$

in a finite number of steps by applying the computation rules: for any $v, v_1, v_2 \in V$, any $w, w_1, w_2 \in W$ and any $c \in \mathbb{C}$ we have

1. $(v_1 + v_2, w) \sim (v_1, w) + (v_2, w)$;

2. $(v, w_1 + w_2) \sim (v, w_1) + (v, w_2)$;

3. $(cv, w) \sim (v, cw)$;

4. $c(v, w) \sim (cv, w)$;

and the substitution rules: for any $X_1, X_2, Y \in VW$ such that $X_1 \sim X_2$ and any $c \in \mathbb{C}$ we have

1. $X_1 + Y \sim X_2 + Y$;

2. $cX_1 \sim cX_2$.

Physicists should note that we use "finite" in the mathematical sense, meaning that the number of steps can be zero or any natural number.

Proposition 2.12 *The relation* $\sim$ *of Definition 2.13 is an equivalence relation.*

Experienced mathematicians may wish to bypass this proposition by defining $\sim$ to be the smallest equivalence relation containing $(v_1 + v_2, w) \sim (v_1, w) + (v_2, w)$, $c(v, w) \sim (cv, w)$, $(cv, w) \sim (v, cw)$, etc. This is all right as long as one knows how to show that there is such a smallest relation, and it is unique. But note that the proof of Proposition 2.12 is not hard, and our definition has the virtue of showing the relationship of the mathematical concept of equivalence with the physics tradition of understanding through computation.

Proof. The relation is reflexive, since we can get from any linear combination to itself by applying zero rules, i.e., no rules at all.[5] The relation is symmetric, since $X \sim Y$ implies that there is a finite number of steps taking X to Y; by reversing the steps we can take Y to X, and hence $Y \sim X$. Transitivity follows from the fact that the sum of two finite numbers is a finite number. $\square$

Definition 2.14 *Suppose V and W are complex vector spaces. The* (complex) tensor product of V and W is

$$V \otimes W := VW/\sim,$$

where VW and $\sim$ are defined as above. If $v \in V$ and $w \in W$ we denote,[6] the equivalence class of vw by $v \otimes w$.

Because of the substitution rules in Definition 2.14, the complex vector space structure of VW descends to $V \otimes W$, so $V \otimes W$ is a vector space.

In practice, if we have bases of V and W, then there is a much easier way to think about the tensor product vector space $V \otimes W$.

Proposition 2.13 *Suppose $\{v_1, \ldots, v_n\}$ is a basis of the vector space V and $\{w_1, \ldots, w_m\}$ is a basis of the vector space W. Then*

$$\{v_i \otimes w_j : i, j \in \mathbb{N}, i \leq n, j \leq m\}$$

is a basis for the vector space $V \otimes W$.

Proof. First we will show that any element of $V \otimes W$ can be written as a linear combination of elements of the form $v_i \otimes w_j$. Because any arbitrary

---

[5]The semantic distinction between "zero rules" and "no rules at all" is deep. An interesting book on this subject is *Signifying Nothing: The Semiotics of Zero* [Rot].

[6]This definition can be applied, *mutatis mutandis* to any two vector spaces over the same scalar field, not just over $\mathbb{C}$. See [Hal58, Section 26].

element of $V \otimes W$ is a linear combination of terms of the form $v \otimes w$, it suffices to show that any $v \otimes w$ can be written as a linear combination of our alleged basis vectors. But because the v_i's and w_j's form bases, we can write any $v \in V$ as $c_1 v_1 + \cdots + c_n v_n$ and any $w \in W$ as $\tilde{c}_1 w_1 + \cdots + \tilde{c}_m w_m$. By definition of the equivalence relation, we have

$$(v, w) \sim \sum_{i=1}^{n} \sum_{j=1}^{m} c_i \tilde{c}_j (v_i, w_j) \in VW,$$

and hence

$$v \otimes w = \sum_{i=1}^{n} \sum_{j=1}^{m} (c_i \tilde{c}_j) v_i \otimes w_j \in V \otimes W.$$

So the set $\{v_i \otimes w_j : i, j \in \mathbb{N}, i \leq n, j \leq m\}$ spans $V \otimes W$.

Next we must show that the elements are linearly independent. For this proof it will be useful to consider an *invariant of the equivalence relation*. For example, a mathematical object that can be calculated from any element of VW is an invariant of the equivalence relation of Definition 2.13 if it is the same when calculated from any two elements related by a computation rule. More generally, given any set S and any equivalence $\sim$, an *invariant of the equivalence relation* is a function J whose domain is S and for which $J(s_1) = J(s_2)$ for any $s_1, s_2 \in S$ such that $s_1 \sim s_2$. Given any element $z = \sum_{j=1}^{N} c_j (x_j, y_j)$, with each x_j in V and y_j in W, we define the *coefficient of v_1 in z* as follows. Expand each x_j as a linear combination of the basis vectors $v_1, \ldots, v_n$ of V. Now let $\tilde{z}$ denote the element obtained from z by replacing each of $v_2, \ldots, v_n$ by 0. Then $\tilde{z}$ takes the form $\sum_{j=1}^{\tilde{N}} c_j (b_j v_1, \tilde{y}_j)$, where b_j and c_j are complex numbers and $y_j \in W$. Define

$$J(z) := \sum_{j=1}^{\tilde{N}} (c_j b_j) \tilde{y}_j.$$

Note that $J(z) \in W$. We call $J(z)$ the *coefficient of v_1 in z*. Since $\{v_1, \ldots, v_2\}$ is a basis, $J(z)$ is well defined as a function of z. Notice that each of the computation rules defining our equivalence relation leaves the coefficient of v_1 unchanged: for example, we have $\alpha(v, w) \sim (\alpha v, w)$, and while making this substitution changes the computation of the c_j's and b_j's, it leaves the products $c_j b_j$ unchanged. The reader should check the other computation rules. Thus if $z_1 \sim z_2$, the coefficient of v_1 in z_1 must equal the coefficient of v_1 in z_2. Similarly, we can define the *coefficient of v_i in z* for any i from 1 to n. Now suppose we have complex numbers c_{ij} such that

$$\sum_{i=1}^{n}\sum_{j=1}^{m} c_{ij} v_i \otimes w_j = 0.$$

Then $\sum_{i=1}^{n}\sum_{j=1}^{m} c_{ij}(v_i, w_j) \sim 0$ in VW. So the coefficient of v_i in the sum $\sum_{i=1}^{n}\sum_{j=1}^{m} c_{ij}(v_i, w_j)$ must be equal to the coefficient of v_i in 0. Hence, for each i we have

$$0 = \sum_{j=1}^{m} c_{ij} w_j$$

in W. But the w_j's form a basis, so this implies that each $c_{ij} = 0$. This proves that the $v_i \otimes w_j$'s form a basis.   □

Let us check Proposition 2.13 in our two examples. A basis for $\mathbb{C}^2$ is $\{(1, 0)^T, (0, 1)^T)\}$, while a basis for $(\mathbb{C}^3)^*$ is $\{(1, 0, 0), (0, 1, 0), (0, 0, 1)\}$. Using the recipe in the proposition, we expect that the set of all six products of basis elements should be a basis for $\mathbb{C}^2 \otimes (\mathbb{C}^3)^*$. And indeed, these are just the six different matrices with a one and five zeroes. Similarly, the basis we exhibited in Formula 2.7 is the set of all products of one element from $\{u, v\}$ (a basis of $\mathcal{P}^1$) with one element from $\{x^2, xy, y^2\}$ (a basis of $\mathcal{P}^2$).

It is often useful to consider the elements of a tensor product that can be expressed without addition. The following definition is useful in the proof of Propositions 5.14 and crucial to the statement and proof of Proposition 11.1.

Definition 2.15 *Suppose $n \in \mathbb{N}$ and for each $j = 1, \ldots, n$ we have a vector space V_j. Then an element x of the tensor product*

$$\bigotimes_{j=1}^{n} V_j = V_1 \otimes \cdots \otimes V_n$$

is an elementary tensor *if there are elements $v_j \in V_j$ such that*

$$x = \bigotimes_{j=1}^{n} v_j = v_1 \otimes \cdots \otimes v_n.$$

Elementary tensors are also known as *decomposable tensors*.

Physicists have a nice trick for visualizing tensor products. For example, to picture a typical element of $\mathbb{C}^2 \otimes (\mathbb{C}^3)^*$, one pictures a typical element of $\mathbb{C}^2$ as a vector, say

$$v = \begin{pmatrix} v_1 \\ v_2 \end{pmatrix},$$

and a typical element of $(\mathbb{C}^3)^*$ as a row vector

$$w = \begin{pmatrix} w_1 & w_2 & w_3 \end{pmatrix}.$$

Now replace each entry of v with that entry times w:

$$v \otimes w = \begin{pmatrix} v_1 \begin{pmatrix} w_1 & w_2 & w_3 \end{pmatrix} \\ v_2 \begin{pmatrix} w_1 & w_2 & w_3 \end{pmatrix} \end{pmatrix}$$

and carry out the suggested multiplications to obtain

$$\begin{pmatrix} v_1 w_1 & v_1 w_2 & v_1 w_3 \\ v_2 w_1 & v_2 w_2 & v_2 w_3 \end{pmatrix}.$$

One nice feature of this visualization is that it generalizes to tensor products of linear transformations. A drawback is that in some situations the answer will differ depending on arbitrary choices.

Again, we remind physicists that tensor products of vector spaces are neither as general nor as powerful as the objects called "tensors" appearing in general relativity. Issues of "covariance" and contravariance" have to do with multiple coordinate systems. Because quantum mechanics is linear, we do not need the more general notion of "tensor" in this book, so we do not stop to introduce it. We do, however, offer our condolences and a few references to physicists searching for clarification. See Footnote 4 in this chapter.

We will use Cartesian sums and tensor products to build and decompose representations in Chapters 5 and 7. Tensor products are useful in combining different aspects of one particle. For instance, when we consider both the mobile and the spin properties of an electron (in Section 11.4) the state space is the tensor product of the mobile state space ($L^2(\mathbb{R}^3)$, defined in Chapter 3) and the spin state space ($\mathbb{C}$).

2.7   Exercises

Exercise 2.1 *Consider the set of homogeneous polynomials in two variables with real coefficients. There is a natural addition of polynomials and a natural scalar multiplication of a polynomial by a complex number. Show that the set of homogeneous polynomials with these two operations is not a complex vector space.*

Exercise 2.2 (Used in Appendix A)   *Suppose that $m_1, \ldots, m_n$ are distinct integers. For each j, let $e^{im_j(\cdot)}$ denote the function $[0, \pi] \to \mathbb{C}, x \mapsto e^{im_jx}$. Show that the set*

$$\left\{ e^{im_j(\cdot)} : j = 1, \ldots, n \right\}$$

is linearly independent.

Exercise 2.3   *Show that $\mathbb{C}$ (with the usual addition and multiplication) is itself a complex vector space of dimension 1. Then show that $\mathbb{C}$ with the usual addition but with scalar multiplication by real numbers only is a real vector space of dimension 2.*

Exercise 2.4   *Show that for any natural number n, the Cartesian product $\mathbb{C}^n$ is a complex vector space of dimension n. Then show that $\mathbb{C}^n$ with the usual addition but with scalar multiplication by real numbers only is a real vector space of dimension 2n.*

Exercise 2.5   *Consider the complex vector space $\mathbb{C}^2$. Is the set*

$$\left\{ \begin{pmatrix} 1 \\ i \end{pmatrix}, \begin{pmatrix} i \\ -1 \end{pmatrix} \right\}$$

linearly independent? Now consider $\mathbb{C}^2$ as a real vector space. Is the same set linearly independent?

Exercise 2.6   *Let V be an arbitrary complex vector space of dimension n. Show that by restricting scalar multiplication to the reals one obtains a real vector space of dimension 2n.*

Exercise 2.7   *Consider the complex plane $\mathbb{C}$ as a real vector space of dimension two. Is complex conjugation a real linear transformation?*

Exercise 2.8   *Consider the complex plane $\mathbb{C}$ as real vector space of dimension two, and the quaternions $\mathbf{Q}$ as a real vector space of dimension four. Show that function $f_\mathbf{i} : \mathbb{C} \to \mathbf{Q}$ defined by*

$$f_\mathbf{i}(a + ib) := a + b\mathbf{i}$$

is a real linear transformation. Similarly, define

$$f_\mathbf{j}(a + ib) := a + b\mathbf{j}, \qquad f_\mathbf{k}(a + ib) := a + b\mathbf{k}$$

and show that they too are linear transformations. Next, show that we can consider $\mathbf{Q}$ as a two-dimensional complex vector space with basis $\{1, \mathbf{j}\}$. Are $f_\mathbf{i}, f_\mathbf{j}$ and $f_\mathbf{k}$ complex linear functions?

Exercise 2.9 (Relevant to Proposition 7.3) *Suppose V is the vector space of all polynomials in three variables. Suppose q is a polynomial in three variables. Show that multiplication by q is a linear transformation. In other words, consider the function taking any $p(x, y, z) \in V$ to $q(x, y, z)p(x, y, z)$. Show that this function is linear. What is its range? (Remark: these statements hold true for polynomials in any number of variables.) Now let $\mathcal{P}_3^\ell$ denote the homogeneous polynomials in three variables of degree ℓ. Let r^2 denote the polynomial $x^2 + y^2 + z^2$. Show that $r^2 : \mathcal{P}_3^\ell \to \mathcal{P}_3^{\ell+2}$. For each ℓ, find an element of $\mathcal{P}_3^{\ell+2}$ that is not in the image of r^2. For each ℓ, find the kernel of r^2 in $\mathcal{P}_3^\ell$.*

Exercise 2.10 *Prove that the function $\mathbb{R} \to \mathbb{R}$, $x \mapsto \sin x$ is not a polynomial. Is $x \mapsto \arcsin(\sin x)$ a polynomial? Is $x \mapsto \sin(\arcsin x)$ a polynomial?*

Exercise 2.11 *Show that the dimension of the vector space of homogeneous polynomials of degree n on $\mathbb{R}^d$ is $(n + d - 1)!/(n!(d - 1)!)$.*

Exercise 2.12 *Show that the set $\mathcal{C}_2$ of twice-differentiable complex-valued functions on $\mathbb{R}^3$ is a complex vector space. Find its dimension. Show that the Laplacian ∇^2 is a linear operator on $\mathcal{C}_2$.*

Exercise 2.13 *Suppose V is a complex vector space of finite dimension. Suppose W is a subspace of V and $\dim W = \dim V$. Show that $W = V$.*

Exercise 2.14 (Used in Section 5.5) *Let V denote a complex vector space. Let V^* denote the set of complex linear transformations from V to $\mathbb{C}$. Show that V^* is a complex vector space. Show that if V is finite dimensional then $\dim V^* = \dim V$. The vector space V^* is called the dual vector space or, more simply, the dual space.*

Exercise 2.15 (Used in Proposition 11.1) *Suppose V is a finite-dimensional complex vector space. Show that $V = (V^*)^*$. (See Exercise 2.14 for a definition of the dual V^*.) Is this true for all complex vector spaces?*

Exercise 2.16 *Consider the kets of a spin-1/2 system. Physicists know that we can express any ket $c_+ |+\mathbf{z}\rangle + c_- |-\mathbf{z}\rangle$ in terms of the x-axis basis. That is, there are complex numbers b_+ and b_- such that $c_+ |+\mathbf{z}\rangle + c_- |-\mathbf{z}\rangle = b_+ |x+\rangle + b_- |x-\rangle$. Is the function taking a pair (c_+, c_-) to a pair (b_+, b_-) a linear transformation?*

Exercise 2.17 (Used in Proposition 8.9) *Suppose T is a linear transformation from a finite-dimensional vector space V to a vector space W. Suppose*

T takes a basis of V to a basis of W. Show that T is an isomorphism of vector spaces.

Exercise 2.18 *Show that the composition of two linear transformations is a linear transformation.*

Exercise 2.19 *Show that the function* $\mathbb{R} \to \mathbb{R}$, $x \mapsto x+1$ *is not a real linear transformation. Why do you think this function is often called "linear" in precalculus and calculus classes?*

Exercise 2.20 (Used in Section 3.4) *Show that if T is a linear transformation with domain V, and W is any linear subspace of V, then the restriction* $T|_W$ *of T to W is a linear transformation.*

Exercise 2.21 *Let* $\mathcal{P}^\ell$ *denote the complex vector space of homogeneous complex-valued polynomials of degree* ℓ *in three real variables. Consider the linear transformation* ∇^2_ℓ *defined as the restriction of the Laplacian* ∇^2 *to* $\mathcal{P}^\ell$. *Show that the image of this linear transformation lies in* $\mathcal{P}^{\ell-2}$.

Exercise 2.22 *Show that the matrix B in Equation 2.5 is invertible, so it makes sense to write* B^{-1}.

Exercise 2.23 *Show that* $\{(x+iy)^2, (x+iy)z, (x+iy)(x-iy), (x-iy)z, (x-iy)^2\}$ *is a basis of the complex vector space* $\mathbb{H}^2$ *of homogeneous harmonic polynomials of degree 2. Find the matrix B that changes this basis into the basis* $\{xy, yz, xz, x^2 - y^2, y^2 - z^2, 2z^2 - x^2 - y^2\}$.

Exercise 2.24 (Used in Exercise 3.20) *Suppose V and W are vector spaces. Define* $\mathrm{Hom}(V, W)$ *to be the set of linear transformations from V to W. Show that* $\mathrm{Hom}(V, W)$ *is a vector space. Express its dimension in terms of the dimensions of V and W.*

Exercise 2.25 *Show that the determinant of a linear transformation is the product of its eigenvalues (with multiplicity). Show that the trace of a linear transformation is the sum of its eigenvalues (with multiplicity).*

Exercise 2.26 *Suppose* $T: V \to V$ *is a linear operator and* λ *is an eigenvalue of T. Show that the set*

$$\{v \in V : Tv = \lambda v\}$$

is a nontrivial vector subspace of V. This set is called the λ-eigenspace (of *T) or, more succinctly, an* eigenspace.

Exercise 2.27 *Show that if $v \in \mathbb{C}^n$ and $w \in (\mathbb{C}^m)^*$, then the $n \times m$ matrix vw has rank at most one. Under what conditions on v and w is the rank of the matrix vw zero? Show that if a matrix M has positive rank k, then one can write it as the sum of k products:*

$$M = \sum_{j=1}^{k} v_j w_j,$$

where each $v_j \in \mathbb{C}^n$ and each $w_j \in (\mathbb{C}^m)^*$). Show that the matrix in Formula 2.6 has rank two.

Exercise 2.28 *Find a nontrivial complex vector space V and a linear operator T from V to V such that T has no eigenvalues. (Hint: consider the space $\bigoplus_{n \in \mathbb{N}} \mathbb{C}$, which is, by definition, the complex vector space of sequences of complex numbers with only a finite number of nonzero entries. Then think about shifting sequences to the left or right.)*

Exercise 2.29 *Suppose V is a finite-dimensional real vector space. Suppose $T : V \to V$ is a linear operator and $\det T = 0$. Show that there is a vector $v \in V$ such that $Tv = 0$. (Readers familiar with fields should prove this statement for finite-dimensional vector spaces over any field.)*

Exercise 2.30 *Define an equivalence of matrices by: $A_1 \sim A_2$ if and only if there is a matrix B such that $A_1 = BA_2B^{-1}$. Show that matrix multiplication is well defined on equivalence classes. Show that trace and determinant are well defined on equivalence classes. Show that eigenvalues are well defined, but eigenvectors are not. Finally, show that given a vector space V, any linear operator on V corresponds to precisely one equivalence class of matrices.*

Exercise 2.31 *Suppose V is a finite-dimensional vector space.*

1. Define an equivalence relation $\sim$ on the tensor product

$$\bigotimes_{k=1}^{n} V$$

in the style of Definition 2.13, using the computation rules

$$v_1 \otimes \cdots \otimes v_n \sim v_{\sigma(1)} \otimes \cdots \otimes v_{\sigma(n)}$$

for each permutation σ of n numbers.

2. *For any natural number n, define the* symmetric tensor product

$$\text{Sym}^n V := \left(\bigotimes_{k=1}^{n} V \right) / \sim .$$

Show that $\text{Sym}^n V$ is a vector space and that its dimension is $\binom{d+n-1}{n}$, where d is the dimension of V.

3. *Suppose W is a finite-dimensional vector space. For each natural number n, construct an isomorphism between the vector space*

$$\bigoplus_{j+k=n} (\text{Sym}^j V \otimes \text{Sym}^k W)$$

and the vector space $\text{Sym}^n (V \oplus W)$.

Exercise 2.32 *Suppose V is a finite-dimensional vector space.*

1. *Define an equivalence relation $\sim$ on the tensor product*

$$\bigotimes_{k=1}^{n} V$$

in the style of Definition 2.13, using the computation rules

$$v_1 \otimes \cdots \otimes v_n \sim \text{sgn}(\sigma) v_{\sigma(1)} \otimes \cdots \otimes v_{\sigma(n)}$$

for each permutation σ of n numbers. Here $\text{sgn}(\sigma)$ is the sign of the permutation σ.

2. *For any natural number n, define the* alternate tensor product

$$\Lambda^n V := \left(\bigotimes_{k=1}^{n} V \right) / \sim .$$

Show that $\Lambda^n V$ is a vector space and that its dimension is $\binom{d}{n}$, where d is the dimension of V.

3. *Suppose W is a finite-dimensional vector space. For each natural number n, construct an isomorphism between the vector space*

$$\bigoplus_{j+k=n} (\Lambda^j V \otimes \Lambda^k W)$$

and the vector space $\Lambda^n (V \oplus W)$.

Exercise 2.33 *Think of a set of computation rules you have used in some other context. Can you define an equivalence relation from them in the style of Definition 2.13?*

3
Complex Scalar Product Spaces (a.k.a. Hilbert Spaces)

Hermione stepped forward.

"Neville," she said, "I'm really, really sorry about this."

She raised her wand.

"Petrificus Totalus!" she cried, pointing it at Neville.

Neville's arms snapped to his sides. His legs sprang together. His whole body rigid, he swayed where he stood and then fell flat on his face, stiff as a board.

—J.K. Rowling, *Harry Potter and the Sorcerer's Stone* [Row, p. 273]

The natural mathematical setting for any quantum mechanical problem is a *complex scalar product space*, defined in Definition 3.2. The primary complex scalar product space used in the study of the motion of a particle in three-space is called $L^2(\mathbb{R}^3)$, pronounced "ell-two-of-are-three." Our analysis of the hydrogen atom (and hence the periodic table) will require a few other complex scalar product spaces as well. Also, the representation theory we will introduce and use depends on the abstract notion of a complex scalar product space. In this chapter we introduce the complex vector space $L^2(\mathbb{R}^3)$, define complex scalar products, discuss and exploit analogies between complex scalar products and the familiar Euclidean dot product[1] and do some of the analysis necessary to apply these analogies to infinite-dimensional complex scalar product spaces.

---

[1] Also known as the *real scalar product* or the *inner product*.

Physicists often refer to complex scalar product spaces as *Hilbert spaces*. The formal mathematical definition of a Hilbert space requires more than just the existence of a complex scalar product: the space must be "closed" a.k.a. "complete" in a certain technical sense. Because every scalar product space is a subset of some Hilbert space, the discrepancy in terminology between mathematicians and physicists does not have dire consequences. However, in this text, to avoid discrepancies with other mathematics textbooks, we will use "complex scalar product."

In Section 3.1 we introduce Lebesgue equivalence and define the complex vector space $L^2(\mathbb{R}^3)$. In Section 3.2 we define complex scalar products in general and on $L^2(\mathbb{R}^3)$ in particular. We show how the complex scalar product helps us to use our orthogonal Euclidean intuition to study complex scalar product spaces in Section 3.3. In particular, we introduce orthogonal projections and complementary subspaces. In Section 3.4 we introduce norms and use them to define approximation. Finally, we give several approximation theorems in Section 3.5.

3.1   Lebesgue Equivalence and $L^2(\mathbb{R}^3)$

Perhaps the reader has noticed some caginess in the introductory paragraph of this chapter. Why did we not say simply that the Hilbert space $L^2(\mathbb{R}^3)$ is the set of wave functions on three-space? Such a statement would be at best vague and at worst false, due to a mathematical subtlety: if we cannot distinguish two functions via integration, we should consider them equivalent. The standard mathematical definition of a function says that in order for two functions to be equal they must take the same value at every point. This requirement is too stringent for us. In quantum mechanical calculations, we never evaluate wave functions at particular points. The most we ever do is multiply two functions together and take an integral, as in Equation 1.3 in Section 1.2. (Note that the other common quantum mechanical calculation, integrating the absolute value squared of a wave function as in Equation 1.2, can be accomplished by multiplication followed by integration.) Thus we would like to consider two functions the same if they cannot be distinguished by multiplication followed by integration.

We make this idea precise by defining an equivalence relation (see Section 1.7): we define $\phi_1 \sim \phi_2$ (and say ϕ_1 *is equivalent to* ϕ_2) if and only if, for all functions $\psi : \mathbb{R}^3 \to \mathbb{C}$ and subsets A of $\mathbb{R}^3$, we have $\int_S \psi \phi_1 = \int_S \psi \phi_2$ whenever both integrals are well defined. This equivalence relationship is not trivial; there are indeed functions that do not agree pointwise yet are equivalent — see Exercise 3.5.

A fully rigorous treatment of this equivalence relation requires the notions of *measurable functions* and the *Lebesgue integral*. This integral is one of the mainstays of modern mathematics, necessary for the proper definition of the Fourier transform. We recommend that budding mathematicians study Lebesgue integration thoroughly at some point. However, it is not a prerequisite for this book. Readers unfamiliar with Lebesgue integration must take it on faith that in calculations the Lebesgue integral behaves just like the Riemann integral taught in most first-year calculus courses. The advantage of the Lebesgue integral is that it applies to a wider class of functions than does the Riemann integral, and that there are a few theorems (such as the *Lebesgue dominated convergence theorem*) that apply to the Lebesgue integral alone. The Lebesgue integral is particularly well suited to situations where one is interested in calculating probabilities. Functions which can be integrated via the Lebesgue integral are called *measurable functions*. Anyone wishing to learn more might consult the intuitive overview by Dym and McKean [DyM, Section 1.1] or the rigorous treatment of Rudin [Ru74, Chapters 1 and 2].

One theorem from the theory of Lebesgue integration will be particularly helpful to us. Fubini's theorem answers the question, "How do we know that we can switch the order of integration?" A physical scientist might answer that she and her colleagues have done it hundreds, if not thousands, of times without ill consequences. A mathematician needs a different kind of justification. In fact, it is possible to construct counterexamples: functions giving different values for different orders of integration. Fubini's theorem assures mathematicians that given one simple condition, one can switch the order of integration without changing the value of the integral. Fubini's theorem has another, more subtle use: it guarantees that certain functions defined by Lebesgue integration are well defined (up to Lebesgue equivalence).

We will need only one special case of Fubini's theorem.

Theorem 3.1 (Fubini's Theorem) *Suppose f is a measurable complex-valued function of three variables. Suppose further that*

$$\int_{\mathbb{R}^3} |f(r, \theta, \phi)| \, r^2 dr \, \sin\theta \, d\theta \, d\phi < \infty.$$

Then the function

$$F_1(r) := \int_{S^2} f(r, \theta, \phi) \sin\theta \, d\theta \, d\phi$$

is a well defined measurable function (possibly taking infinite values) on $\mathbb{R}^{\geq 0}$, the function

$$F_2(\theta, \phi) := \int_0^\infty f(r, \theta, \phi) r^2 dr$$

is a well defined measurable function (possibly taking infinite values) on S^2 and

$$\int_0^\infty F_1(r) r^2 dr = \int_{S^2} F_2(\theta, \phi) \sin\theta \, d\theta \, d\phi$$

$$= \int_{\mathbb{R}^3} |f(r, \theta, \phi)| \, r^2 dr \sin\theta \, d\theta \, d\phi < \infty.$$

We will use Fubini's theorem in the proofs of Propositions 7.7 and A.3, both to define measurable functions and to switch the order of integration. A proof of Fubini's theorem is available in [Hal50, Section 36] or [Ru74, Theorem 7.8].

Next we define the complex vector space $L^2(\mathbb{R}^3)$:

Definition 3.1 *Let $L^2(\mathbb{R}^3)$ denote the set*

$$\left\{ f : f \text{ is a measurable function from } \mathbb{R}^3 \text{ to } \mathbb{C}, \int_{\mathbb{R}^3} |f|^2 \leq \infty \right\} / \sim,$$

i.e., $L^2(\mathbb{R}^3)$ is the set of equivalence classes of square-integrable complex-valued functions on $\mathbb{R}^3$, under the equivalence relation $\sim$ defined above.

It may not be immediately obvious that $L^2(\mathbb{R}^3)$ is indeed a vector space. The trickiest part is to show that if $f \in L^2(\mathbb{R}^3)$ and $g \in L^2(\mathbb{R}^3)$, then $f + g \in L^2(\mathbb{R}^3)$. Because the usual rules of integration hold for Lebesgue integrals, the result follows from the observation that for any two numbers a and b in $\mathbb{C}$ we have $|a + b|^2 \leq 2|a|^2 + 2|b|^2$. Thus $\int_{\mathbb{R}^3} |f + g|^2 \leq 2\int_{\mathbb{R}^3} |f|^2 + 2\int_{\mathbb{R}^3} |g|^2 < \infty$, so $f + g \in L^2(\mathbb{R}^3)$. The reader should check that the other criteria of Definition 2.1 are satisfied.

A second bit of caginess in the introduction is our statement that $L^2(\mathbb{R}^3)$ is "the primary complex scalar product space used in the study of a particle in three-space." Beware the passive voice! We used it here to gloss over the fact that $L^2(\mathbb{R}^3)$ is not the set of all states of the particle. The fact that we want $\int_{\mathbb{R}^3} |\phi|^2 = 1$ is only part of the story. Because the only numbers we can measure physically are of the form $\left| \int_A \psi^* \phi \right|$, we cannot distinguish between two wave functions ϕ_1 and ϕ_2 such that

$$\left| \int_A \psi^* \phi_1 \right| = \left| \int_A \psi^* \phi_2 \right| \tag{3.1}$$

for all suitable functions ψ and sets A. For example, if we take any function $\phi_1(x, y, z)$ and any real number u and define $\phi_2(x, y, z) := e^{iu}\phi_1(x, y, z)$ for all (x, y, z), then the constant *phase factor* e^{iu} will not affect the absolute value of the integral and Equation 3.1 will be satisfied for all suitable functions ψ and sets A.

To be absolutely precise, a one-dimensional subspace of $L^2(\mathbb{R}^3)$ describes the state of a particle moving in $\mathbb{R}^3$ — that is, each one-dimensional subspace can be used to predict the outcome of any quantum mechanical experiment involving the particle's position. Physicists call these subspaces *rays*. Just as the familiar rays of Euclidean geometry (such as the positive x-axis) are closed under multiplication by a positive real number, these subspaces are closed under multiplication by a complex scalar. Note that these quantum-mechanical rays are one-dimensional as complex vector spaces. See Exercise 2.6. Many people find it easier to think of vectors rather than rays, and in many, many situations (including the first eight chapters of this book) there is no harm done by thinking of quantum states as vectors. The natural mathematical way to deal with the issue of different wave functions labelling the same state is, as before, to introduce an equivalence relation. Physicists sometimes refer to this equivalence as *ray equivalence*. This leads to the notion of a *projective vector space*. We introduce projective vector spaces formally in Section 10.1. Readers who wish to understand spin rigorously must study projective vector spaces and rays; readers who are willing to fudge some of the details can save effort by pretending that states correspond to single vectors and by keeping in mind that the phase factor sometimes introduces some complications.

We hope that this section has made clear the precise relationship between the space $L^2(\mathbb{R}^3)$ and the state space of a mobile quantum mechanical particle in $\mathbb{R}^3$. Although $L^2(\mathbb{R}^3)$ is not, strictly speaking, the state space in question, it is close enough to provide a reasonable model.

3.2   Complex Scalar Products

We start with the definition of a *complex scalar product* (also known as a *Hermitian inner product*, a *complex inner product* or a *unitary structure*) on a complex vector space. Then we present several examples of complex scalar product spaces.

Definition 3.2 *Let V be a complex vector space. An operation*

$$\langle \cdot, \cdot \rangle : V \times V \to \mathbb{C}$$

is a complex scalar product *if and only if it satisfies:*

1. *The operation* $\langle \cdot, \cdot \rangle$ is linear in the second argument. *In other words, for all* $c \in \mathbb{C}$ *and all* $v, w_1, w_2 \in V$ *we have* $\langle v, w_1 + w_2 \rangle = \langle v, w_1 \rangle + \langle v, w_2 \rangle$ *and* $\langle v, cw_1 \rangle = c \langle v, w_1 \rangle$.

2. *The bracket is* Hermitian symmetric. *In other words, for all* $v, w \in V$ *we have* $\langle v, w \rangle = \langle w, v \rangle^*$, *where the* * *denotes complex conjugation.*

3. *The bracket is* positive definite: *for all* $v \in V$ *we have* $\langle v, v \rangle \geq 0$. *Also, it is* nondegenerate: $\langle v, v \rangle = 0$ *if and only if* $v = 0$.

Mathematicians should note that we have taken the physicists' convention in criterion 1; in many mathematics texts, the definition requires linearity in the first argument. See Exercise 3.4. A complex vector space with a complex scalar product defined on it is known as a *complex scalar product space*. The complex scalar product is sometimes called a *unitary structure* on the space.

Physicists should take special note of the positive definiteness. Although there are many useful examples of brackets that satisfy all but the positive definiteness requirement (such as the Minkowski metric on spacetime in special relativity), we are concerned here with positive definite brackets.

For example, for any natural number n there is a natural complex scalar product on the n-dimensional complex vector space $\mathbb{C}^n$ defined by

$$\left\langle \begin{pmatrix} v_1 \\ \vdots \\ v_n \end{pmatrix}, \begin{pmatrix} w_1 \\ \vdots \\ w_n \end{pmatrix} \right\rangle := \sum_{j=1}^{n} v_j^* w_j = v^* w,$$

where the last expression is matrix multiplication of a row n-vector (v^*) and a column n-vector (w). It is not hard to check that this operation satisfies the three requirements of a complex scalar product. For instance, to check the last criterion, note that

$$\langle v, v \rangle = \sum_{j=1}^{n} |v_j|^2 \geq 0,$$

with equality only if $v = 0$. The space $\mathbb{C}^n$ may be familiar to physicists from the analysis of spin-$(n-1)/2$ systems. In particular, if $n = 2$, this is the complex vector space for spin states of the electron, as we discuss in Section 10.2.

There are other complex scalar products on $\mathbb{C}^n$ as well. In fact, for any set of strictly positive real numbers $\lambda_1, \ldots, \lambda_n$, there is a complex scalar product defined by

$$\left\langle \begin{pmatrix} v_1 \\ \vdots \\ v_n \end{pmatrix}, \begin{pmatrix} w_1 \\ \vdots \\ w_n \end{pmatrix} \right\rangle := \sum_{j=1}^{n} \lambda_j v_j^* w_j = v^* \begin{pmatrix} \lambda_1 & & \\ & \ddots & \\ & & \lambda_n \end{pmatrix} w.$$

Again, the proof is straightforward; for instance,

$$\langle v, v \rangle = \sum_{j=1}^{n} \lambda_j \left| v_j \right|^2 \geq 0,$$

with equality only if $v = 0$. More generally, any Hermitian-symmetric matrix with positive eigenvalues corresponds to a complex scalar product on $\mathbb{C}^n$, and vice versa. See Exercise 3.25.

Recall the vector space $\mathcal{P}^n$ of homogeneous polynomials of degree n in two variables defined in Section 2.2. We will find it useful (see Proposition 4.7) to define the following complex scalar product on $\mathcal{P}^n$:

$$\left\langle a_0 x^n + a_1 x^{n-1} y + \cdots + a_n y^n, b_0 x^n + b_1 x^{n-1} y + \cdots + b_n y^n \right\rangle$$

$$:= \sum_{k=0}^{n} a_k^* b_k k! (n - k)!.$$

Because $k!(n - k)! > 0$ for each $k = 0, \ldots, n$, this bracket satisfies Definition 3.2.

One complex scalar product on $C[-1, 1]$, the complex vector space of continuous functions on $[-1, 1]$, is

$$\langle f, g \rangle := \frac{1}{2} \int_{-1}^{1} f^*(t) g(t) dt.$$

The verification of Definition 3.2 follows from the basic properties of integration of continuous functions. The hardest part is to show that if

$$\frac{1}{2} \int_{-1}^{1} |f(t)|^2 dt = 0,$$

then $f(t) = 0$ for all $t \in [-1, 1]$. But if $f(t_0) \neq 0$ then there is an interval J of strictly positive length containing t_0 such that $|f(t)| > 0$ for all $t \in J$. See Figure 3.1. So we have

$$\int_{-1}^{1} |f(t)|^2 dt \geq \int_{J} |f(t)|^2 dt > 0.$$

So the proposed scalar product satisfies Definition 3.2.

Does our main example, the bracket on $L^2(\mathbb{R}^3)$, satisfy the definition?

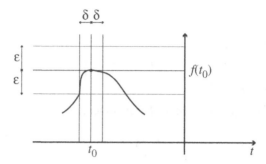

Figure 3.1. If a continuous function is nonzero at a point, then it is nonzero over an interval. The definition of continuity ensures that for any $\epsilon > 0$ there is a $\delta > 0$ such that if $|t - t_0| < \delta$ then $|f(t) - f(t_0)| < \epsilon$.

Proposition 3.1 *For any two functions $f, g \in L^2(\mathbb{R}^3)$, define*

$$\langle f, g \rangle := \int_{\mathbb{R}^3} f^* g.$$

This bracket is a complex scalar product.

Proof. [Sketch] We leave it to the reader to check the first two criteria of Definition 3.2. As for Criterion 3, positive definiteness follows directly from the definition of the integral, while nondegeneracy can be deduced from the theory of Lebesgue integration, using the first equivalence relation defined in Section 3.1. The interested reader can work out the details in Exercise 3.9 or consult Rudin [Ru74, Theorem 1.39]. □

Finally, we introduce another complex scalar product space necessary to our analysis.

Definition 3.3 *Suppose S is a set on which integration is well defined. Let $L^2(S)$ denote the complex vector space*

$$\left\{ f : f \text{ is a measurable function from } S \text{ to } \mathbb{C}, \int_S |f|^2 \leq \infty \right\} / \sim,$$

where the equivalence relation $\sim$ is defined as for $L^2(\mathbb{R}^3)$, mutatis mutandis. Define

$$\langle f, g \rangle := \int_S f^* g.$$

The verification that $L^2(S)$ is a vector space and that $\langle \cdot, \cdot \rangle$ is a complex scalar product resembles the corresponding verifications for $L^2(\mathbb{R}^3)$. For instance, we will eventually consider the spaces $L^2(S^2)$, where S^2 is the unit sphere

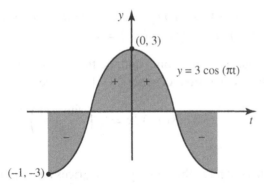

Figure 3.2. The signed area under the graph of $3\cos(\pi\cdot)$ on the interval $[-1, 1]$ is zero.

in $\mathbb{R}^3$, as well as $L^2(B_R)$, where B_R is a ball of radius R around the origin in $\mathbb{R}^3$.

Complex scalar products arise naturally in quantum mechanics because there is an experimental interpretation for the complex scalar product of two wave functions (as we saw in Section 1.2). Students of physics should note that the traditional "brac-ket" notation[2] is consistent with our complex scalar product notation — just put a bar in place of the comma. The physical importance of the bracket will allow us to apply our intuition about Euclidean geometry (such as orthogonality) to states of quantum systems.

3.3 Euclidean-style Geometry in Complex Scalar Product Spaces

Since a complex scalar product resembles the Euclidean dot product in its form and definition, we can use our intuition about perpendicularity in the Euclidean three-space we inhabit to study complex scalar product spaces. However, we must be aware of two important differences. First, we are dealing with complex scalars rather than real scalars. Second, we are often dealing with infinite-dimensional spaces. It is easy to underestimate the trouble that infinite dimensions can cause. If this section seems unduly technical (especially the introduction to orthogonal projections), it is because we are careful to avoid the infinite-dimensional traps.

By analogy to the geometry of Euclidean space we define perpendicularity.

---

[2]We propose that "brac-ket" be used in place of the more popular but orthographically inferior "bra-ket".

Definition 3.4 *Suppose V is a complex scalar product. Two vectors v_1, v_2 in V are* perpendicular *if and only if $\langle v_1, v_2 \rangle = 0$.*

For example, the constant function 3 and the function $\cos \pi x$ are perpendicular in the complex scalar product space $C[-1, 1]$ since

$$\langle 3, \cos(\pi \cdot) \rangle = \frac{1}{2} \int_{-1}^{1} (3^*) \cos(\pi x) dx = \frac{3}{2} \left(\frac{\sin(\pi x)}{\pi} \right) \Big|_{x=-1}^{x=1} = 0.$$

See Figure 3.2.

Recall from linear algebra that orthogonal matrices have columns that form an orthonormal basis. Orthogonal linear operators *preserve the Euclidean structure*, i.e., if we let a dot denote the Euclidean dot product we have

$$(T v_1) \cdot (T v_2) = v_1 \cdot v_2$$

for any Euclidean vectors v_1 and v_2. By analogy we define a *unitary* operator to be one that preserves the complex scalar product.

Definition 3.5 *Suppose V is a complex scalar product space and $T : V \to V$ is a linear transformation. We say that T is a* unitary *operator if and only if for all $v_1, v_2 \in V$ we have*

$$\langle v_1, v_2 \rangle = \langle T v_1, T v_2 \rangle .$$

Unitary operators are also known as *complex orthogonal* operators. If we use the standard basis and the standard complex scalar product on $\mathbb{C}^n$, then the columns of the matrix of a unitary operator are all mutually perpendicular and have length one.[3] In other words, a transformation $T : \mathbb{C}^n \to \mathbb{C}^n$ is unitary if and only if $T^*T = I$.

In Euclidean space we sometimes talk of complementary subspaces. For example, the z-axis is the complementary subspace to the xy-plane inside $\mathbb{R}^3$. We define complementary subspaces of complex scalar product spaces.

Definition 3.6 *Suppose B is an arbitrary subset of a complex scalar product space V. Then the* perpendicular space *to B in V is*

$$B^\perp := \{x \in V : \forall y \in B \ \langle x, y \rangle = 0\} .$$

If W is a subspace of V, then the perpendicular space $W^\perp$ is called the complementary subspace *of W in V.*

---

[3] A vector v in a complex scalar product space has length one if and only if $\langle v, v \rangle = 1$. See Definition 3.12.

Often the ambient space V is clear from context, so the notation does not reflect the dependence of the perpendicular space on V. The issue is the same in Euclidean space: the space perpendicular to the x-axis might be the y-axis (in the plane) or the yz-plane (in three-space).

In Euclidean space, orthonormal bases help both to simplify calculations and to prove theorems. *Unitary bases*, also called *complex orthonormal bases*, play the same role in complex scalar product spaces. To define a unitary basis for arbitrary (including infinite-dimensional) complex scalar product spaces, we first define spanning.

Definition 3.7 *Suppose B is a subset of a complex scalar product space V. If $B^\perp = \{0\}$ in V, then we say that B spans V.*

If V is finite dimensional, then Definition 3.7 is consistent with Definition 2.2 (Exercise 3.13). In infinite-dimensional complex scalar product spaces, Definition 3.7 is usually simpler than an infinite-dimensional version of Definition 2.2. To make sense of an infinite linear combination of functions, one must address issues of convergence; however, arguments involving perpendicular subspaces are often relatively simple. We can now define unitary bases.

Definition 3.8 *Suppose V is a complex scalar product space and B is a subset of V. Suppose that B satisfies the following:*

1. *For all $b_1, b_2 \in B$ with $b_1 \neq b_2$, we have $\langle b_1, b_2 \rangle = 0$;*

2. *For all $b \in B$ we have $\|b\| = 1$.*

3. *The perpendicular space to B inside V contains only the zero element, i.e., $B^\perp = \{0\}$;*

Then B is a unitary basis *of V.*

For example, if we consider $\mathbb{C}^n$ with the standard complex scalar product, then the set $\{e_k: k = 1, \ldots, n\}$, where e_k denotes the vector whose kth entry is 1 and all of whose other entries are 0, is a unitary basis of V. A more sophisticated example (left to readers in Exercise 3.14) is that the set of functions

$$\left\{ \frac{1}{\sqrt{2}} e^{ik(\cdot)} : k \in \mathbb{Z} \right\}$$

is a unitary basis of $L^2[-1, 1]$.

The next proposition gives a convenient way to recognize unitary transformations and construct unitary bases.

Proposition 3.2 *Suppose V is a finite-dimensional complex scalar product space. Suppose $T : V \rightarrow V$ is a linear operator. Then T is unitary if and only if the columns of its matrix in any unitary basis form a unitary basis.*

We will use this proposition in Section 4.4.

Proof. First suppose T is unitary and suppose that $B = \{b_1, \ldots, b_n\}$ is a unitary basis of V. Then the kth column of the matrix of T in the basis B consists of the coefficients of the vector Tb_k in the basis B. In other words,

$$Tb_k = \sum_{j=1}^{n} T_{jk} b_j.$$

We must show that the set $\{Tb_1, \ldots, Tb_n\}$ is a unitary basis for V. If $k \neq j$ we have

$$\langle Tb_j, Tb_k \rangle = \langle b_j, b_k \rangle = 0,$$

where the first equality follows from the hypothesis that T is unitary and the second from the hypothesis that B is a basis. Similarly, for any $b_k \in B$, we have

$$\|Tb_k\| = \|b_k\| = 1.$$

It follows that the Tb_k's are linearly independent; since there are n of them, the set $\{Tb_1, \ldots, Tb_n\}^{\perp} = 0$.

On the other hand, suppose that the columns of T in any unitary basis form a unitary basis. Suppose $B = \{b_1, \ldots, b_n\}$ is a unitary basis of V. Then for any complex n-tuples $(a_1, \ldots, a_n)$ and $(c_1, \ldots, c_n)$ we have

$$\left\langle \sum_{k=1}^{n} a_k b_k, \sum_{j=1}^{n} c_j b_j \right\rangle = \sum_{k=1}^{n} a_k^* c_k$$

$$= \left\langle \sum_{k=1}^{n} a_k T b_k, \sum_{j=1}^{n} c_j T b_j \right\rangle = \left\langle T \sum_{k=1}^{n} a_k b_k, T \sum_{j=1}^{n} c_j b_j \right\rangle.$$

Hence T is unitary.   □

For the proof of Proposition 11.1 in Section 11.3 we will need *adjoint linear transformations*, also known more briefly as *adjoints*, defined below. Adjoints arise in many fields of mathematics. Although, with appropriate care, adjoints can be defined in infinite-dimensional complex scalar product spaces, we will limit ourselves to the finite-dimensional case.

Definition 3.9 *Suppose V and W are finite-dimensional complex scalar product spaces, and let $\langle \cdot, \cdot \rangle_V$ and $\langle \cdot, \cdot \rangle_W$ denote their complex scalar products. Suppose $T : V \to W$ is a linear transformation, that is, suppose $T \in$ Hom (V, W). Then the* adjoint *of T is the unique linear transformation $T^* : W \to V$ such that for all $v \in V$ and all $w \in W$ we have*

$$\langle w, Tv \rangle_W = \langle T^*w, v \rangle_V .$$

To justify this definition we must show that T^* exists and is unique. We show uniqueness first. Suppose U satisfies the definition as well as T^*. Then for any $w \in W$ and $v \in V$ we have

$$\langle Uw, v \rangle_V = \langle w, Tv \rangle_W = \langle T^*w, v \rangle_V .$$

Because the complex scalar product on V is nondegenerate (by condition 3 of Definition 3.2), we conclude that for any $w \in W$ we have $Uw - T^*w = 0$. Hence $U = T^*$, completing the proof of uniqueness. To show existence, let $\{v_1, \ldots, v_m\}$ be a unitary basis for V and let $\{w_1, \ldots, w_n\}$ be a basis for W. Let A denote the matrix of the transformation T in these bases, i.e., for any $j = 1, \ldots, m$ and any $k = 1, \ldots, n$,

$$A_{kj} = \langle w_k, Tv_j \rangle_W .$$

Then define T^* to be the linear transformation from W to V whose matrix in the given basis is the conjugate transpose A^* of A: for each j and k we have matrix entries

$$A^*_{jk} := (A_{kj})^* .$$

Does this T^* have the desired property? By the bilinearity of the complex scalar products (condition 1 of Definition 3.2), it suffices to check the condition on basis elements. For any j and any k we have

$$\langle T^*w_k, v_j \rangle_V = \left(A^*_{jk} \right)^* = A_{kj} = \langle w_k, Tv_j \rangle_W .$$

So T^* has the desired property, completing the proof of existence.

For example, consider the linear transformation $T : \mathbb{C}^3 \to \mathbb{C}^2$ defined by the matrix

$$\begin{pmatrix} 1 & 0 & 0 \\ 0 & 0 & i \end{pmatrix}$$

in the standard bases of $\mathbb{C}^3$ and $\mathbb{C}^2$. The matrix of the adjoint transformation $T^* : \mathbb{C}^2 \to \mathbb{C}^3$ in these bases is

$$\begin{pmatrix} 1 & 0 \\ 0 & 0 \\ 0 & -i \end{pmatrix},$$

as we will now check. For any $v \in \mathbb{C}^3$ and any $w \in \mathbb{C}^2$, we have

$$\langle w, Tv \rangle_2 = \left\langle \begin{pmatrix} w_1 \\ w_2 \end{pmatrix}, \begin{pmatrix} 1 & 0 & 0 \\ 0 & 0 & i \end{pmatrix} \begin{pmatrix} v_1 \\ v_2 \\ v_3 \end{pmatrix} \right\rangle_2 = w_1^* v_1 + i w_2^* v_3$$

$$= \left\langle \begin{pmatrix} 1 & 0 \\ 0 & 0 \\ 0 & -i \end{pmatrix} \begin{pmatrix} w_1 \\ w_2 \end{pmatrix}, \begin{pmatrix} v_1 \\ v_2 \\ v_3 \end{pmatrix} \right\rangle_3 = \langle T^* w, v \rangle_3 .$$

Although our definition of adjoint applies only to finite-dimensional vector spaces, we cannot resist giving an infinite-dimensional example. The proof of uniqueness works for infinite-dimensional spaces as well, but our proof of existence fails.[4] Fix an element $\alpha \in L^2(\mathbb{R}^3)$ and consider the linear transformation $T : L^2(\mathbb{R}^3) \to \mathbb{C}$ defined by

$$Tf := \langle \alpha, f \rangle .$$

The adjoint of T is the linear transformation $T^* : \mathbb{C} \to L^2(\mathbb{R}^3)$ defined by

$$T^* c := c\alpha.$$

Indeed, for any $c \in \mathbb{C}$ and any function $f \in L^2(\mathbb{R}^3)$, we have

$$\langle c, Tf \rangle_{\mathbb{C}} = c^* \langle \alpha, f \rangle_{L^2(\mathbb{R}^3)} = \langle c\alpha, f \rangle_{L^2(\mathbb{R}^3)} .$$

For another infinite-dimensional example, see Exercise 3.24.

The complex scalar product lets us define an analog of Euclidean orthogonal projections. First we need to define *Hermitian* operators. These are analogous to symmetric operators on $\mathbb{R}^n$.

Definition 3.10 *Suppose V is a complex scalar product space. A* Hermitian *linear operator (also known as a* Hermitian symmetric *operator or self-adjoint operator) on V is a linear operator $T : V \to V$ such that for all $v_1, v_2 \in V$ we have*

$$\langle v_1, Tv_2 \rangle = \langle Tv_1, v_2 \rangle . \tag{3.2}$$

---

[4] If the infinite-dimensional complex scalar product space is a Hilbert space, in the strict mathematical sense, then there is a proof of existence. The main tool in the proof is the Riesz representation theorem. See any text on Hilbert spaces or functional analysis, such as [RS, Theorem II.4].

Note that on a finite-dimensional vector space V, a linear operator is Hermitian if and only if $T = T^*$. More concretely, in $\mathbb{C}^n$, a linear operator is Hermitian-symmetric if and only if its matrix M in the standard basis satisfies $M = M^*$, where M^* denotes the conjugate transpose matrix. To check that a linear operator is Hermitian, it suffices to check Equation 3.2 on basis vectors. Physics textbooks often contain expressions such as $\langle +\mathbf{z}|\, H\, |-\mathbf{z}\rangle$. These expressions are well defined only if H is a Hermitian operator. If H were not Hermitian, the value of the expression would depend on where one applies the H.

Now we can define projections.

Definition 3.11 *Suppose V is a complex scalar product space. An* orthogonal projection $\Pi: V \to V$ *is a Hermitian linear operator Π such that $\Pi^2 = \Pi$.*

To see that this algebraic definition corresponds to the geometric notion of projection, consider, for example, the projection onto the x-axis in $\mathbb{R}^3$. Because the projection acts like the identity on the x-axis itself, projecting twice yields the same result as projecting once. Furthermore, letting Π denote orthogonal projection onto the x-axis, the dot product of a vector v_1 with a vector Πv_2 parallel to the x-axis depends only on the x-components of v_1 and v_2, which is the geometric content of the second condition in Definition 3.11.

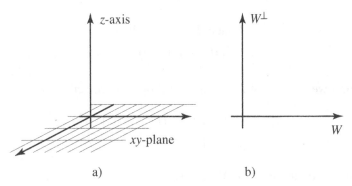

a)   *b)*

Figure 3.3. Complementary subspaces. a.) A literal picture of a real example. b.) A schematic picture of the general situation.

Next we prove a few technical propositions that will be useful to us later. These may seem obvious because their finite-dimensional real analogs are geometrically obvious; however, infinite-dimensional vector spaces are tricky and one must proceed carefully.

Proposition 3.3 *Suppose Π is an orthogonal projection. Let W denote the image of Π. Then if $w \in W$ we have $\Pi w = w$. Also, the kernel of Π is $W^{\perp}$.*

Proof. If w lies in W then, by the definition of the image, there is a $v \in V$ such that $w = Pv$. Then

$$\Pi w = \Pi^2 v = \Pi v = w.$$

To show that $W^\perp$ is the kernel of Π, note first that if $v_1 \in W^\perp$, then for any $v_2 \in V$ we have

$$\langle \Pi v_1, v_2 \rangle = \langle v_1, \Pi v_2 \rangle = 0,$$

since $\Pi v_2 \in W$ and $v_1 \in W^\perp$. By the nondegeneracy of the complex scalar product, it follows that $\Pi v_1 = 0$. Hence $W^\perp$ is a subset of the kernel of Π.

On the other hand, if v lies in the kernel of Π and $w \in W$ we have

$$\langle v, w \rangle = \langle v, \Pi w \rangle = \langle \Pi v, w \rangle = \langle 0, w \rangle = 0,$$

so $v \in W^\perp$. Hence the kernel of Π is a subset of $W^\perp$. Combining this conclusion with the conclusion of the previous paragraph we find that $W^\perp$ is equal to the kernel of Π. $\qquad\square$

Proposition 3.4 *Suppose Π is an orthogonal projection. Let W denote the image of Π. Then the function $I - \Pi$ is an orthogonal projection. Furthermore,*

$$\text{Image}(I - \Pi) = W^\perp \tag{3.3}$$
$$\ker(I - \Pi) = W. \tag{3.4}$$

Proof. First we verify that $I - \Pi$ is an orthogonal projection. We calculate

$$(I - \Pi)(I - \Pi) = I^2 - \Pi I - I\Pi + \Pi^2$$
$$= I - \Pi - \Pi + \Pi = I - \Pi.$$

Furthermore, for any $v_1, v_2 \in V$ we have

$$\langle v_1, (I - \Pi)v_2 \rangle = \langle v_1, v_2 \rangle - \langle v_1, \Pi v_2 \rangle = \langle I v_1, v_2 \rangle - \langle \Pi v_1, v_2 \rangle$$
$$= \langle (I - \Pi)v_1, v_2 \rangle .$$

So $I - \Pi$ is indeed an orthogonal projection.

Next we show that W is the kernel of $I - \Pi$. If $w \in W$ then

$$(I - \Pi)w = w - w = 0,$$

so W is a subset of the kernel of $I - \Pi$. On the other hand, if $(I - \Pi)v = 0$ then $v = \Pi v \in W$, so the kernel of $I - \Pi$ is a subset of W. Putting these two assertions together we find that the kernel of $I - \Pi$ is equal to W.

Finally, we show that the image of $I - \Pi$ is $W^\perp$. Suppose v lies in the image of $I - \Pi$. Then there is a u such that $v = (I - \Pi)u$ and hence for any $w \in W$ we have

$$\langle v, w \rangle = \langle u, w \rangle - \langle \Pi u, w \rangle = \langle u, w \rangle - \langle u, \Pi w \rangle = 0.$$

Hence $v \in W^\perp$. On the other hand, suppose that $v \in W^\perp$. Then $(I - \Pi)v = v$, so v lies in the image of $I - \Pi$. $\qquad\square$

Not every subspace W of V can be the image of an orthogonal projection (see Exercise 3.29). However, any finite-dimensional subspace can be the image of an orthogonal projection. In our investigation of the structure of the hydrogen atom we will want to construct orthogonal projections with finite-dimensional images. See Propositions 6.6, 6.7 and 7.6.

Proposition 3.5 *Suppose W is a finite-dimensional subspace of a complex scalar product space V. Then there is an orthogonal projection Π_W whose image is W. Also, there is an orthogonal projection $\Pi_{W^\perp}$ onto the subspace $W^\perp$ orthogonal to W.*

Proof. We will prove the first conclusion of this proposition by induction on the dimension of W. We start with the subspace of dimension zero, i.e., the trivial subspace $\{0\}$. It is easy to check that the linear transformation taking every vector of V to the zero vector is an orthogonal projection onto $\{0\}$.

Next we must prove the inductive step. Fix any natural number n. Suppose that there exists an orthogonal projection onto any subspace of dimension n. Consider a subspace W of dimension $n + 1$. Fix an element $w \in W$ such that $\langle w, w \rangle = 1$. Let $\tilde{W}$ denote the subspace of W perpendicular to w. Then $\tilde{W}$ has dimension n, so there is a well-defined orthogonal projection $\Pi_{\tilde{W}}$ onto $\tilde{W}$. Note that $\Pi_{\tilde{W}} w = 0$. Define $P : V \to V$ by

$$Pv := \Pi_{\tilde{W}}v + \langle w, v \rangle\, w.$$

We must verify that P is an orthogonal projection. First, we note that $\Pi_{\tilde{W}}(w) = 0$ and $\langle \Pi_{\tilde{W}}v, w \rangle = \langle v, \Pi_{\tilde{W}}w \rangle = 0$ and we calculate

$$\begin{aligned}
P^2 v &= \Pi_{\tilde{W}}\left(\Pi_{\tilde{W}}v + \langle w, v \rangle\, w\right) + \left\langle w, \Pi_{\tilde{W}}v + \langle w, v \rangle\, w\right\rangle w \\
&= \Pi_{\tilde{W}}v + \left\langle w, \Pi_{\tilde{W}}v\right\rangle w + \langle w, v \rangle \langle w, w \rangle\, w \\
&= \Pi_{\tilde{W}}v + \langle w, v \rangle\, w = Pv.
\end{aligned}$$

Next, let v_1 and v_2 denote arbitrary elements of V.

$$\begin{aligned}
\langle v_1, P v_2 \rangle &= \langle v_1, \Pi_{\tilde{W}}(v_2) + \langle w, v_2 \rangle\, w \rangle \\
&= \langle v_1, \Pi_{\tilde{W}} v_2 \rangle + \langle v_1, \langle w, v_2 \rangle\, w \rangle \\
&= \langle \Pi_{\tilde{W}} v_1, v_2 \rangle + \langle v_1, w \rangle \langle w, v_2 \rangle \\
&= \langle \Pi_{\tilde{W}} v_1, v_2 \rangle + \langle \langle w, v_1 \rangle\, w, v_2 \rangle \\
&= \langle P v_1, v_2 \rangle.
\end{aligned}$$

Hence P is an orthogonal projection with image W, and the inductive step is complete.

Finally, note that the existence of the orthogonal projection $\Pi_{W^\perp}$ follows from the first conclusion and Proposition 3.4. $\square$

In this section we have extended perpendicularity and orthogonal projections to the context of complex scalar product spaces. In the next section we extend another Euclidean idea — distance.

3.4   Norms and Approximations

In this section we define distance in complex scalar product spaces and apply the idea to a space of functions. We show how distance lets us make precise statements about approximating functions by other functions.

In order to exploit our intuition about distance in Euclidean geometry, we distill some of the most important properties of distance into a definition.

Definition 3.12 *Suppose V is a complex vector space and $\|\cdot\| : V \to \mathbb{R}$ satisfying the following:*

1. *If $x \in V$ then $x = 0$ if and only if $\|x\| = 0$.*

2. *If $x \in V$ and $c \in \mathbb{C}$ then $\|cx\| = |c|\, \|x\|$.*

3. *(Triangle inequality) if $x, y \in V$ then $\|x + y\| \leq \|x\| + \|y\|$.*

Then the function $\|\cdot\|$ is a norm.

It is helpful to think of the norm $\|v\|$ as the length of the vector v, i.e., the distance from the point v to the point 0. As in Euclidean geometry, we think of the norm of a difference $\|v - w\|$ as the distance from v to w.

For example, the absolute value, also known as the *modulus*, is a norm on the one-dimensional complex vector space $\mathbb{C}$. More generally, for any natural number n, the function

$$\|(x_1, \ldots, x_n)\| := \sqrt{|x_1|^2 + \cdots + |x_n|^2}$$

is a norm on $\mathbb{C}^n$. We leave the proof to the reader — three of the conditions of Definition 3.12 follow from the properties of distance in $\mathbb{R}^{2n}$, and the other property follows from a straightforward algebraic calculation.

Proposition 3.6 *Suppose $\langle \cdot, \cdot \rangle$ is a complex scalar product on a complex vector space V. Define $\|\cdot\| : V \to \mathbb{R}$ by*

$$\|v\| := \sqrt{\langle v, v \rangle}.$$

Then $\|\cdot\|$ is a norm. It is often called the norm associated to the scalar product $\langle \cdot, \cdot \rangle$. *Furthermore, we have the so-called* Schwarz inequality:[5] *if $x, y \in V$ then $|\langle x, y \rangle| \leq \|x\| \|y\|$.*

Proof. [Sketch][6] Except for the Schwarz inequality, unimaginative calculations suffice for the proof. The Schwarz inequality follows from

$$0 \leq \langle (x \|y\| - y \|x\|), (x \|y\| - y \|x\|) \rangle.$$

The triangle inequality follows from the Schwarz inequality.   □

An example of particular interest to us is a norm on $L^2[-1, 1]$, the complex vector space of square-integrable complex-valued functions on the interval $[-1, 1]$.

Definition 3.13 *Let $L^2[-1, 1]$ denote the set*

$$\left\{ f : f \text{ measurable, from } [-1, 1] \text{ to } \mathbb{C} \text{ and } \int_{-1}^{1} |f|^2 \leq \infty \right\} / \sim,$$

where "measurable" and $\sim$ have the same meanings as in Definition 3.1.
For any function $f \in L^2[-1, 1]$ we define

$$\|f\| := \left(\int_{-1}^{1} |f|^2 \leq \infty \right)^{1/2}.$$

---

[5] Also known as the Cauchy–Bunyakovskii–Schwarz inequality.
[6] For details see Bartle [Bart, Section 8]. Although the proof there is for a real scalar product, the same calculations work in the case of a complex scalar product.

It follows from Propositions 3.1 and 3.6 that this is indeed a norm. In fact, this is the norm associated to the standard scalar product on $L^2[-1, 1]$.

The beauty of the norm is that it allows us to make rigorous mathematical sense of the idea of approximation.

Definition 3.14 *Given a vector space V with a norm, an element $v \in V$ and a set $S \subset V$, we say that we can approximate v by elements of S if and only if, for every $\epsilon > 0$ there is an element $s \in S$ such that $\|s - v\| < \epsilon$.*

It may help to think of ϵ as a desired precision or allowable error. In physics problems or other applications, there is usually a particular precision, determined by experimental constraints. For instance, if the best ruler one has is marked in tenths of a centimeter, one could not expect the precision of measurement to be much less than one-hundredth of a centimeter ($\epsilon = 0.001$ centimeters). In this case, two lengths that differ by less than 0.001 centimeters are indistinguishable. In mathematics, we are interested in truths that transcend the limitations of any one particular experimental setup; hence our Definition 3.14 applies only if we can use elements of S to approximate v to any precision, no matter how small. Approximation is closely related to mathematical limits;[7] see Exercise 3.33.

Any function in $L^2[-1, 1]$ can be approximated by *trigonometric polynomials (of period 2)*. A trigonometric polynomial is a finite (complex) linear combination of the functions

$$\ldots, e^{-2\pi i x}, e^{-\pi i x}, 1, e^{\pi i x}, e^{2\pi i x}, \ldots.$$

For instance, $\sin(\pi x) = \frac{i}{2}e^{-\pi i x} - \frac{i}{2}e^{\pi i x}$ is a trigonometric polynomial. Let $\mathcal{T}_2$ denote the set of trigonometric polynomials of period 2. Because the sum of a finite number of trigonometric polynomials is a trigonometric polynomial and the product of a complex number with a trigonometric polynomial is also a trigonometric polynomial, the set $\mathcal{T}_2$ is a linear subspace of $L^2[-1, 1]$. Hence by Exercise 2.20, $\mathcal{T}_2$ is a complex scalar product space.

In the language of Definition 3.14 the claim that any function in $L^2[-1, 1]$ can be approximated by trigonometric polynomials means that given any function $f \in L^2[-1, 1]$ and any real number $\epsilon > 0$, there is a trigonometric polynomial $T \in \mathcal{T}_2$ such that $\|T - f\| < \epsilon$. We will not prove this claim (however, see Exercise 3.32) but we hope that our brief exploration of it will help the reader understand our definition of approximation. As an example,

---

[7] Students of topology will recognize that approximation is also closely related to the notion of density in the topology whose basic open sets are open balls defined in terms of the norm.

consider the function $f \in L^2[-1, 1]$ defined by

$$f(x) := \begin{cases} -1, & -1 \le x < 0, \\ 0, & x = 0, \\ 1, & 0 < x \le 1. \end{cases} \qquad (3.5)$$

See Figure 3.4. This function is often, legitimately, denoted $x/|x|$.

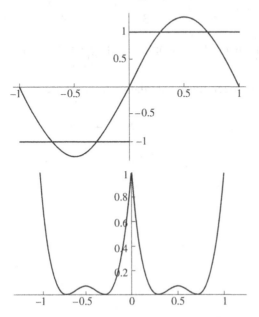

Figure 3.4. (a) graphs of $f(x)$ and $\frac{4}{\pi} \sin \pi x$. (b) graph of $\left| f(x) - \frac{4}{\pi} \sin \pi x \right|^2$.

Sticklers might object that although $f(x) = x/|x|$ for any nonzero x, division by zero is undefined. This is true, but the objection is overruled: in $L^2[-1, 1]$ functions whose values differ at a finite number of points are equivalent, so we can omit a finite number of points from the definition of the function. See Definition 3.13.

The theory of Fourier series gives a method to find approximations of f by trigonometric polynomials. We will not delve into the theory here, but we will report some results. We hope that readers will, at the very least, appreciate these results and put Fourier series on their list of interesting topics for future study; at the other extreme, readers well versed in the theory might find it satisfying to derive the results in this paragraph as an exercise. In any case, according to the theory, one trigonometric polynomial worth considering as an approximation for $f(x)$ is $T_1(x) := \frac{2i}{\pi}(e^{-\pi i x} - e^{\pi i x}) = \frac{4}{\pi} \sin \pi x$. See Figure 3.4(a). It turns out that $\|f\| = 2$ and $\|f - T_1\| = \sqrt{2 - 16/\pi^2} \approx 0.62$. To put it another way, the norm of the error in this approximation is about $\frac{.62}{2} = 32\%$ of the norm of the function f. To get the error down to 5% one can use the 162-term trigonometric polynomial

$$T_{81}(x) := \frac{2i}{\pi} \left(\frac{1}{81} e^{-81\pi ix} + \frac{1}{79} e^{-79\pi ix} + \cdots \right.$$

$$+ \frac{1}{3} e^{-3\pi ix} + e^{-\pi ix} - e^{\pi ix} - \frac{1}{3} e^{3\pi ix} \cdots - \frac{1}{79} e^{79\pi ix} - \frac{1}{81} e^{81\pi ix} \left. \right)$$

$$= \frac{4}{\pi} \left(\sin \pi x + \frac{1}{3} \sin 3\pi x \cdots + \frac{1}{79} \sin 79\pi x + \frac{1}{81} \sin 81\pi x \right).$$

See Figure 3.5. To check these calculations, see Exercise 3.34.

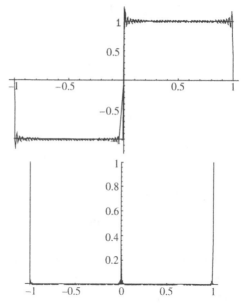

Figure 3.5. (a) graphs of $f(x)$ and $T_{81}(x)$. (b) graph of $|f(x) - T_{81}(x)|^2$.

Notice that we have approximated a discontinuous function by a continuous one. It turns out that any function in $L^2[-1, 1]$ can be approximated by trigonometric polynomials — this is one of the important results of the theory of Fourier series.[8]

We will use approximation of functions to prove the crucial spanning results in the next section.

---

[8]For more detail on Fourier series, see Rudin's book [Ru76] (Chapter 8, especially Theorem 8.15) or Section 1.4 of Dym and McKean's book [DyM]. See also Exercise 3.32.

3.5 Useful Spanning Subspaces

The goal of this section is to find useful spanning subspaces of $C[-1, 1]$ and $L^2(S^2)$. Recall from Definition 3.7 that a subspace spans if the perpendicular subspace is trivial. In a finite-dimensional space V, there are no proper spanning subspaces: any subspace that spans must have the same dimension as V and hence is equal to V. However, for an infinite-dimensional complex scalar product space the situation is more complicated. There are often proper subspaces that span. We will see that polynomials span both $C[-1, 1]$ and $L^2(S^2)$ in Propositions 3.8 and 3.9, respectively. In the process, we will appeal to the Stone–Weierstrass theorem (Theorem 3.2) without giving its proof.

The Stone–Weierstrass theorem uses another notion of approximation: *uniform approximation*.

Definition 3.15 *Suppose $\mathcal{A}$ is a set of complex-valued functions on a set S and suppose that $f : S \to \mathbb{C}$. (Note that f is not necessarily an element of $\mathcal{A}$.) We say that f can be uniformly approximated by elements of $\mathcal{A}$ if and only if, for every $\epsilon > 0$ there is a function $\phi \in \mathcal{A}$ such that $|f - \phi| < \epsilon$.*

With the help of Exercise 3.1 we can see that uniform approximation can be applied to Lebesgue equivalence classes of functions.

Note that our previous notion of approximation (which we here call L^2-*approximation* to distinguish it from uniform approximation) applies to points in normed vector spaces, while uniform approximation applies to functions. As we have seen, many sets of functions are indeed vector spaces, so it is useful to know how these two different notions of approximation relate to one another when both apply. We will find the following proposition useful.

Proposition 3.7 *Suppose S is a set on which integration is well defined. Suppose that S has finite volume, i.e., that $\int_S 1 < \infty$. Consider the norm*

$$\|f\| := \left(\int_S |f|^2 \right)^{1/2}$$

and complex scalar product

$$\langle f, g \rangle := \int_S f^* g$$

on $L^2(S)$. If a function f can be uniformly approximated by a set $\mathcal{A}$ of complex-valued functions, then we can approximate f by $\mathcal{A}$ in L^2 (i.e., in the sense of Definition 3.14).

Furthermore we have

$$\mathcal{A}^\perp = 0.$$

Proof. Suppose f can be uniformly approximated by $\mathcal{A}$. We want to show that f can be L^2-approximated by $\mathcal{A}$. Suppose that $\epsilon > 0$. We must find a function $\phi \in \mathcal{A}$ such that $\| f - \phi \| < \epsilon$. Let K denote the *total volume* of S, i.e.,

$$K := \int_S 1 < \infty.$$

Because f is uniformly approximated by $\mathcal{A}$, there is a $\phi \in \mathcal{A}$ such that $|f - \phi| < \frac{\epsilon}{\sqrt{K}}$. Then we have

$$\| f - \phi \| = \left(\int_S |f - \phi|^2 \right)^{1/2} < \left(\int_S \frac{\epsilon^2}{K} \right)^{1/2} = \epsilon.$$

So f can be approximated by $\mathcal{A}$ in L^2.

Next we must show that $\mathcal{A}$ spans $L^2(S)$. Let $\epsilon > 0$ be given. For any $f \in \mathcal{A}^\perp$ we can choose a $q \in \mathcal{A}$ such that $\| f - q \| < \epsilon$. We have

$$\| f \|^2 = |\langle f, q \rangle + \langle f, f - q \rangle| = |\langle f, f - q \rangle| \leq \| f \| \, \| f - q \|,$$

where the inequality is a consequence of the Schwarz Inequality (Proposition 3.6). It follows that $\| f \| \leq \| f - q \| < \epsilon$. Since ϵ was arbitrary, we conclude that $\| f \| = 0$. Hence

$$\mathcal{A}^\perp = 0$$

inside $L^2(S^2)$.   □

Propositions 3.8 and 3.9 below are both consequences of Proposition 3.7 and the Stone–Weierstrass theorem. Before stating the Stone–Weierstrass theorem, we must define compactness[9] for subsets of $\mathbb{R}^n$.

Definition 3.16 *A subset S of $\mathbb{R}^n$ (respectively, $\mathbb{C}^n$) is* bounded *if there is a real number R such that $\| s \| < R$ for every $s \in S$. A subset S of $\mathbb{R}^n$ (respectively, $\mathbb{C}^n$) is* closed *if, for every point $x \in \mathbb{R}^n \setminus S$ there is an $\epsilon > 0$ such that the open ball (of radius ϵ around x) $\{y \in \mathbb{R}^n : \| x - y \| < \epsilon\}$ lies in $\mathbb{R}^n \setminus S$ (respectively, $\mathbb{C}^n \setminus S$). A subset of $\mathbb{R}^n$ (respectively, $\mathbb{C}^n$) is* compact *if and only if it is closed and bounded.*

---

[9]Compactness is usually defined in terms of open covers, and the characterization we give as a definition is usually the statement of the Heine–Borel theorem [Ru76, Theorem 2.41]. In infinite-dimensional spaces (such as $L^2(\mathbb{R}^3)$) one can have closed, bounded sets that are not compact. See Exercise 3.31.

The definition of a closed set can be restated in terms of approximations: a set S is closed if any point that can be approximated by S (in the sense of Definition 3.14) must lie in S itself. See Exercise 3.36.

For example, the set $[-1, 1] \subset \mathbb{R}$ is compact. It is bounded: we have

$$|x| \le 1 < 2$$

for any $x \in [-1, 1]$. Also, $[-1, 1]$ is closed: if $x \notin [-1, 1]$ then $|x| > 1$, so $\epsilon := (|x| - 1)/2$ is strictly positive and the open interval $(x - \epsilon, x + \epsilon)$ lies entirely in $\mathbb{R} \setminus [-1, 1]$. See Figure 3.6.

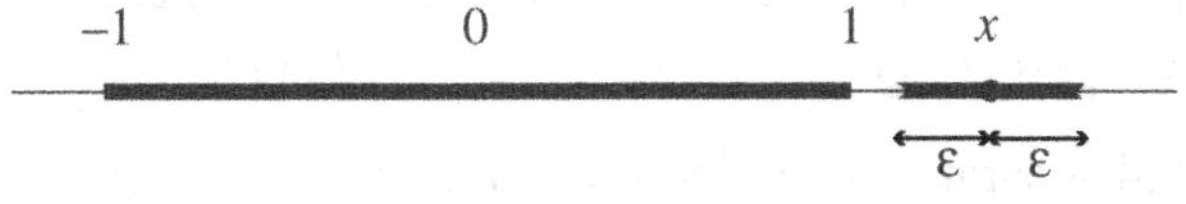

$$\begin{array}{cccc} -1 & 0 & 1 & x \end{array}$$

Figure 3.6. The interval $[-1, 1]$ is closed, since every point x outside the interval lies in an open ball (i.e., open interval of strictly positive length ϵ) outside the interval.

Another important compact set in our story is the unit two-sphere S^2 in $\mathbb{R}^3$. We have $S^2 := \left\{ v \in \mathbb{R}^3 : |v| = 1 \right\}$. This set is clearly bounded, as for every $v \in S^2$ we have $|v| = 1 < 2$. This set is also closed: if $v \notin S^2$, then $|v| \ne 1$, so

$$\epsilon := \left| \frac{|v| - 1}{2} \right| > 0$$

(the number ϵ is half the distance from v to the sphere) and the open ball of radius ϵ around v lies either entirely inside the unit sphere or entirely outside the unit sphere. In either case, the open ball lies entirely in the set $\mathbb{R}^3 \setminus S^2$. So S^2 is compact. See Figure 3.7.

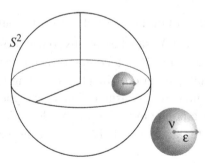

Figure 3.7. The sphere S^2 is closed, since every point x not on S^2 lies in an open ball that does not intersect S^2.

Finally, consider the set $B_R := \left\{ v \in \mathbb{R}^3 : |v| \le R \right\}$, where R is a strictly positive real number. This set is compact, by an argument similar to the one given above for S^2 and left to the reader in Exercise 3.30. We will use the compactness of B_R in Proposition 7.5.

We can now state the Stone–Weierstrass theorem.

Theorem 3.2 (Stone–Weierstrass) *Suppose $\mathcal{A}$ is a set of complex-valued functions on a compact set S with the following properties:*

1. *The functions in $\mathcal{A}$ form a complex vector space under the usual addition and scalar multiplication of functions.*

2. *The set $\mathcal{A}$ is closed under multiplication of functions, i.e., if $f, g \in \mathcal{A}$ then $fg \in \mathcal{A}$.*

3. *The set $\mathcal{A}$ is closed under complex conjugation, i.e., if $f \in \mathcal{A}$ then $f^* \in \mathcal{A}$, where f^* is defined by $f^* : x \mapsto (f(x))^*$.*

4. *The set $\mathcal{A}$ separates points in S, i.e., if $x, y \in S$ and $x \neq y$ then there is a function $f \in \mathcal{A}$ such that $f(x) \neq f(y)$.*

5. *For every $x \in S$ there is at least one $f \in \mathcal{A}$ such that $f(x) \neq 0$.*

Then any continuous function on S can be uniformly approximated by functions in $\mathcal{A}$.

For a proof, see [Ru76, Theorem 7.33].

In Proposition 6.14 we will need a particular application of the Stone–Weierstrass Theorem. Recall the complex scalar product space $C[-1, 1]$ of continuous functions on $[-1, 1]$, introduced in Section 2.1.

Proposition 3.8 *Let V denote the complex vector space of polynomials in one variable (restricted to the interval $[-1, 1]$). Then $V^\perp = 0$ in the complex scalar product space $C[-1, 1]$.*

This fact will be at the heart of the proof of our main result in Section 6.5.

Proof. First, we show that V satisfies the hypotheses of the Stone–Weierstrass theorem. We know that V is a complex vector space under the usual addition and scalar multiplication of functions: adding two polynomials or multiplying a polynomial by a constant yields a polynomial. The product of two polynomials is a polynomial. To see that V is closed under complex conjugation, note that for any $x \in [-1, 1]$ and any constant complex numbers $a_0, \ldots, a_n$ we have

$$\left(a_0 + a_1 x + \cdots + a_n x^n\right)^* = a_0^* + a_1^* x + \cdots + a_n^* x^n \in V,$$

since x is real. To see that V separates points on $[-1, 1]$, consider the polynomial $x \mapsto x$, which takes a different value on each point in $[-1, 1]$. Finally, the constant function 1 is a polynomial taking a nonzero value at each

$x \in [-1, 1]$. So V is a set of complex-valued functions on the compact set $[-1, 1]$ and satisfies all the conditions of the Stone–Weierstrass theorem. We conclude that any continuous function on $[-1, 1]$ is a uniform limit of polynomials on $[-1, 1]$.

Second, we apply Proposition 3.7 to conclude that

$$V^{\perp} = 0 \in L^2([-1, 1]).$$

Since $C[-1, 1]$ is a subspace of $L^2([-1, 1])$, it follows that $V^{\perp} = 0$ inside $C[-1, 1]$ as well.   □

Proposition 3.9 *Let V denote the subspace of $L^2(S^2)$ consisting of restrictions of complex-coefficient polynomials in three variables to the sphere. In the complex scalar product space $L^2(S^2)$ we have*

$$V^{\perp} = 0.$$

Proof. We start by showing that V satisfies the hypotheses of the Stone–Weierstrass theorem. Most of the hypotheses follow easily from the fact that polynomials form a vector space closed under multiplication and complex conjugation. It remains to show that the restrictions of polynomials separate points on the two-sphere S^2. Suppose we have two points, (x_1, y_1, z_2) and (x_2, y_2, z_2) such that $(x_1, y_1, z_2) \neq (x_2, y_2, z_2)$. Then either $x_1 \neq x_2$ or $y_1 \neq y_2$ or $z_1 \neq z_2$. In the first case, the polynomial x takes different values at the two points. In the second case y does and in the third case z does. So V separates points on the two-sphere S^2. Hence by Proposition 3.7 we have $V^{\perp} = 0$.   □

In this section we have shown that polynomials span both $C[-1, 1]$ and $L^2(S^2)$. This fact is the mathematical justification for the physicists' habit of using polynomials instead of arbitrary functions in certain calculations. To put it vaguely, all of $C[-1, 1]$ and $L^2(S^2)$ can be reached by polynomials. The same ideas apply in many other contexts, such as $L^2(\mathbb{R}^3)$. Different problems find resolution via different sets of functions — some require polynomials, but others require spherical harmonics, "Bessel functions" or even fancier "special functions." Behind each applications of special functions in physics is a mathematical proposition that the special functions span the space in question.

Note that in order to prove the spanning propositions we have appealed to a theorem of analysis whose proof is beyond the scope of this text. The Stone–Weierstrass theorem will allow us to be sure from our armchairs, without stepping into a laboratory or consulting a history book, that our lists (of spherical harmonic functions or of irreducible representations) are complete. We hope that even the most skeptical readers will appreciate the power of this kind of result. Even more, we hope that if any reader-experimentalist works in the future on a quantum system with symmetry, she will remember to consult the mathematicians for the appropriate classification corresponding to the symmetry.

3.6   Exercises

Exercise 3.1 (Used in Section 3.5) *In this exercise we show how to make sense of inequalities on Lebesgue equivalence classes of functions. Suppose S is a set with an integral defined on it and ϕ is a real-valued functions on S. Let $[\phi]$ denote the Lebesgue equivalence class of f. We say that $[\phi]$ is strictly positive $(0 < [\phi])$ if, for every function ψ such that $0 < \psi(x)$ for all $x \in S$, we have*

$$0 < \int_S \phi\psi.$$

Show that the truth of this statement depends only on the equivalence class of ϕ. Show that any inequality (such as $[\phi] < \epsilon$) can be rewritten in the form $0 < something$. Thus we can make sense of inequalities of Lebesgue equivalence classes of functions.

Exercise 3.2 (For students of Lebesgue measure) *Show that $0 < [\phi]$ if and only if ϕ is strictly positive on the complement of a set of measure zero.*

Exercise 3.3 *Suppose V is a complex scalar product space. Suppose W is a subspace of V. Show that the restriction of the complex scalar product to W makes W a complex scalar product space.*

Exercise 3.4 *Show that no nontrivial complex scalar product $\langle \cdot, \cdot \rangle$ is linear in the first argument.*

Exercise 3.5 *Let ϕ be any function from $\mathbb{R}^3$ to $\mathbb{C}$ such that $\phi(0, 0, 0) \neq 0$. Define a second function by*

$$\tilde{\phi} : \mathbb{R}^3 \to \mathbb{C}$$

$$(x, y, z) \mapsto \begin{cases} \phi(x, y, z) & (x, y, z) \neq (0, 0, 0) \\ 0 & (x, y, z) = (0, 0, 0). \end{cases}$$

Note that these two functions are not equal in the usual sense. Using either Riemann or Lebesgue integration, show that for any function $\psi : \mathbb{R}^3 \to \mathbb{C}$ and any set S such that $\int_S \psi\phi$ is well defined, one finds

$$\int_S \psi\phi = \int_S \psi\tilde{\phi}.$$

Exercise 3.6 *Suppose $\phi \sim \psi$ and both ϕ and ψ are continuous functions. Show that the functions ϕ and ψ must be equal, i.e., $\phi(x, y, z) = \psi(x, y, z)$ for every (x, y, z).*

Exercise 3.7 *Let V_E denote the subset of even functions in $C[-1, 1]$, i.e., the set of all functions $f \in C[-1, 1]$ satisfying $f(-x) = f(x)$ for every $x \in [-1, 1]$. Let V_O denote the subset of odd functions, i.e., the set of all functions $f \in C[-1, 1]$ satisfying $f(-x) = -f(x)$ for all $x \in [-1, 1]$. Show that V_E and V_O are subspaces, that $V_E = (V_O)^\perp$, and that*

$$C[-1, 1] = V_E \oplus V_O.$$

Exercise 3.8 (For students of Lebesgue measure) *Show that $\phi \sim \psi$ if and only if they differ only on a set of measure zero. In other words, show that $\phi \sim \psi$ if and only if the set*

$$\{(x, y, z) \in \mathbb{R}^3 : \phi(x, y, z) \neq \psi(x, y, z)\}$$

has measure zero.

Exercise 3.9 (For students of Lebesgue measure) *Prove rigorously that the bracket relation on $L^2(\mathbb{R}^3)$ satisfies the definition of a complex scalar product.*

Exercise 3.10 *Is the real-valued function on $\mathbb{C}^2$ defined by*

$$(x_1, x_2) \mapsto \sqrt{2|x_1|^2 + |x_2|^2}$$

a norm? Is the function on $\mathbb{C}^3$ defined by

$$(x_1, x_2, x_3) \mapsto \sqrt{|x_1|^2 + |x_3|^2}$$

a norm? Is the real-valued function on $\mathbb{C}^2$ defined by

$$(x_1, x_2) \mapsto \sqrt{|x_1|^3 + |x_2|^3}$$

a norm? Is the real-valued function on $\mathbb{C}^2$ defined by

$$(x_1, x_2) \mapsto \left(|x_1|^3 + |x_2|^3\right)^{1/3}$$

a norm?

Exercise 3.11 *Define a complex vector subspace of $L^2(\mathbb{R})$ by*

$$V := \left\{ f \in L^2(\mathbb{R}) : f^{(n)} \in L^2(\mathbb{R}) \text{ for any } n \in \mathbb{N} \right\}.$$

In other words, V consists of functions in $L^2(\mathbb{R})$ whose derivatives all exist and lie in $L^2(\mathbb{R})$. Show that V is not trivial by finding a nonzero function in V. Now consider the Laplacian $\nabla^2 \colon V \to V$, $x \mapsto \partial_x^2 f(x)$ for any $x \in \mathbb{R}$. Show that for any $\lambda \geq 0$, there is a complex-valued function f_λ such that $\nabla^2 f_\lambda = \lambda f_\lambda$. However, show also that ∇^2 has no eigenvalues and no eigenvectors in V. (Budding analysts should prove that any element of $L^2(\mathbb{R})$ can be approximated by elements of V.)

Exercise 3.12 *Suppose U_1 and U_2 are both unitary operators on a complex scalar product space V. Show that $U_1 \circ U_2$ is a unitary operator on V. Also, show that every unitary operator on V has an inverse that is a unitary operator on V.*

Exercise 3.13 *In this exercise V is a finite-dimensional complex scalar product space and W is a subspace of V. Show that $W^\perp = 0$ in V if and only if W spans V in the sense given in Definition 2.2.*

Exercise 3.14 *Show that the set*

$$\left\{ \frac{1}{\sqrt{2}} e^{ik(\cdot)} : k \in \mathbb{Z} \right\}$$

is a unitary basis of $L^2[-1, 1]$. (Hint: to show that this set spans, use the fact that the Fourier series of any function in $L^2[-1, 1]$ converges in the norm to the function.)

Exercise 3.15 *Consider the set $B := \{ie_1, ie_2, ie_3\}$ in $\mathbb{C}^3$, where e_1, e_2 and e_3 are the standard basis vectors. Show that B is a unitary basis. Show that*

$$v = \langle ie_1, v \rangle \, ie_1 + \langle ie_2, v \rangle \, ie_2 + \langle ie_3, v \rangle \, ie_3$$

for arbitrary $v \in \mathbb{C}^3$. Next, let $\tilde{B} := \{b_1, b_2, b_3\}$ be any unitary basis of $\mathbb{C}^3$. Show that for any $v \in \mathbb{C}^3$ we have

$$v = \langle b_1, v \rangle \, b_1 + \langle b_2, v \rangle \, b_2 + \langle b_3, v \rangle \, b_3.$$

(Hint: calculate the scalar product of each basis vector with the difference of the two sides of the equation.)

Exercise 3.16 *Suppose that V is a complex scalar product space and $\Pi: V \to V$ is an orthogonal projection. Show that the only possible eigenvalues for Π are 0 and 1. Show that Π is diagonalizable, i.e., show that there is a basis of V composed of eigenvectors of Π.*

Exercise 3.17 *Show that any Cartesian sum $V_1 \oplus \cdots \oplus V_n$ of complex scalar product spaces has a complex scalar product defined by*

$$\langle (v_1, \ldots, v_n), (w_1, \ldots, w_n) \rangle := \sum_{k=1}^{n} \langle v_k, w_k \rangle_k ,$$

where $\langle \cdot, \cdot \rangle_k$ denotes the complex scalar product on V_k. Show that the function Π_k defined in Definition 2.12 is an orthogonal projection. What is the matrix of this projection?

Exercise 3.18 *Any linear transformation $T: V \to V$ on a vector space V, satisfying $T^2 = T$ is called a* projection. *Find a complex scalar product space V and a linear transformation $T: V \to V$ such that T is a projection but not an orthogonal projection.*

Exercise 3.19 (Used in Exercises 5.21 and 5.22) *Suppose V is a finite-dimensional complex scalar product space. Recall the dual vector space V^* from Exercise 2.14. Consider the function $\tau: V \to V^*$ defined by*

$$(\tau v)w := \langle v, w \rangle$$

for any $v, w \in V$. Show that τ is an isomorphism of vector spaces. Then show that the operation $\langle \cdot, \cdot \rangle_*$ on V^* defined by

$$\langle \alpha, \beta \rangle_* := \left\langle \tau^{-1}\alpha, \tau^{-1}\beta \right\rangle$$

for each $\alpha, \beta \in V^*$ is a complex scalar product on V^*. (This operation on V^* is called the natural complex scalar product induced on V^*.)

Exercise 3.20 (Used in Exercises 5.21 and 5.22) *Suppose V and W are complex vector spaces with complex scalar products $\langle \cdot, \cdot \rangle_V$ and $\langle \cdot, \cdot \rangle_W$, respectively. Recall the vector space $\mathrm{Hom}(V, W)$ of linear transformations from V to W. Show that there is a complex scalar product on $\mathrm{Hom}(V, W)$ defined by*

$$\langle A, B \rangle_{\mathrm{Hom}} := \mathrm{Tr}(A^*B),$$

where $A^* \in \mathrm{Hom}(W, V)$ denotes the adjoint of the linear transformation A.

Exercise 3.21 *Suppose V is a complex scalar product space and* $\Pi: V \to V$ *is an orthogonal projection. Show that V and* (ker Π) $\oplus$ (Image V) *are isomorphic as complex scalar product spaces.*

Exercise 3.22 *Show that the set of harmonic polynomials on* $\mathbb{R}^3$ *is not closed under multiplication. (The point of this exercise is that in Chapter 7, when we wish to show that the restrictions of harmonic polynomials to* S^2 *span* S^2, *we will not be able to appeal directly to the Stone–Weierstrass theorem. Rather, we will relate restrictions of harmonic functions to restrictions of polynomial functions and then appeal to the results of Section 3.5.)*

Exercise 3.23 *Show that C* $([-1, 1])$ *is a complex vector space. Show that the set of complex-valued polynomials in one variable is a vector subspace. Show that the bracket* $\langle \cdot, \cdot \rangle$ *(defined as in Section 3.2) is a complex scalar product on C* $([-1, 1])$.

Exercise 3.24 *Consider the linear transformation* $T: L^2(\mathbb{R}) \to L^2(\mathbb{R})$ *defined by*
$$(Tf)(x) := f(x+1)$$
for any $x \in \mathbb{R}$. *Find* T^*.

Exercise 3.25 *Suppose M is an* $n \times n$ Hermitian-symmetric *matrix, i.e., suppose* $M^* = M$, *where* M^* *denotes the conjugate transpose of M. Suppose every eigenvalue of M is strictly positive. Define*
$$\langle v, w \rangle := v^* M w,$$
where v^* *is the conjugate transpose of v, i.e., a row vector whose entries are the conjugates of entries of v. Show that this bracket is a complex scalar product on* $\mathbb{C}^n$.

Conversely, suppose $\langle \cdot, \cdot \rangle$ *is a complex scalar product in* $\mathbb{C}^n$. *Show that there is a Hermitian-symmetric matrix M such that* $\langle v, w \rangle = v^* M w$ *for any* $v, w \in \mathbb{C}^n$.

Exercise 3.26 (Used in Proposition A.3) *Consider the Laplacian in spherical coordinates (see Exercise 1.12):*
$$\partial_r^2 + \frac{2}{r}\partial_r + \frac{1}{r^2}\partial_\theta^2 + \frac{\cos\theta}{r^2\sin\theta}\partial_\theta + \frac{1}{r^2\sin^2\theta}\partial_\phi^2.$$
Show that for any fixed nonzero value of r the angular part
$$\nabla_{\theta,\phi}^2 := \frac{1}{r^2}\left(\partial_\theta^2 + \frac{\cos\theta}{\sin\theta}\partial_\theta + \frac{1}{\sin^2\theta}\partial_\phi\right)$$

is Hermitian-symmetric with respect to the complex scalar product on the subspace V of $L^2(S^2)$ consisting of infinitely differentiable functions of θ and ϕ.

Exercise 3.27 *Show that the operator Π_+ defined in Section 2.3 satisfies Definition 3.11.*

Exercise 3.28 *True or false? "No orthogonal projection is unitary."*

Exercise 3.29 *Show that if W is a finite-dimensional subspace of a complex scalar product space V, then $(W^\perp)^\perp = W$. Note that V need not be finite dimensional. Find a counterexample in infinite dimensions, i.e., find an infinite-dimensional subspace W of a complex scalar product space V such that $(W^\perp)^\perp \neq W$.*

Exercise 3.30 (Used in Proposition 7.5) *Suppose R is a strictly positive real number. Show that the set*

$$B_R := \left\{ v \in \mathbb{R}^3 : |v| \leq R \right\}$$

is compact.

Exercise 3.31 (For readers familiar with open sets) *Here is the standard definition of compactness: A set S is compact if every open cover of S has a finite subcover. More explicitly, if $\{G_\alpha\}$ is a collection of open sets such that $S \subset \bigcup_\alpha G_\alpha$, then a finite subcover is a finite set $\{\alpha_1, \ldots, \alpha_n\}$ such that $S \subset \bigcup_{k=1}^n G_{\alpha_k}$.*
Show that the unit ball in $L^2(\mathbb{R}^3)$, i.e., the set

$$\left\{ f \in L^2(\mathbb{R}^3) : \|f\| \leq 1 \right\},$$

where $\|f\|$ is defined to be

$$\left(\int_{\mathbb{R}^3} |f|^2 \right)^{1/2},$$

is closed and bounded but not compact (by the definition of compactness given in this exercise). (Remark: this does not contradict the Heine–Borel theorem, as the unit ball in $L^2(\mathbb{R}^3)$ is not a subset of $\mathbb{R}^n$ for any n.)

Exercise 3.32 *Show that the set T_2 of trigonometric polynomials of period 2 is closed under addition, scalar multiplication and multiplication. Use the Stone–Weierstrass theorem to conclude that any function $f \in L^2[-1, 1]$ can be approximated (in $L^2[-1, 1]$) by trigonometric polynomials.*

Exercise 3.33 (For students of analysis) *Consider an arbitrary vector space V with a norm. Use Definition 3.14 to show that if* $\lim_{n \to \infty} a_n = \ell$, *then the set* $S := \{a_n : n \in \mathbb{N}\}$ *approximates the point* ℓ. *On the other hand, given a point* $\ell \in V$ *and a subset S of V approximating* ℓ, *find a sequence* $\{a_1, a_2, \dots\}$ *of elements of S such that* $\lim_{n \to \infty} a_n = \ell$.

Can you relate our mathematical definition of approximation to the standard definition of the limit of a function at a point in its domain?

Exercise 3.34 (For students of Fourier series) *Check the Fourier series calculations about the function f in Section 3.4.*

Exercise 3.35 *Suppose* $\mathcal{A}$ *is a complex vector space of bounded, complex-valued functions on a set S. For any* $f \in \mathcal{A}$, *define*

$$\| f \|_\infty := \sup \{|f(s)| : s \in S\},$$

where sup *denotes the* supremum, *i.e., the least upper bound. Show that a function g can be approximated by* $\mathcal{A}$ *(in the sense of Definition 3.14) if and only if g can be uniformly approximated by elements of* $\mathcal{A}$ *(in the sense of Definition 3.15).*

Exercise 3.36 *Suppose that* $S \subset \mathbb{R}^n$ *is closed. Suppose* $x \in \mathbb{R}^n$ *and x can be approximated by S. Show that* $x \in S$. *Conversely, suppose that* $S \subset \mathbb{R}^n$ *and any* $x \in \mathbb{R}^n$ *that can be approximated by S lies in S. Show that S is closed.*

4

Lie Groups and Lie Group Representations

Presently she began again. "I wonder if I shall fall right *through* the earth! How funny it'll seem to come out among the people that walk with their heads downwards! The Antipathies, I think—" (she was rather glad there *was* no one listening, this time, as it didn't sound at all the right word) "—but I shall have to ask them what the name of the country is, you know. Please, Ma'am, is this New Zealand or Australia?" (and she tried to curtsey as she spoke— fancy *curtseying* as you're falling through the air! Do you think you could manage it?)

— Lewis Carroll, *Alice's Adventures in Wonderland* [Car, pp. 27–8]

The notion of a group is a natural mathematical abstraction of physical symmetry. Because quantum mechanical state spaces are linear, symmetries in quantum mechanics have the additional structure of group representations. Formally, a group is a set with a binary operation that satisfies certain criteria, and a representation is a natural function from a group to a set of linear operators.

In this chapter we introduce groups, representations and characters. We discuss the structure of a few particular groups in detail and introduce an important family of representations of the group $SU(2)$.

4.1   Groups and Lie Groups

In this section we define and discuss groups and group homomorphisms, including "differentiable" group homomorphisms, otherwise known as *Lie group homomorphisms*.

Definition 4.1 *A group* $(G, \cdot)$ *is a set G with an operation $G \times G \to G$ denoted by juxtaposition and satisfying*

1. *Associativity: for all g_1, g_2 and g_3 in G we have $(g_1 g_2)g_3 = g_1(g_2 g_3)$.*

2. *Existence of Identity Element: there is an element I in G such that for all $g \in G$ we have $Ig = gI = g$. The element I is called the* identity element of the group G.

3. *Existence of Inverses: for all $g \in G$ there is an element, denoted g^{-1}, such that $gg^{-1} = g^{-1}g = I$. The group element g^{-1} is called the* inverse of g.

It is useful to know that the inverse is unique.

Proposition 4.1 *Suppose G is a group. Then there is a unique identity element. If g is an element of G then the inverse g^{-1} is unique.*

Proof. Suppose I and $\tilde{I}$ both satisfy the definition of the identity element in Definition 4.1. Then $I = I\tilde{I} = \tilde{I}$. So the identity element is uniquely defined. Next suppose that h and $\tilde{h}$ both satisfy the definition of the inverse of g. Then $h = hI = hg\tilde{h} = I\tilde{h} = \tilde{h}$. So the inverse of g is uniquely defined.   □

One of the easiest groups to understand is the *circle group*:

$$\mathbb{T} := \{\lambda \in \mathbb{C} \colon |\lambda| = 1\};$$

the group operation is complex multiplication. The reader should check that the group axioms are satisfied. It is useful to note that $\lambda^{-1} = \lambda^*$ for any $\lambda \in \mathbb{T}$.

Another group is the set of two-by-two real matrices of the form

$$M_\theta := \begin{pmatrix} \cos\theta & -\sin\theta \\ \sin\theta & \cos\theta \end{pmatrix}.$$

This group is called $SO(2)$. Each M_θ is a rotation of the real two-dimensional plane through the angle θ. Thanks to various trigonometrical identities, multiplying two rotations yields a rotation; more precisely,

$$M_\theta M_\phi = M_{\theta+\phi}.$$

The identity element is

$$M_0 = \begin{pmatrix} 1 & 0 \\ 0 & 1 \end{pmatrix}$$

and $M_{-\theta}$ is the inverse of M_θ.

A slightly more complicated example is the set of unit quaternions (defined in Section 1.5) with its usual multiplication. By Exercise 1.14, the multiplication is associative. The quaternion $1 \in \mathbf{Q}$ is the identity element. Also, for any unit quaternion $u + x\mathbf{i} + y\mathbf{j} + z\mathbf{k}$ we have $u^2 + x^2 + y^2 + z^2 = 1$ and hence

$$(u + x\mathbf{i} + y\mathbf{j} + z\mathbf{k})(u - x\mathbf{i} - y\mathbf{j} - z\mathbf{k})$$
$$= (u^2 + x^2 + y^2 + z^2) + (-ux + ux - yz + yz)\mathbf{i}$$
$$+ (-uy + xz + uy - xz)\mathbf{j} + (-uz - xy + xy + uz)\mathbf{j}$$
$$= 1.$$

It follows from this calculation that $u - x\mathbf{i} - y\mathbf{j} - z\mathbf{k}$ is the inverse of $u + x\mathbf{i} + y\mathbf{j} + z\mathbf{k}$. We are almost done proving that the unit quaternions form a group. (Any reader puzzled to find that we are not completely done should pause to think about what might be left to do.) Note that Definition 4.1 requires that the product of two elements of G should itself lie in G. We know that the product of any two quaternions is a quaternion, but to be complete we must show that the product of any two unit quaternions is a unit quaternion. See Exercise 1.15.

Given any set S, the set $\mathcal{T}(S, S)$ of all invertible transformations from S to itself forms a group under composition of transformations. Often a set S will have some kind of extra structure we are interested in preserving. For example, a vector space V has a *linear structure*, i.e., it is a vector space. It is often useful to consider invertible transformations that preserve the structure. For example, given any vector space V (over any scalar field, not necessarily $\mathbb{C}$), the set $\mathcal{GL}(V)$ of invertible linear transformations from V to itself forms a group. The group operation is composition of transformations, with the transformation $T_1 T_2 \colon V \to V$ defined by $v \mapsto T_1(T_2(v))$. If we have chosen a particular basis for V, then we can write each element of $\mathcal{GL}(V)$ as a matrix. For instance, because there is a standard basis of $\mathbb{C}^n$, we can always think of $\mathcal{GL}(\mathbb{C}^n)$ as the set of $n \times n$ invertible matrices with complex entries. Whenever we write a group as a set of matrices, we tacitly assume that the group multiplication is matrix multiplication. For another example, consider a complex scalar product space V. Such a space has a linear structure as well as a *unitary structure*, i.e., a complex scalar product. Recall the notion (Definition 3.5) of a unitary operator on V. By Exercise 3.12, every unitary operator on V has an inverse unitary on V and the composition of two unitary operators is unitary. In other words, unitary operators form a group.

Definition 4.2 *Suppose V is a complex scalar product space. The group of unitary operators on V is called the* unitary group *(of V) and denoted $\mathcal{U}(V)$. If V is finite dimensional, then we also define*

$$SU(V) := \{A \in \mathcal{U}(V) : \det A = 1\}.$$

It is often useful to think of relationships between various groups. To this end we define *group homomorphisms* and *group isomorphisms*.

Definition 4.3 *Suppose $\Phi : G \to \tilde{G}$, where G and $\tilde{G}$ are groups. Suppose that for any $g, h \in G$ we have*

$$\Phi(gh) = \Phi(g)\Phi(h).$$

Then the function Φ is a group homomorphism. *Let $\tilde{I}$ denote the identity element of the group $\tilde{G}$. If Φ is a group homomorphism, then the set $\Phi^{-1}[\tilde{I}]$ is called the* kernel *of Φ.*

The definition requires only that the function preserve multiplication. Preservation of inversion and the identity element follow.

Proposition 4.2 *Suppose G and $\tilde{G}$ are groups and $\Phi : G \to \tilde{G}$ is a group homomorphism. If I is the identity element of G and $\tilde{I}$ is the identity element of $\tilde{G}$, then $\Phi(I) = \tilde{I}$. For any $g \in G$ we have $\Phi\left(g^{-1}\right) = \Phi(g)^{-1}$.*

Proof. To show that Φ preserves the identity, note that

$$\Phi(I) = \tilde{I}\Phi(I) = \left(\Phi(I)^{-1}\Phi(I)\right)\Phi(I) = \Phi(I)^{-1}\left(\Phi(I)\Phi(I)\right)$$
$$= \Phi(I)^{-1}\Phi(II) = \Phi(I)^{-1}\Phi(I) = \tilde{I}.$$

To show that Φ preserves inversion, note that $\Phi\left(g^{-1}\right)\Phi(g) = \Phi\left(g^{-1}g\right) = \Phi(I) = \tilde{I}$ and, similarly, $\Phi(g)\Phi\left(g^{-1}\right) = \Phi\left(gg^{-1}\right) = \Phi(I) = \tilde{I}$. So $\Phi(g^{-1}) = \Phi(g)^{-1}$. $\qquad\square$

As an example, consider the determinant. It is a standard result in linear algebra that if A and B are square matrices of the same size, then $\det(AB) = (\det A)(\det B)$. In other words, for each natural number n, the function $\det : \mathcal{GL}(\mathbb{C}^n) \to \mathbb{C} \setminus \{0\}$ is a group homomorphism. The kernel of the determinant is the set of matrices of determinant one. The kernel is itself a group, in this example and in general. See Exercise 4.4. A composition of

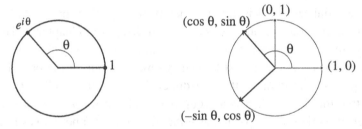

Figure 4.1. a) The point $e^{i\theta}$ lies θ units along the unit circle. b) The matrix M_θ is rotation through an angle θ.

group homomorphisms is a group homomorphism. This result will be used in the proof of Proposition 5.15, the construction of "pullback representations."

Proposition 4.3 *Suppose G_1, G_2 and G_3 are groups. Suppose $\Phi: G_1 \to G_2$ and $\Psi: G_2 \to G_3$ are group homomorphisms. Then $\Psi \circ \Phi: G_1 \to G_3$ is a group homomorphism.*

Proof. It suffices to check that $\Psi \circ \Phi$ preserves multiplication. Let g_1 and g_2 be arbitrary elements of G_1. Then

$$\Psi \circ \Phi(g_1 g_2) = \Psi(\Phi(g_1 g_2)) = \Psi(\Phi(g_1)\Phi(g_2)) = \Psi(\Phi(g_1))\Psi(\Phi(g_2))$$
$$= (\Psi \circ \Phi(g_1))(\Psi \circ \Phi(g_2)),$$

where the second equality follows from the fact that Φ preserves multiplication and the third from the fact that Ψ preserves multiplication. We have shown that $\Psi \circ \Phi$ is a group homomorphism. Note further that because the domain of Ψ is all of G_2, the domain of Φ is all of G_1. Also, the range of $\Psi \circ \Phi$ lies in the range of Ψ, namely, G_3.   □

We can use the notion of a *group isomorphism* to describe the relationship between two groups that are the same as far as multiplication goes.

Definition 4.4 *An injective group homomorphism $\Psi: G_1 \to G_2$ whose inverse Φ is a group homomorphism from G_2 to G_1 is a group isomorphism. If there is a group isomorphism from a group G_1 to another group G_2, we say that the groups G_1 and G_2 are isomorphic.*

Intuitively, two groups that are isomorphic are essentially the same, although they may arise in different contexts and consist of different types of mathematical objects. For example, the unit circle in the complex plane is isomorphic as a group to the set of 2×2 rotation matrices. See Figure 4.1. One is a set of complex numbers, and one is a set of matrices with real entries, but if we strip away their contexts and consider only how the multiplication operation works, they have identical mathematical structure.

This essential sameness is at play when people speak of the "$SO(4)$ symmetry of the hydrogen atom," which we will discuss in Chapter 8. The hydrogen atom is not a four-dimensional system, much less a system rotating in four dimensions. Yet the largest known symmetry group of the bound states of hydrogen is isomorphic to the four-dimensional rotation group $SO(4)$.

Note that the determinant is a group isomorphism for $n = 1$, but not for any other n; while for any particular n the determinant function is surjective (any real number is the determinant of some $n \times n$ matrix), it is not injective for $n \geq 2$. Only when $n = 1$ does the determinant determine every entry of the matrix.

Each of the groups we introduce in this text is a *Lie group*. We give the formal definition in terms of "manifolds"; however, readers unfamiliar with differential geometry may think of a manifold as analogous to a nicely parametrized surface embedded in $\mathbb{R}^3$. More to the point for our purposes, a manifold is a set on which differentiability is well defined. Since all the manifolds we will consider are nicely parameterized, we can define differentiability in terms of the parameters.

Definition 4.5 *A* Lie group *is a group whose set of elements is a* differentiable manifold *such that multiplication and inversion are differentiable functions.*

Each group we discuss is a Lie group because products and inverses are differentiable functions of the parameters. For example, the circle group $\mathbb{T}$ is parameterized by θ (see Figure 4.1, part a). Because $e^{-i\theta}$ is a differentiable function of θ, inversion is differentiable. Because $e^{i(\theta_1 + \theta_2)}$ is differentiable with respect to both θ_1 and θ_2, multiplication is a differentiable function.

For a gentle introduction to manifolds and Lie groups, see the author's previous work [Si]; for a more standard approach, see Warner [Wa]. Understanding these general concepts is not required for our work here; however, we urge readers familiar with these concepts to make explicit connections between the particular calculations in this book and the more abstract or general theorems they may already know.

Definition 4.6 *Suppose G_1 and G_2 are Lie groups. Suppose $\Psi: G_1 \to G_2$ is a group homomorphism. If Ψ is differentiable, then Ψ is a* Lie group homomorphism. *If Ψ is a also a group isomorphism and Ψ^{-1} is differentiable then Ψ is a* Lie group isomorphism.

For this definition to be valid, we must know what we mean by "differentiable" in this context. Since we will give explicit parameterizations of all the groups we encounter, and all of these groups are in matrix form, we can think of "differentiable" as meaning that the entries of the matrix $\Psi(g)$ should be differentiable functions of the parameters on the group G_1. All of the group homomorphisms we discuss in this text are Lie group homomorphisms.

In Section 4.5 we will explain how groups arise in physical systems with symmetry. This idea has myriad applications in classical and quantum mechanics, as the reader might see by glancing at the tables of contents of *Foundations of Mechanics* [AM] and *Lie Groups and Physics* [St].

4.2 The Key Players: $SO(3)$, $SU(2)$ and $SO(4)$

In this section we introduce the three-dimensional rotation group $SO(3)$ and the special unitary group $SU(2)$. Along with the circle group, these are the groups we need to understand the spatial symmetry of the hydrogen atom. Through Chapter 7 we will need no other groups. We also introduce the group $SO(4)$ (rotations of $\mathbb{R}^4$), which appears only in later chapters.

The *special orthogonal group* $SO(3)$ is the group of rotations of three-dimensional Euclidean space $\mathbb{R}^3$. We use the standard basis of Euclidean space, $\{(1, 0, 0)^T, (0, 1, 0)^T, (0, 0, 1)^T)\}$, to write elements of the group as matrices. Because rotations are linear transformations, we can think of this group as a group of matrices. It is helpful to recall (or realize) that the first column of a 3×3 matrix is the image of the vector $(1, 0, 0)^T$, the second column is the image of the vector $(0, 1, 0)^T$, and the third column is the image of the vector $(0, 0, 1)^T$. Because a rotation carries the standard orthonormal basis in $\mathbb{R}^3$ into another orthonormal basis, each matrix in $SO(3)$ has a set of three mutually orthogonal, length one columns. Because rotations preserve orientation, the three columns must obey the right-hand rule. We can express these conditions mathematically by defining

$$SO(3) := \left\{ M \in \mathcal{GL}\left(\mathbb{R}^3\right) \colon M^T M = I \text{ and } \det M = 1 \right\}.$$

Note that the condition $M^T M = I$ is equivalent to the condition that M should preserve lengths in $\mathbb{R}^3$ (Exercise 1.25).

An explicit parameterization of the group $SO(3)$ is often useful. The most common parameters are called *Euler angles*. Euler angles arise from the observation that any rotation of xyz-space can be expressed as a rotation around the z-axis, followed by a rotation around the x-axis, followed by a rotation around the z-axis. For example, rotating through an angle θ around the y-axis is the same as rotating $\frac{3\pi}{2}$ around the z-axis, followed by rotating θ around the x-axis, followed by rotating $\frac{\pi}{2}$ around the z-axis. We can say this more

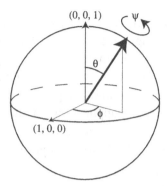

Figure 4.2. Euler angles: The dark arrow is the image of the north pole $(0, 0, 1)$ under the transformation $\mathbf{Z}_\phi \mathbf{X}_\theta \mathbf{Z}_\psi$. The angles (ϕ, θ) are the spherical angle coordinates of the image of the north pole, while ψ measures the amount of rotation around that axis.

formally if we introduce some notation. Let $\mathbf{X}_\theta$, $\mathbf{Y}_\theta$ and $\mathbf{Z}_\theta$ denote rotations through an angle of θ around the x-, y- and z-axes, respectively. More concretely, we have

$$\mathbf{X}_\theta = \begin{pmatrix} 1 & 0 & 0 \\ 0 & \cos\theta & -\sin\theta \\ 0 & \sin\theta & \cos\theta \end{pmatrix}, \quad \mathbf{Y}_\theta = \begin{pmatrix} \cos\theta & 0 & -\sin\theta \\ 0 & 1 & 0 \\ \sin\theta & 0 & \cos\theta \end{pmatrix}$$

$$\text{and} \quad \mathbf{Z}_\theta = \begin{pmatrix} \cos\theta & -\sin\theta & 0 \\ \sin\theta & \cos\theta & 0 \\ 0 & 0 & 1 \end{pmatrix}.$$

The reader can easily check that, as claimed above, $\mathbf{Z}_{\frac{3\pi}{2}} \mathbf{X}_\theta \mathbf{Z}_{\frac{\pi}{2}} = \mathbf{Y}_\theta$. Note that the first rotation performed is the right-hand factor in the matrix multiplication. The proof that any element of $SO(3)$ can be written as $\mathbf{Z}_\phi \mathbf{X}_\theta \mathbf{Z}_\psi$ for some real ϕ, θ and ψ is harder and is left as Exercise 4.24. The angles ϕ, θ and ψ are known in the physics literature as Euler angles. For their geometric interpretation, see Figure 4.2.

Finally, we must introduce the *special unitary group* $SU(2)$. The "unitary" in the name is analogous to the "orthogonal" in the group $SO(3)$. We set

$$SU(2) := \{ M \in \mathcal{GL}(\mathbb{C}) : M^*M = I \text{ and } \det M = 1 \}.$$

This is the group of determinant-one linear operators on the complex vector space $\mathbb{C}^2$ preserving the natural complex scalar product $\langle \cdot, \cdot \rangle$ defined in Section 3.2. Note that

$$\langle Mv, Mw \rangle = \langle v, w \rangle$$

for all $v, w \in \mathbb{C}^2$ if and only if

$$v^* M^* M w = (Mv)^* Mw = v^* w$$

for all $v, w \in \mathbb{C}$ if and only if

$$M^*M = I.$$

In other words, a linear operator U on $\mathbb{C}^2$ is in $SU(2)$ if and only if $\langle v, w \rangle = \langle Uv, Uw \rangle$ for every v and w in $\mathbb{C}^2$ and $\det(U) = 1$. If we choose the standard basis $\{(1, 0)^T, (0, 1)^T\}$ of $\mathbb{C}^2$, then (as the reader is invited to check in Exercise 4.18) we obtain a convenient way of writing matrices in $SU(2)$:

$$\begin{pmatrix} \alpha & -\beta^* \\ \beta & \alpha^* \end{pmatrix},$$

where $\alpha, \beta \in \mathbb{C}$ and $|\alpha|^2 + |\beta|^2 = 1$.

There is a Lie group isomorphism between unit quaternions and the special unitary group $SU(2)$. Define a function Ψ from the unit quaternions to $SU(2)$ by

$$\Psi(u + x\mathbf{i} + y\mathbf{j} + z\mathbf{k}) := \begin{pmatrix} u + ix & -y + iz \\ y + iz & u - ix \end{pmatrix}.$$

To see that this is a group homomorphism, consider any two unit quaternions $q = u + x\mathbf{i} + y\mathbf{j} + z\mathbf{k}$ and $\tilde{q} = \tilde{u} + \tilde{x}\mathbf{i} + \tilde{y}\mathbf{j} + \tilde{z}\mathbf{k}$. Then from Formula 1.6 in Section 1.5 we have

$$\Psi(q\tilde{q}) = \begin{pmatrix} \begin{array}{l} u\tilde{u} - x\tilde{x} - y\tilde{y} - z\tilde{z} \\ +(u\tilde{x} + x\tilde{u} + y\tilde{z} - z\tilde{y})i \end{array} & \begin{array}{l} -(u\tilde{y} + y\tilde{u} + z\tilde{x} - x\tilde{z}) \\ +(u\tilde{z} + z\tilde{u} + x\tilde{y} - y\tilde{x})i \end{array} \\ \\ \begin{array}{l} (u\tilde{y} + y\tilde{u} + z\tilde{x} - x\tilde{z}) \\ +(u\tilde{z} + z\tilde{u} + x\tilde{y} - y\tilde{x})i \end{array} & \begin{array}{l} u\tilde{u} - x\tilde{x} - y\tilde{y} - z\tilde{z} \\ -(u\tilde{x} + x\tilde{u} + y\tilde{z} - z\tilde{y})i \end{array} \end{pmatrix}$$

$$= \begin{pmatrix} u + ix & -y + iz \\ y + iz & u - ix \end{pmatrix} \begin{pmatrix} \tilde{u} + i\tilde{x} & -\tilde{y} + i\tilde{z} \\ \tilde{y} + i\tilde{z} & \tilde{u} - i\tilde{x} \end{pmatrix} = \Psi(q)\Psi(\tilde{q}).$$

To show that Ψ is a group isomorphism, we must check injectivity and surjectivity. By Exercise 4.9 we can prove Ψ injective by showing that $\Psi^{-1}[I]$ contains only the identity element of the unit quaternions. But $\Psi(q) = I$ if and only if $u = 1$ and $x = y = z = 0$, which holds only if $q = 1$, the identity quaternion. So Ψ^{-1} exists. It is easy to see that Ψ^{-1} has domain $SU(2)$: for any element of $SU(2)$ we have

$$\begin{pmatrix} \alpha & -\beta^* \\ \beta & \alpha^* \end{pmatrix} = \Psi^{-1}\left(\Re\alpha + \Im\alpha\mathbf{i} + \Re\beta\mathbf{j} + \Im\beta\mathbf{k}\right).$$

To show differentiability of Ψ, we must parameterize the unit quaternions and $SU(2)$ with open sets in $\mathbb{R}^3$ and write Ψ in terms of the parameterizations. Away from the set $z = 0$, for example, we can parameterize the unit quaternions by u, x and y. Then applying Ψ to a unit quaternion $\left(u + x\mathbf{i} + y\mathbf{j} \pm \sqrt{1 - u^2 - x^2 - y^2}\mathbf{k}\right)$ we obtain the matrix

$$\begin{pmatrix} u + ix & -y \pm i\sqrt{1 - u^2 - x^2 - y^2} \\ y \pm i\sqrt{1 - u^2 - x^2 - y^2} & u - ix \end{pmatrix}.$$

Where $z \neq 0$ we have $1 - u^2 - x^2 - y^2 \neq 0$, so the expression on the right-hand side is a differentiable function of u, x and y. Hence Ψ is differentiable at unit quaternions with $z \neq 0$. A similar argument shows that Ψ is differentiable at any unit quaternion with at least one nonzero coefficient. But each unit quaternion has at least one nonzero coefficient, so we have shown Ψ to be differentiable on its domain. An almost identical argument shows that Ψ^{-1} is differentiable. Hence Ψ is an isomorphism of Lie groups.

We will also encounter the group of four-dimensional rotations:

$$SO(4) := \left\{M \in \mathcal{GL}\left(\mathbb{R}^4\right): M^T M = I \text{ and } \det M = 1\right\}.$$

The four columns of a matrix in $SO(4)$ are mutually perpendicular and each has length one. The ordering of the columns is restricted by the determinant-one condition.

Each of the groups we have introduced so far is compact, i.e., satisfies Definition 3.16. Note that an $n \times n$ matrix with complex entries can be thought of as a subset of $\mathbb{C}^{n \times n}$. We leave the verification to the reader in Exercise 4.5.

The groups in this section are the key players in our drama. The rotation group $SO(3)$ is the physical symmetry group of the hydrogen atom (from the lone electron's point of view). There is a close relationship between $SU(2)$ and $SO(3)$, made explicit in Section 4.3, that will allow us to use $SU(2)$ to deduce important facts about $SO(3)$. Finally, the group $SO(4)$ appears (miraculously) as a symmetry of the hydrogen atom. We will explore this symmetry in Chapters 8 and 9. However, the importance of these groups is by no means limited to our application. On the contrary, these groups are indispensable first examples of the phenomena and techniques of the theory of compact Lie groups. Even a reader with no particular interest in the hydrogen atom would be well advised to master this section.

4.3   The Spectral Theorem for $SU(2)$ and the Double Cover of $SO(3)$

This section presents two crucial results about particular groups. The first, the Spectral Theorem for $SU(2)$, shows that any element of $SU(2)$ can be written in a particular convenient form. The second result is the existence of a two-to-one group homomorphism from $SU(2)$ to $SO(3)$, known as a *double cover*.

We begin with the Spectral Theorem.

Proposition 4.4 (The Spectral Theorem for $SU(2)$) *Consider an element U of $SU(2)$. Then there is a complex number λ of modulus one (i.e., $\lambda \in \mathbb{T}$) and a matrix $M \in SU(2)$ such that*

$$M^*UM = \begin{pmatrix} \lambda & 0 \\ 0 & \lambda^* \end{pmatrix}, \tag{4.1}$$

where M^* denotes the conjugate transpose of M. Furthermore, if we write

$$U = \begin{pmatrix} \alpha & -\beta^* \\ \beta & \alpha^* \end{pmatrix},$$

then we can choose $\lambda = \Re(\alpha) + i\sqrt{1 - (\Re(\alpha))^2}$.

As its extravagantly capitalized name[1] indicates, this proposition is a special case of an important theorem of linear algebra. With suitable care, one can generalize this theorem to unitary operators[2] on Hilbert spaces — this is one of the fundamental results of the mathematical field called *functional analysis*. Physicists implicitly use this theorem (or one of its relatives) every time they make a calculation using what they call a "complete basis" of eigenfunctions of a particular operator. Readers who have done such calculations might ask themselves how one knows that the eigenfunctions actually form a basis, i.e., why there are enough eigenfunctions to go around. Some operators do not have any eigenfunctions at all. See Exercises 2.28 and 3.11 for examples. In its more advanced forms, the Spectral Theorem gives a sophisticated generalization of the notions of eigenfunctions and eigenvectors.[3]

Proof. To find the eigenvalues, we calculate the characteristic polynomial explicitly. We have $\det(zI - U) = z^2 - 2\Re(\alpha)z + 1$. By the quadratic formula, we find that the eigenvalues are $\Re(\alpha) \pm i\sqrt{1 - \Re(\alpha)^2}$. Note that because

---

[1] The words "spectrum" for eigenvalues and its associated adjective "spectral" come from the Latin word *spectrum*, which means appearance. Astronomers observing light from distant stars find a characteristic set of lines appearing in their data; these lines were found to correspond to differences of energy eigenvalues of the hydrogen atom. This is evidence for the claim that distant stars are composed largely of hydrogen.

[2] And to various other kinds of operators, including "self-adjoint" operators.

[3] See, for example, Reed and Simon [RS, Part VII]).

$\Re(\alpha)^2 \leq |\alpha|^2 \leq |\alpha|^2 + |\beta|^2 = 1$, the argument of the square root is nonnegative. We are free to choose

$$\lambda := \Re(\alpha) + i\sqrt{1 - \Re(\alpha)^2}.$$

It is easy to calculate that $|\lambda| = 1$ and the two eigenvalues are λ and λ^*.

We will build the matrix M out of eigenvectors of U. To find the desired eigenvectors, we take two cases. If $\lambda^2 = 1$, it follows that $(\Re(\alpha))^2 = 1$ and hence $U = \pm I$. In this case we can take $M := I \in SU(2)$. If $\lambda^2 \neq 1$, then we must work a little harder. Note that $\lambda^2 \neq 1$ implies that $\lambda \neq \lambda^*$. By the definition of eigenvalues, there are nonzero vectors $v, w \in \mathbb{C}^2$ such that $Uv = \lambda v$ and $Uw = \lambda^* w$. Without loss of generality we may assume that $\|v\| = \|w\| = 1$. Because $\lambda^2 \neq 1$, it follows from

$$\langle w, v \rangle = \langle Uw, Uv \rangle = \langle \lambda^* w, \lambda v \rangle = \lambda^2 \langle w, v \rangle,$$

that $\langle w, v \rangle = 0$. Define a two-by-two matrix whose columns are v and w:

$$\tilde{M} := \begin{pmatrix} v & w \end{pmatrix}.$$

The matrix $\tilde{M}$ is almost, but not quite, the matrix we need. We have

$$\tilde{M}^* U \tilde{M} = \begin{pmatrix} v^* \\ w^* \end{pmatrix} U \begin{pmatrix} v & w \end{pmatrix} = \begin{pmatrix} v^* \\ w^* \end{pmatrix} \begin{pmatrix} \lambda v & \lambda^* w \end{pmatrix} = \begin{pmatrix} \lambda & 0 \\ 0 & \lambda^* \end{pmatrix},$$

as desired, but it is possible that $\tilde{M}$ is not in $SU(2)$. We do have

$$\tilde{M}^* \tilde{M} = \begin{pmatrix} \langle v, v \rangle & \langle v, w \rangle \\ \langle w, v \rangle & \langle w, w \rangle \end{pmatrix} = I,$$

but there is no guarantee that $\det \tilde{M} = 1$. A slight modification yields a matrix in $SU(2)$. The calculation just above shows that $\tilde{M}$ is invertible and that $|\det \tilde{M}|^2 = \det \tilde{M}^* \det \tilde{M} = 1$. Hence there must be a complex number γ such that $|\gamma| = 1$ and $\gamma^2 \det \tilde{M} = 1$. Set $M := \gamma \tilde{M}$. Then M satisfies all our conditions:

$$M^* U M = \gamma^* \tilde{M}^* U \tilde{M} \gamma = \tilde{M}^* U \tilde{M} = \begin{pmatrix} \lambda & 0 \\ 0 & \lambda^* \end{pmatrix},$$

$M^* M = \tilde{M}^* \gamma^* \gamma \tilde{M} = \tilde{M}^* \tilde{M} = I$ and $\det M = \gamma^2 \det \tilde{M} = 1$. So M is an element of $SU(2)$ and satisfies the requirements of the theorem. $\qquad \square$

There is an important surjective group homomorphism Φ from $SU(2)$ to $SO(3)$. We will find the homomorphism useful in Section 6.6 for deriving the list of irreducible representations of $SO(3)$ from the list of irreducible representations of $SU(2)$. There is no *a priori* reason to expect such a homomorphism between two arbitrary groups, so the fact that $SU(2)$ and $SO(3)$ are related in this way is quite special. Here is the definition of Φ:

$$\Phi\begin{pmatrix} \alpha & -\beta^* \\ \beta & \alpha^* \end{pmatrix} := \begin{pmatrix} |\alpha|^2 - |\beta|^2 & -2\Re(\alpha\beta) & -2\Im(\alpha\beta) \\ 2\Re(\alpha^*\beta) & \Re(\alpha^2 - \beta^2) & \Im(\alpha^2 - \beta^2) \\ 2\Im(\alpha^*\beta) & -\Im(\alpha^2 + \beta^2) & \Re(\alpha^2 + \beta^2) \end{pmatrix}. \quad (4.2)$$

Readers should take a little time to familiarize themselves with this homomorphism by concrete calculations such as those in Exercise 4.38. Readers who wish to check by brute calculation that Φ is indeed a homomorphism should consult Exercise 4.32. We will take another approach, one that is more appealing geometrically (because we will see how an element of $SU(2)$ can rotate an actual geometric object) and theoretically (because it uses concepts that generalize to other Lie groups).

There is a geometric way to construct this homomorphism.[4] We pull out of our hat a certain set of matrices:

$$S := \left\{ \begin{pmatrix} x & y - iz \\ y + iz & -x \end{pmatrix} : (x, y, z) \in \mathbb{R}^3 \right\}.$$

Note that every matrix M of S is *Hermitian symmetric*, i.e., writing M^* to denote the conjugate transpose of M, we have $M^* = M$. Note also that the trace of each $M \in S$ is zero and

$$\det\begin{pmatrix} x & y - iz \\ y + iz & -x \end{pmatrix} = -x^2 - y^2 - z^2. \quad (4.3)$$

In other words, we can think of this set of matrices as $\mathbb{R}^3$, and the negative of the determinant gives the square of the distance from the origin. To be precise, we can define a correspondence (i.e., a one-to-one function)

$$F: \mathbb{R}^3 \to \text{Hermitian symmetric } 2 \times 2 \text{ matrices}$$

$$\begin{pmatrix} x \\ y \\ z \end{pmatrix} \mapsto \begin{pmatrix} x & y - iz \\ y + iz & -x \end{pmatrix}.$$

---

[4]Readers familiar with the theory of Lie groups may recognize this construction as the adjoint action of $SU(2)$. In general, one can always use the adjoint action to construct a homomorphism from a Lie group G to G/Z, where Z is the *center* of the group, i.e., the set of group elements that commute with every element of G.

Notice that this correspondence is linear (as a function between real vector spaces) and invertible (i.e., injective and surjective).

Now consider any particular element g of $SU(2)$. We can use g and the correspondence F to define a linear transformation of $\mathbb{R}^3$:

$$T_g : \mathbb{R}^3 \to \mathbb{R}^3$$
$$v \mapsto F^{-1}(gF(v)g^{-1}).$$

The reader should verify that $T_g(v + w) = T_g(v) + T_g(w)$ and $T_g(rv) = rT_g(v)$ for all v, w in $\mathbb{R}^3$ and for all $r \in \mathbb{R}$.

Matrix calculations show that the 3×3 real matrix corresponding to T_g in the standard basis is $\Phi(g)$. To show how this kind of calculation goes, we will find the first column of the matrix of T_g. The first column of the matrix will be the image of the vector $(1, 0, 0)^T$. Writing $g = \begin{pmatrix} \alpha & -\beta^* \\ \beta & \alpha^* \end{pmatrix}$, we find

$$F \begin{pmatrix} 1 \\ 0 \\ 0 \end{pmatrix} = \begin{pmatrix} 1 & 0 \\ 0 & -1 \end{pmatrix}$$

and so

$$gF \begin{pmatrix} 1 \\ 0 \\ 0 \end{pmatrix} g^{-1} = \begin{pmatrix} \alpha & -\beta^* \\ \beta & \alpha^* \end{pmatrix} \begin{pmatrix} 1 & 0 \\ 0 & -1 \end{pmatrix} \begin{pmatrix} \alpha^* & \beta^* \\ -\beta & \alpha \end{pmatrix}$$
$$= \begin{pmatrix} |\alpha|^2 - |\beta|^2 & 2\alpha\beta^* \\ 2\alpha^*\beta & |\beta|^2 - |\alpha|^2 \end{pmatrix}.$$

Applying F^{-1}, we see that the first column of T_g is indeed

$$T_g \begin{pmatrix} 1 \\ 0 \\ 0 \end{pmatrix} = \begin{pmatrix} |\alpha|^2 - |\beta|^2 \\ 2\Re(\alpha^*\beta) \\ 2\Im(\alpha^*\beta) \end{pmatrix},$$

which is the first column on the right-hand side of Equation 4.2. Similar calculations work for the second and third columns. See Exercise 4.31.

Proposition 4.5 *The function* $\Phi : SU(2) \to SO(3)$ *is a surjective, two-to-one Lie group homomorphism. The kernel of this homomorphism is* $\{I, -I\} \subset SU(2)$; *i.e., if* $\Phi(x) = I \in SO(3)$ *then* $x = \pm I \in SU(2)$.

Proof. We must show each of the statements of the proposition, and we must check that for each $g \in SU(2)$ we have $\Phi(g) \in SO(3)$.

First we will show that Φ is a Lie group homomorphism. For any $g_1, g_2 \in SU(2)$ and any $v \in \mathbb{R}^3$ we have

$$
\begin{aligned}
\Phi(g_1 g_2) v = T_{g_1 g_2}(v) &= F^{-1} \left(g_1 g_2 F(v)(g_1 g_2)^{-1} \right) \\
&= F^{-1} \left(g_1 g_2 F(v) g_2^{-1} g_1^{-1} \right) \\
&= F^{-1} \left(g_1 F \left(F^{-1} \left(g_2 F(v) g_2^{-1} \right) \right) g_1^{-1} \right) \\
&= T_{g_1} \left(T_{g_2}(v) \right) \\
&= \Phi(g_1) \Phi(g_2) v.
\end{aligned}
$$

Hence Φ is a group homomorphism. In the defining formula for Φ given in Equation 4.2, every matrix entry is a differentiable function of the real parameters $\Re(\alpha)$, $\Im(\alpha)$, $\Re(\beta)$ and $\Im(\beta)$. Because these parameters are differentiable functions on $SU(2)$, the function Φ is differentiable. Hence Φ is a Lie group homomorphism.

Next we show that for each $g \in SU(2)$ we have $\Phi(g) \in SO(3)$. For any $(x, y, z)^T \in \mathbb{R}^3$ we have, by Equation 4.3,

$$
\begin{aligned}
\left\| T_g \begin{pmatrix} x \\ y \\ z \end{pmatrix} \right\| &= -\det \left(g \begin{pmatrix} x & y - iz \\ y + iz & -x \end{pmatrix} g^{-1} \right) \\
&= -\det \begin{pmatrix} x & y - iz \\ y + iz & -x \end{pmatrix} = \left\| \begin{pmatrix} x \\ y \\ z \end{pmatrix} \right\|.
\end{aligned}
$$

Hence $\Phi(g)$ preserves the length on $\mathbb{R}^3$, so by Exercise 1.25, we have $\Phi(g)^T \Phi(g) = I$. It remains to show that $\det \Phi(g) = 1$. If g is a diagonal element of $SU(2)$, then we can make a direct calculation:

$$
\det \left(\Phi \begin{pmatrix} \lambda & 0 \\ 0 & \lambda^* \end{pmatrix} \right) = \det \begin{pmatrix} 1 & 0 & 0 \\ 0 & \Re(\lambda^2) & \Im(\lambda^2) \\ 0 & -\Im(\lambda^2) & \Re(\lambda^2) \end{pmatrix} = |\lambda|^4 = 1.
$$

If g is an arbitrary element of $SU(2)$, then we can apply the Spectral Theorem to find a λ and a matrix $M \in SU(2)$ such that

$$
M^* g M = \begin{pmatrix} \lambda & 0 \\ 0 & \lambda^* \end{pmatrix}
$$

and $M^* = M^{-1}$. We have

$$\det \Phi(g) = (\det \Phi(M))^{-1} \det \Phi(g) \det \Phi(M) = \det \Phi(M^{-1}gM)$$

$$= \det \Phi(M^*gM) = \det \left(\Phi \begin{pmatrix} \lambda & 0 \\ 0 & \lambda^* \end{pmatrix} \right) = 1.$$

Hence for any $g \in SU(2)$ we have $\Phi(g) \in SO(3)$.

Next we show that Φ is surjective onto $SO(3)$. By Exercise 4.24, it suffices to show that for any $\theta \in \mathbb{R}$, $\mathbf{X}_\theta$ and $\mathbf{Z}_\theta$ lie in the image of Φ. But according to Exercise 4.38,

$$\mathbf{X}_\theta = \Phi \begin{pmatrix} e^{-i\theta/2} & 0 \\ 0 & e^{i\theta/2} \end{pmatrix}.$$

Also, note that

$$\mathbf{Z}_\theta = \begin{pmatrix} 0 & 0 & 1 \\ 0 & -1 & 0 \\ 1 & 0 & 0 \end{pmatrix} \mathbf{X}_\theta \begin{pmatrix} 0 & 0 & 1 \\ 0 & -1 & 0 \\ 1 & 0 & 0 \end{pmatrix}$$

and, again by Exercise 4.38, the permutation matrix in this equation is equal to

$$\Phi \left(\frac{1}{\sqrt{2}} \begin{pmatrix} -i & -1 \\ 1 & i \end{pmatrix} \right).$$

Hence, since Φ is a group homomorphism, we have

$$\Phi \left(\frac{1}{\sqrt{2}} \begin{pmatrix} -i & -1 \\ 1 & i \end{pmatrix} \begin{pmatrix} e^{-i\theta/2} & 0 \\ 0 & e^{i\theta/2} \end{pmatrix} \left(\frac{1}{\sqrt{2}} \begin{pmatrix} -i & -1 \\ 1 & i \end{pmatrix} \right) \right)$$

$$= \Phi \left(\frac{1}{\sqrt{2}} \begin{pmatrix} -i & -1 \\ 1 & i \end{pmatrix} \right) \Phi \begin{pmatrix} e^{-i\theta/2} & 0 \\ 0 & e^{i\theta/2} \end{pmatrix} \Phi \left(\frac{1}{\sqrt{2}} \begin{pmatrix} -i & -1 \\ 1 & i \end{pmatrix} \right)$$

$$= \begin{pmatrix} 0 & 0 & 1 \\ 0 & -1 & 0 \\ 1 & 0 & 0 \end{pmatrix} \mathbf{X}_\theta \begin{pmatrix} 0 & 0 & 1 \\ 0 & -1 & 0 \\ 1 & 0 & 0 \end{pmatrix} = \mathbf{Z}_\theta.$$

Thus we have shown that any rotation around the z- or x-axis is in the image of the group homomorphism Φ. Because any element of $SO(3)$ can be written as a product of three such rotations (by Exercise 4.24), and because Φ is a group homomorphism, it follows that any element of $SO(3)$ is in the image of Φ. It remains only to show that Φ is two-to-one. Note first that

$$\Phi \begin{pmatrix} \alpha & -\beta^* \\ \beta & \alpha^* \end{pmatrix} = \begin{pmatrix} 1 & 0 & 0 \\ 0 & 1 & 0 \\ 0 & 0 & 1 \end{pmatrix} \qquad (4.4)$$

only if $|\alpha|^2 - |\beta|^2 = 1$ and $\Re(\alpha^2 - \beta^2) = 1$. Recalling that $|\alpha|^2 + |\beta|^2 = 1$, it follows from the first equation that $|\alpha| = 1$ and $\beta = 0$. Then the second equation implies that $(\Re(\alpha))^2 = 1$ and hence $\alpha = \pm1$. Hence there are at most two solutions to Equation 4.4. But both of the candidate solutions are in fact solutions:

$$\Phi \begin{pmatrix} 1 & 0 \\ 0 & 1 \end{pmatrix} = \Phi \begin{pmatrix} -1 & 0 \\ 0 & -1 \end{pmatrix} = \begin{pmatrix} 1 & 0 & 0 \\ 0 & 1 & 0 \\ 0 & 0 & 1 \end{pmatrix}.$$

So there are precisely two elements of $SU(2)$ in the preimage of the identity in $SO(3)$ under Φ, namely I and $-I$. From Exercise 4.9 we conclude that Φ is a two-to-one function. □

In Proposition 4.8 and in Section 6.3 we will use the Spectral Theorem to simplify calculations in $SU(2)$. In Section 6.6 we will use the homomorphism Φ between $SU(2)$ and $SO(3)$ to make some calculations about $SO(3)$ that would be harder to make directly.

4.4   Representations: Definition and Examples

In this section we define representations and give examples. We also define homomorphisms and isomorphisms of representations, as well as unitary representations and isomorphisms.

A representation of a group G is an interpretation of the group in terms of linear operators. We fix a vector space V and assign to each element of the group a linear transformation of V in such a way that group multiplication corresponds to composition of linear transformations. Recall the group $\mathcal{GL}(V)$ of invertible linear operators on a vector space V and, from Section 4.1, the notion of a group homomorphism.

Definition 4.7 *A representation is a triple (G, V, ρ) where G is a group, V is a vector space and $\rho \colon G \to \mathcal{GL}(V)$ is a group homomorphism. We often write that ρ is a representation of G on V. If G is a Lie group and ρ is a Lie group homomorphism, then the representation (G, V, ρ) is a Lie group representation.*

If G and ρ are clear from context, one can call V "the representation." This slight abuse of language is quite common in the literature.

For example, the following function ρ is a representation of $\mathbb{T}$ on $\mathbb{C}^2$:

$$\rho : \mathbb{T} \to \mathcal{GL}\left(\mathbb{C}^2\right)$$

$$e^{i\theta} \mapsto \begin{pmatrix} 1 & 0 \\ 0 & e^{i\theta} \end{pmatrix}.$$

See Figure 4.3. Clearly $\mathbb{T}$ is a group and $\mathbb{C}^2$ is a vector space. We check

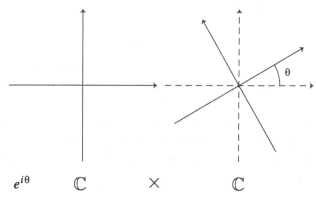

$$e^{i\theta} \qquad \mathbb{C} \qquad \times \qquad \mathbb{C}$$

Figure 4.3. A representation of the circle group $\mathbb{T}$ on $\mathbb{C}^2$.

explicitly that ρ is a group homomorphism:

$$\rho\left(e^{i\theta_1}\right) \rho\left(e^{i\theta_2}\right) = \begin{pmatrix} 1 & 0 \\ 0 & e^{i\theta_1} \end{pmatrix} \begin{pmatrix} 1 & 0 \\ 0 & e^{i\theta_2} \end{pmatrix}$$

$$= \begin{pmatrix} 1 & 0 \\ 0 & e^{i(\theta_1+\theta_2)} \end{pmatrix} = \rho\left(e^{i(\theta_1+\theta_2)}\right) = \rho\left(e^{i\theta_1} e^{i\theta_2}\right).$$

Representations are powerful mathematical tools because they allow us to use the concepts of linear algebra to study a group. For instance, we cannot talk about the eigenvalues of an element of an abstract group G, but given a representation (G, V, ρ) we can talk about the eigenvalues of $\rho(g)$ for any $g \in G$.

One common and useful technique for constructing representations is to use an *action* of a group G on a set S to induce a representation of G on the vector space V of complex-valued functions on S. Recall the group $T(S, S)$ of invertible function from S onto S defined in Section 4.1.

Definition 4.8 *An* action *of a group on a set is a triple* (G, S, σ) *where G is a group, S is a set, and σ is a group homomorphism from G to $T(S, S)$.*

In many texts an alternative (but equivalent) definition is given for an action: an action should be a function $G \times S \to S$ satisfying certain conditions with respect to the group. Readers familiar with this alternative definition

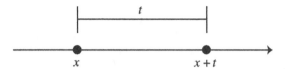

Figure 4.4. Translation by a positive number t.

should take the trouble to prove that it is equivalent to Definition 4.8; see Exercise 4.42.

Notice that every representation is an example of an action, but the notion of an action is far more general. For example, there is an action of the group $\mathbb{R}$ (with addition as the group "multiplication") on the real line given by $(\mathbb{R}, \mathbb{R}, \sigma)$, where for each t in the group $\mathbb{R}$ we define $\sigma(t) \colon \mathbb{R} \to \mathbb{R}, x \mapsto x + t$. This action is called the *translation action*. See Figure 4.4. However, the transformation $\sigma(t)$ is a linear transformation if and only if $t = 0$. So the action $(\mathbb{R}, \mathbb{R}, \sigma)$ is not a representation.

Given any action (G, S, σ), there is a representation (G, V, ρ), where V is defined to be the complex vector space of complex-valued functions on S and ρ is given by the formula

$$\rho(g) \cdot f \colon S \to V$$
$$s \mapsto f(\sigma(g^{-1})s)$$

for each $g \in G$ and each $f \in V$. We say that the action (G, S, σ) is *induced* by the representation (G, V, ρ). Alternatively, we say that ρ is the representation *corresponding to* the action (G, S, σ). Let us check that ρ satisfies the definition of a representation (Definition 4.7). The group is G, and the vector space is V. We must check that ρ is a group homomorphism from G to $\mathcal{GL}(V)$. Certainly the domain of ρ is G. Also, for any $g \in G$ and any $f \in V$ we have $\rho(g)f \in V$, i.e., $\rho(g)f$ is a complex-valued function on S. Finally, we check that for any $s \in S$, any $f \in V$ and any $g_1, g_2 \in G$ we have

$$(\rho(g_1)\rho(g_2)f)(s) = \rho(g_1)f(\sigma(g_2^{-1})s) = f(\sigma(g_2^{-1})\sigma(g_1^{-1})s)$$
$$= f(\sigma((g_1g_2)^{-1}s) = (\rho(g_1g_2)f)(s).$$

Let us verify the second equality more explicitly. Let h denote the function $\rho(g_2)f$; i.e., define $h(s) := f(\sigma(g_2^{-1})s)$. Then we have

$$\rho(g_1)f(\sigma(g_2^{-1})s) = \rho(g_1)h(s) = h(\sigma(g_1^{-1})s) = f(\sigma(g_2^{-1})\sigma(g_1^{-1})s).$$

Note the role the inverse plays: it undoes the order reversal introduced by the passage from the ρ's to the σ's. So ρ is indeed a group homomorphism, and hence (G, V, ρ) is a representation.

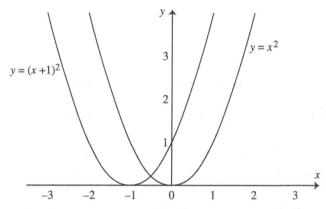

Figure 4.5. The graphs of f and $\rho(-1) \cdot f$ with $f(x) := x^2$.

For example, consider the translation action $(\mathbb{R}, \mathbb{R}, \sigma)$ defined above. Let V denote the complex vector space of complex-valued functions of one real variable. The induced representation of the additive group $\mathbb{R}$ on V is given by

$$\rho(t): V \to V$$
$$f \to f(\cdot - t)$$

for each group element $t \in \mathbb{R}$. So, for example, if $f(x) := x^2$, then

$$(\rho(-1) \cdot f)(x) = (x+1)^2.$$

Here $\sigma(-1)$ moves points on $\mathbb{R}$ one unit to the left and $\rho(-1)$ moves graphs one unit to the left. More generally, if $\sigma(t)$ moves points t units to the left (resp., right), $\rho(t)$ moves graphs of functions t units to the left (resp., right). See Figure 4.5.

For another example, let $S = \mathbb{R}^2$, let $G = SO(2)$ and let σ be the natural action. That is, if we fix the standard basis on $\mathbb{R}^2$ and write a typical group element

$$R_\theta := \begin{pmatrix} \cos\theta & -\sin\theta \\ \sin\theta & \cos\theta \end{pmatrix}$$

we have

$$\sigma(R_\theta) \cdot \begin{pmatrix} r_1 \\ r_2 \end{pmatrix} := \begin{pmatrix} \cos\theta & -\sin\theta \\ \sin\theta & \cos\theta \end{pmatrix} \begin{pmatrix} r_1 \\ r_2 \end{pmatrix}.$$

In words, the group element R_θ rotates the real plane counterclockwise around 0 through an angle of θ. Let V denote the set of complex-valued functions on $\mathbb{R}^2$. Then in the corresponding representation on V the group element R_θ rotates the graph of each function f counterclockwise through an angle of θ. For example, consider the functions $f_i: \mathbb{R}^2 \to \mathbb{C}$, $(r_1, r_2)^T \mapsto r_i$ for $i = 1$ or $i = 2$. The graph of f_1 is a plane containing the r_2 axis and making an angle of $\pi/4$ with the positive r_1 axis. If we rotate this plane through an angle of $\pi/2$ (parallel to the r_1r_2-plane) we get a plane containing the r_1 axis and making an angle of $\pi/4$ with the r_2 axis: the graph of f_2. Algebraically, we find that

$$R_{\pi/2} \cdot f_1(r_1, r_2) = f_1\left(\sigma(R_{-\pi/2})(r_1, r_2)\right) = f_1(r_2, -r_1) = r_2 = f_2(r_1, r_2).$$

In other words, $R_{\pi/2} \cdot f_1 = f_2$.

The representation that arises first in the study of the hydrogen atom is the *natural representation of $SO(3)$ on $L^2(\mathbb{R}^3)$*. Recall that we defined the group $SO(3)$ as a group of rotations of real Euclidean three-space. This gives a natural action of the group on the space $\mathbb{R}^3$. Hence there is a natural representation of $SO(3)$ on the space of complex-valued functions on $\mathbb{R}^3$. Sticklers will recall that elements of $L^2(\mathbb{R}^3)$ are equivalence classes of complex-valued functions; because rotating two equivalent functions yields two equivalent functions (as the reader can check in Exercise 4.44), the representation is well defined on equivalence classes. Also, if a function is square-integrable, then rotating it yields a square-integrable function. So we have a *bona fide* representation of $SO(3)$ on the Hilbert space $L^2(\mathbb{R}^3)$. More explicitly, for any $g \in SO(3)$ and any $f \in L^2(\mathbb{R}^3)$, we define the function $g \cdot f$ by

$$(g \cdot f)\begin{pmatrix} x \\ y \\ z \end{pmatrix} := f\left(g^{-1}\begin{pmatrix} x \\ y \\ z \end{pmatrix}\right).$$

Just as the same group can arise in different guises, two different-looking representations can be essentially the same. Hence it is useful to define isomorphisms of representations. Homomorphisms of representations play an important role in the critical technical tools developed in Chapter 6. We will also use them in the proof of Proposition 11.1.

Definition 4.9 *Suppose (G, V, ρ) and $(G, W, \tilde{\rho})$ are representations of the same group G. Suppose $T : V \to W$ is a linear transformation. Then T is a homomorphism of the representations (G, V, ρ) and $(G, W, \tilde{\rho})$ if and only if, for every $g \in G$ we have $\tilde{\rho}(g) \circ T = T \circ \rho(g)$.*

If T is a homomorphism of representations, then T is said to *intertwine* the two representations. Because the condition for T to be a homomorphism is linear in T, it follows that the set of homomorphisms of representations from V to W forms a vector space.

Next we introduce isomorphisms of representations. As usual, isomorphisms are homomorphisms that are injective and surjective.

Definition 4.10 *Suppose $T: V \to W$ is a homomorphism of the representations (G, V, ρ) and $(G, W, \tilde{\rho})$. If T is injective and $T^{-1}: W \to V$ is a homomorphism of representations, then we say that T is an isomorphism of representations. In this case we say that the representations (G, V, ρ) and $(G, W, \tilde{\rho})$ are* isomorphic.

It is useful to have a shorthand for the statement that (G, V, ρ) is isomorphic to $(G, W, \tilde{\rho})$. One can write $\rho \cong \tilde{\rho}$. A notation common in the literature is $V \cong W$. This last shorthand puts the burden on the reader to determine from context what the group and the representations are.

The next proposition is an important tool for telling representations apart.

Proposition 4.6 *Suppose (G, V, ρ) and $(G, W, \tilde{\rho})$ are isomorphic representations of the group G. Then either both V and W are infinite-dimensional, or both are finite dimensional and the dimension of V is equal to the dimension of W.*

Thus if two representations are of different dimensions, they cannot be isomorphic.

Proof. By the definition of "isomorphic," there must be an invertible, surjective linear transformation T from V to W. If W is finite-dimensional, then we can apply the Fundamental Theorem of Linear Algebra (Proposition 2.5) to find

$$\dim V = \dim(\ker T) + \dim(\text{Image } T) = 0 + \dim W = \dim W.$$

Likewise, if V is finite-dimensional, we can apply Proposition 2.5 to T^{-1}. The only other possibility is that V and W are both infinite-dimensional. □

In all the important quantum mechanical applications, the representations are *unitary*. Recall the unitary group $\mathcal{U}(V)$ from Definition 4.2.

Definition 4.11 *Suppose V is a complex scalar product space. A representation (or Lie group representation) (G, V, ρ) is* unitary *if and only if, for each $g \in G$ the linear transformation $\rho(g)$ is a unitary transformation. In other words, a representation is* unitary *if the image of ρ lies in $\mathcal{U}(V) \subset \mathcal{GL}(V)$.*

For example, consider the representation $\rho: \mathbb{T} \to \mathcal{U}\left(\mathbb{C}^2\right)$ defined by

$$e^{i\theta} \mapsto \begin{pmatrix} \cos\theta & -\sin\theta \\ \sin\theta & \cos\theta \end{pmatrix}.$$

By Proposition 3.2, this is a unitary representation with respect to the natural scalar product on $\mathbb{C}^2$, since the columns of the matrix form a unitary basis of

$\mathbb{C}^2$ for any θ. On the other hand, the representation $\rho: \mathbb{T} \to \mathcal{GL}\left(\mathbb{C}^2\right)$ defined by

$$e^{i\theta} \mapsto \begin{pmatrix} \cos\theta & -2\sin\theta \\ \frac{1}{2}\sin\theta & \cos\theta \end{pmatrix}$$

is not unitary in the natural complex scalar product on $\mathbb{C}^2$, since the second column of the matrix does not have norm 1 for every value of θ:

$$\left\| \begin{pmatrix} -2\sin\frac{\pi}{2} \\ \cos\frac{\pi}{2} \end{pmatrix} \right\| = \left\| \begin{pmatrix} -2 \\ 0 \end{pmatrix} \right\| = 2 \neq 1.$$

Isomorphisms of unitary representations ought to preserve the unitary structure. When they do, they are called *unitary isomorphisms of representations*.

Definition 4.12 *Suppose (G, V, ρ) and $(G, W, \tilde{\rho})$ are two representations of the same group. Suppose V and W are complex scalar product spaces. Suppose $T: V \to W$ is an homomorphism of representations. If T respects the complex scalar products, i.e., if for all $v, w \in V$ we have $\langle v, w \rangle = \langle Tv, Tw \rangle_{\sim}$, where $\langle \, , \rangle$ denotes the complex scalar product on V and $\langle \, , \rangle_{\sim}$ denotes the one on W, then we call T a* unitary homomorphism. *If T is an isomorphism as well we call it a* unitary isomorphism *and we say that the representations are* isomorphic as unitary representations.

Note that every unitary homomorphism T of representations is injective: if $v \neq 0 \in V$, then

$$\langle Tv, Tv \rangle = \langle v, v \rangle \neq 0,$$

so the kernel of T is trivial.

Representations are the primary object of our mathematical analysis. In particular, the natural representation of $SO(3)$ on $L^2(\mathbb{R}^3)$ introduced in this section is the mathematical model for the states of the electron in the hydrogen atom, along with its symmetries. We will analyze this representation in detail in Chapter 7. In our preparatory work in the intervening chapters, we will use homomorphisms, isomorphisms and unitary representations many times.

4.5 Representations in Quantum Mechanics

Representations arise naturally in quantum physics, where there is a homomorphism from the symmetry group of the physical space to the group of linear transformations of the Hilbert space of states of the quantum system.

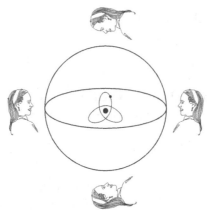

Figure 4.6. Sphere of observers and their hydrogen atom.

The symmetry group of the physical space is the abstraction of the empirical fact that many different observers see the same laws of physics. Let us explain in detail the case that interests us most. Suppose one studies hydrogen atoms in the laboratory, and one discovers that laws of physics governing the hydrogen atom show no directional bias. In other words, the results of experiments do not depend on the angle of the observational equipment (from the vertical, or from any other reference direction). The results might depend on the angle between the observational equipment and other equipment involved in the experiment, but if one rotates the whole setup, one gets the same results. Also, the result of one particular experimental trial might involve some angular data, but statistically (looking at the aggregate of many trials) there is no favored direction.

To see how a group arises, imagine many observers, all at the same distance from one hydrogen atom. All these observers sit at points on an abstract sphere, whose center is the hydrogen atom under study. See Figure 4.6. If we secretly rotate the sphere to a different position, none of the observers would be able to tell the difference. Thus any rotation of the sphere is a symmetry of the system. In other words, the symmetry group of the hydrogen atom contains the group $SO(3)$.

If we model the hydrogen atom as a stable nucleus with one particle moving around it, then the space of states in our model is $L^2(\mathbb{R}^3)$, as discussed in Section 1.2. To create a representation, we find a group homomorphism ρ from $SO(3)$ to the group of linear operators on $L^2(\mathbb{R}^3)$. Fix an arbitrary rotation $g \in SO(3)$. Consider two observers (A and B) whose positions differ by g. In other words, suppose that if we apply the rotation g to the imaginary sphere in Figure 4.6, observer A ends up precisely where observer B was, and facing the same direction as observer B faced. To understand the corresponding linear transformation $\rho(g)$ of $L^2(\mathbb{R}^3)$, consider an arbitrary vector $f \in L^2(\mathbb{R}^3)$. Imagine the hydrogen atom is in the state described by the vector f from observer A's point of view. Now define $\rho(g)f$ to be the vector in $L^2(\mathbb{R}^3)$ that observer B would use to describe that same state. (Astute readers may object that this last vector is not well defined, since the state determines a line in $L^2(\mathbb{R}^3)$, not a single vector. Such readers should see Chapter 10 for the true story; for our purposes here, it is not misleading to assume that each state corresponds to one single vector.) In other words, if we asked each observer to write down the vector describing the state of the mobile particle in the hydrogen atom, observer A would write down f and observer B would write down a state $\tilde{f}$, and then we would define $\rho(g)f := \tilde{f}$. Note that the definition of $\rho(g)$ does not depend on which pair of observers we chose — another pair in the same relative position would yield the same $\rho(g)$.

Of course, we need to check that the $\rho(g)$ we defined is actually a linear transformation. Here the physics helps us. Recall from Section 1.2 that linear combinations of vectors can be interpreted physically: if a beam of particles contains a mixture of orthogonal states, then the probabilities governing experiments with that beam can be predicted from a linear combination of those states. Thus observer A's and observer B's linear combinations must be compatible. In other words, if observer A takes a linear combination $c_1 f_1 + c_2 f_2$, while observer B takes the same linear combination of the corresponding states $c_1 \rho(g) f_1 + c_2 \rho(g) f_2$, the answers should be compatible, i.e.,

$$\rho(g)(c_1 f_1 + c_2 f_2) = c_1 \rho(g) f_1 + c_2 \rho(g) f_2.$$

But this is equivalent to the definition of a linear transformation.

Finally, we need to check that ρ is a group homomorphism. It is not hard to see that ρ respects group multiplication: if we have observers A, B and C, a rotation g_{AB} that takes A to B's position and a rotation g_{BC} that takes B to C's position, then $g_{BC} g_{AB}$ takes A to C's position and hence $\rho(g_{BC} g_{AB})$ is the linear transformation taking states in A's perspective to states in C's perspective. This yields the same as taking states from A's to B's perspective, followed by taking states from B's to C's perspective. So $\rho(g_{BC} g_{AB}) = \rho(g_{BC}) \rho(g_{AB})$. Hence ρ is indeed a group homomorphism.

In addition, ρ is a unitary representation. Because complex inner products of states yield physically measurable quantities, the value of the complex scalar product cannot depend on the angular position of the measurer. More explicitly, the value $\langle \phi, \psi \rangle$ measured by A must be equal to the value $\langle \rho(g)\phi, \rho(g)\psi \rangle$ measured by B. (Again, this is a bit of a lie, harmless for now: we can measure only $|\langle \cdot, \cdot \rangle|$. For the true story see Chapter 10.) So each $\rho(g)$ must be a unitary operator on the state space.

Any physical representation ρ must also be a Lie group homomorphism; i.e., it must be differentiable. This follows from the experimental observation that data observed changes smoothly as an observer changes position smoothly. All the representations we discuss in this book are Lie group representations. In our study of the hydrogen atom, we have an experimental model for the particular state space, namely, $L^2(\mathbb{R}^3)$. In Section 7.3 we will get physical predictions by studying the representation of the group $SO(3)$ on that state space. In other situations, one may know only the group and not the particular state space. For example, one might ask what quantum mechanical systems one might expect in a physical space obeying the rules of special relativity. Such a space is called *Minkowski space* and its group of symmetries is called the *Poincaré group*. Any quantum system must correspond to a representation of the Poincaré group. Therefore, if we can find a way (mathematically) to classify the representations of the Poincaré group, then we can predict something about quantum systems in Minkowski space. With this goal, E. Wigner worked out the classification and predicted that elementary particles should have mass and spin. For more detail, see [St, Section 3.9].

Representation theory encompasses more than just group representations. Because we can add, compose and take *commutators* $(T_1 T_2 - T_2 T_1)$ of linear transformations, we can represent any algebraic structure whose operations are limited to these operations. We will see an important example in Chapter 8, where we introduce and use the representation theory of *Lie algebras* to find more symmetry in and make finer predictions for the hydrogen atom.

Note that the application of representation theory to quantum mechanics depends heavily on the *linear nature of quantum mechanics*, that is, on the fact that we can successfully model states of quantum systems by vector spaces. (By contrast, note that the states of many classical systems cannot be modeled with a linear space; consider for example a pendulum, whose motion is limited to a sphere on which one cannot define a natural addition.) The linearity of quantum mechanics is miraculous enough to beg the question: is quantum mechanics truly linear? There has been some investigation of nonlinear quantum mechanical models but by and large the success of linear models has been enormous and long-lived.

In summary, a set of equivalent observers of a quantum mechanical system gives a unitary representation (G, V, ρ), where G is the symmetry group for the equivalent observers and V is the state space of the system.

4.6 Representations of $SU(2)$ on Homogeneous Polynomials in Two Variables

A family of representations important in our analysis of the hydrogen atom consists of the representations of $SU(2)$ on spaces of homogeneous polynomials. These representations play a major role in our classification of representations in Chapter 6.

Recall from Section 2.2 that the complex vector space $\mathcal{P}^n$ of homogeneous complex polynomials of degree n in two variables has dimension $n + 1$ and has a basis of the form $\{x^n, x^{n-1}y, x^{n-2}y^2, \dots, xy^{n-1}, y^n\}$. The action of the group $SU(2)$ on the complex vector space $\mathbb{C}^2$ (via matrix multiplication) defines a representation $(SU(2), \mathcal{P}^n, R_n)$ as described in Section 4.4: for any $g \in SU(2)$, any $(x, y)^T \in \mathbb{C}^2$ and any polynomial p,

$$(R_n(g)) \, p \begin{pmatrix} x \\ y \end{pmatrix} := p \left(g^{-1} \begin{pmatrix} x \\ y \end{pmatrix} \right).$$

Note that because the action of $SU(2)$ on $\mathbb{C}^2$ is linear and invertible, the transformation $R_n(g)$ preserves polynomial degree. These representations are related to the *spin* of elementary particles as we will see in Section 10.4; in particular, $\mathcal{P}^n$ corresponds to a particle of spin-$n/2$. (*Spin* is a quality of particles that physicists introduced into their equations to model certain mysterious experimental results; we will see in Chapter 10 that spin arises naturally from the spherical symmetry of space.)

Let us calculate some of these R_n's explicitly. Recall that each element of $SU(2)$ has the form $\begin{pmatrix} \alpha & -\beta^* \\ \beta & \alpha^* \end{pmatrix}$, where α and β are complex numbers such that $|\alpha|^2 + |\beta|^2 = 1$. It will help to note that for any $(x, y)^T \in \mathbb{C}^2$ and any $\begin{pmatrix} \alpha & -\beta^* \\ \beta & \alpha^* \end{pmatrix} \in SU(2)$ we have

$$\begin{pmatrix} \alpha & -\beta^* \\ \beta & \alpha^* \end{pmatrix}^{-1} \begin{pmatrix} x \\ y \end{pmatrix} = \begin{pmatrix} \alpha^* & \beta^* \\ -\beta & \alpha \end{pmatrix} \begin{pmatrix} x \\ y \end{pmatrix} = \begin{pmatrix} \alpha^* x + \beta^* y \\ -\beta x + \alpha y \end{pmatrix}.$$

We start with the case $n = 0$. Here our vector space is one-dimensional and the basis consists of the constant function 1. Because $R_0(g) \cdot 1 = 1$ for each $g \in SU(2)$, the matrix of the representation R_0 is (1) in the basis $\{1\}$ of the constant polynomials in two variables. Next we tackle the case $n = 1$. The vector space is two-dimensional with basis $\{x, y\}$. Note that here we are using x to denote the function taking a point in $\mathbb{C}^2$ to its first coordinate. So, for instance,

$$\begin{pmatrix} \alpha & -\beta^* \\ \beta & \alpha^* \end{pmatrix} \cdot x$$

is the function taking the point $(X, Y)^T \in \mathbb{C}^2$ to the first coordinate of

$$\begin{pmatrix} \alpha^* & \beta^* \\ -\beta & \alpha \end{pmatrix} \begin{pmatrix} X \\ Y \end{pmatrix},$$

which is $\alpha^* X + \beta^* Y$. In other words, we have

$$\begin{pmatrix} \alpha & -\beta^* \\ \beta & \alpha^* \end{pmatrix} \cdot x = \alpha^* x + \beta^* y.$$

Similarly, we can calculate that

$$\begin{pmatrix} \alpha & -\beta^* \\ \beta & \alpha^* \end{pmatrix} \cdot y = -\beta x + \alpha y.$$

Hence the matrix of the representation R_1 in the given basis is

$$\begin{pmatrix} \alpha^* & \beta^* \\ -\beta & \alpha \end{pmatrix}.$$

For $n = 2$ we have the three-dimensional vector space spanned by the basis $\{x^2, xy, y^2\}$. We have

$$\begin{pmatrix} \alpha & -\beta^* \\ \beta & \alpha^* \end{pmatrix} \cdot x^2 = (\alpha^* x + \beta^* y)^2 = (\alpha^*)^2 x^2 + \text{other terms},$$

$$\begin{pmatrix} \alpha & -\beta^* \\ \beta & \alpha^* \end{pmatrix} \cdot xy = (\alpha^* x + \beta^* y)(-\beta x + \alpha y)$$

$$= (\alpha^* \alpha - \beta^* \beta)xy + \text{other terms},$$

$$\begin{pmatrix} \alpha & -\beta^* \\ \beta & \alpha^* \end{pmatrix} \cdot y^2 = (-\beta x + \alpha y)^2 = \alpha^2 y^2 + \text{other terms}.$$

In fact, as the reader may show in Exercise 4.45, the matrix of the representation R_2 in the given basis is

$$\begin{pmatrix} (\alpha^*)^2 & 2\alpha^* \beta^* & (\beta^*)^2 \\ -\alpha^* \beta & |\alpha|^2 - |\beta|^2 & \alpha\beta^* \\ \beta^2 & -2\alpha\beta & \alpha^2 \end{pmatrix}.$$

Are the representations R_n unitary? That is, do they satisfy the conditions of Definition 4.11? The question does not make sense until we specify complex scalar products. There are many different choices; we will find it useful to define complex scalar products in which the representations are unitary.

Proposition 4.7 *Fix a nonnegative integer n and consider the complex vector space* $\mathcal{P}^n$. *Define a complex scalar product on* $\mathcal{P}^n$ *by setting*

$$\langle x^{n-j}y^j, x^{n-k}y^k \rangle := \begin{cases} k!(n-k)! & j=k \\ 0 & j \neq k \end{cases}$$

and extending linearly to arbitrary elements of $\mathcal{P}^n$. *With this complex scalar product on* $\mathcal{P}^n$, *the representation* $(SU(2), \mathcal{P}^n, R_n)$ *is unitary.*

Note that it suffices to define the scalar product on basis elements. This proposition plays a crucial role in the proof of Proposition 6.14, the classification of the unitary irreducible representations of the group $SU(2)$.

Proof. We will find it helpful to define a function[5] F of three variables by

$$F(t, x, y) := (x + ty)^n = \sum_{k=0}^{n} t^k \binom{n}{k} x^{n-k} y^k.$$

We will see below that the function $\langle F(s, x, y), F(t, x, y) \rangle$ has important properties. Specifically, $\langle F(s, x, y), F(t, x, y) \rangle$ is a polynomial in s and t; its coefficients contain complete information about the complex scalar product on $\mathcal{P}^n$; and $\langle F(s, x, y), F(t, x, y) \rangle$ is invariant under the action of $SU(2)$ on $\mathbb{C}^2$. Finally, we will show that these properties imply that the complex scalar product is invariant under the representation R_n, and thus the representation is unitary.

To see that $\langle F(s, x, y), F(t, x, y) \rangle$ is a polynomial in s^* and t, note that

$$\langle F(s, x, y), F(t, x, y) \rangle = \sum_{k=0}^{n} \sum_{j=0}^{n} (s^*)^k t^j \binom{n}{k} \binom{n}{j} \langle x^{n-k} y^k, x^{n-j} y^j \rangle.$$

We will find it useful to simplify this expression (using the particular complex scalar product defined in the statement of the proposition) to

$$\langle F(s, x, y), F(t, x, y) \rangle = \sum_{k=0}^{n} (s^*t)^k \binom{n}{k}^2 k!(n-k)! = n!(1 + s^*t)^n. \quad (4.5)$$

---

[5]The function F is an example of a *generating function*, i.e., a power series in an extra variable (in this case, t) whose analysis yields information about its coefficients.

How does the representation of $SU(2)$ affect $\langle F(s, x, y), F(t, x, y)\rangle$? We consider

$$F\left(t, \begin{pmatrix} \alpha & -\beta^* \\ \beta & \alpha^* \end{pmatrix} \cdot (x, y)\right) = F(t, \alpha^* x + \beta^* y, -\beta x + \alpha y)$$

$$= (\alpha^* x + \beta^* y + t(-\beta x + \alpha y))^n$$

$$= (\alpha^* - t\beta)^n \left(x + \left(\frac{t\alpha + \beta^*}{\alpha^* - t\beta}\right) y\right)^n$$

$$= (\alpha^* - t\beta)^n F\left(\frac{t\alpha + \beta^*}{\alpha^* - t\beta}, x, y\right).$$

It follows that

$$\left\langle F\left(s, \begin{pmatrix} \alpha & -\beta^* \\ \beta & \alpha^* \end{pmatrix} \cdot (x, y)\right), F\left(t, \begin{pmatrix} \alpha & -\beta^* \\ \beta & \alpha^* \end{pmatrix} \cdot (x, y)\right)\right\rangle$$

$$= (\alpha - s^*\beta^*)^n (\alpha^* - t\beta)^n \left\langle F\left(\frac{s\alpha + \beta^*}{\alpha^* - s\beta}, x, y\right), F\left(\frac{t\alpha + \beta^*}{\alpha^* - t\beta}, x, y\right)\right\rangle$$

$$= n!(\alpha - s^*\beta^*)^n (\alpha^* - t\beta)^n \left(1 + \left(\frac{s\alpha + \beta^*}{\alpha^* - s\beta}\right)^* \left(\frac{t\alpha + \beta^*}{\alpha^* - t\beta}\right)\right)^n$$

$$= n! (1 + s^* t)^n = \langle F(s, x, y), F(t, x, y)\rangle,$$

where we have used Equation 4.5 and the fact that $|\alpha|^2 + |\beta|^2 = 1$. Hence we conclude that, for any $g \in SU(2)$,

$$\sum_{k=0}^{n} \sum_{j=0}^{n} (s^*)^k t^j \binom{n}{k}\binom{n}{j} \langle x^{n-k} y^k, x^{n-j} y^j\rangle = \langle F(s, x, y), F(t, x, y)\rangle$$

$$= \langle F(s, g \cdot (x, y)), F(t, g \cdot (x, y))\rangle$$

$$= \sum_{k=0}^{n} \sum_{j=0}^{n} (s^*)^k t^j \binom{n}{k}\binom{n}{j} \langle g \cdot x^{n-k} y^k, g \cdot x^{n-j} y^j\rangle.$$

But two polynomials in s and t are equal if and only if the coefficients of monomials in s and t are equal. Hence for any j and any k we have

$$\langle g \cdot x^{n-k} y^k, g \cdot x^{n-j} y^j\rangle = \langle x^{n-k} y^k, x^{n-j} y^j\rangle.$$

We conclude that the representation R_n is unitary with respect to the given complex scalar product.   □

The unitary representations R_n in this section turn out to be the building blocks for all representations of $SU(2)$. They will help us (in Chapters 6 and 7 to identify the representations of $SO(3)$ that occur in $L^2(\mathbb{R}^3)$, and they will show up again in the study of arbitrary spins in Section 10.4.

4.7   Characters of Representations

Let me see: four times five is twelve, and four times six is thirteen, and four
times seven is — oh dear! I shall never get to twenty at that rate!
— Lewis Carroll, *Alice's Adventures in Wonderland* [Car, pp. 38]

In this section we define characters. Associated to each finite-dimensional
representation (G, V, ρ) is a complex-valued function on the group G, called
the *character of the representation*.[6] Recall the *trace* of an operator (Definition 2.8): the sum of the diagonal elements of the corresponding matrix,
expressed in any basis.

Definition 4.13 *Suppose* (G, V, ρ) *is a representation with finite-dimensional vector space* V. *Define the* character $\chi_\rho : G \to \mathbb{C}$ *of the representation*
by

$$\chi_\rho(g) := \operatorname{Tr} \rho(g),$$

for each $g \in G$.

For example, consider the representation of $(SU(2), \mathbb{C}^2, \rho)$, where ρ is
given by matrix multiplication:

$$\rho\left(\begin{pmatrix} \alpha & -\beta^* \\ \beta & \alpha^* \end{pmatrix}\right)\begin{pmatrix} c_1 \\ c_2 \end{pmatrix} := \begin{pmatrix} \alpha & -\beta^* \\ \beta & \alpha^* \end{pmatrix}\begin{pmatrix} c_1 \\ c_2 \end{pmatrix}.$$

The character of this representation is given by the formula

$$\chi_\rho\left(\begin{pmatrix} \alpha & -\beta^* \\ \beta & \alpha^* \end{pmatrix}\right) = 2\Re(\alpha).$$

Because each representation function ρ is a group homomorphism and because the trace function is invariant under conjugation, we have

$$\chi_\rho(hgh^{-1}) = \operatorname{Tr}\left(\rho(hgh^{-1})\right) = \operatorname{Tr}\left(\rho(h)\rho(g)\rho(h)^{-1}\right) = \operatorname{Tr}\left(\rho(g)\right) = \chi_\rho(g)$$

for any group elements g, h. Hence the character of any representation is
invariant under conjugation.

We would like to find the character of each representation of $SU(2)$ on
homogeneous polynomials in two variables, introduced in Section 4.6. For
each nonnegative integer n it suffices to find the diagonal entries of the matrix
form of the transformation $R_n(g)$ in the familiar basis. We calculated some of

---

[6]This is *not* the same as the "characteristic function of a set."

these matrices in Section 4.6. For $n = 0$, the character is $\chi_0(g) = \text{Tr}(1) = 1$. (Note that we are using the shorthand notation $\chi_n := \chi_{R_n}$.) For $n = 1$ the character is

$$\chi_1\left(\begin{pmatrix} \alpha & -\beta^* \\ \beta & \alpha^* \end{pmatrix}\right) = \text{Tr}\begin{pmatrix} \alpha^* & \beta^* \\ -\beta & \alpha \end{pmatrix} = \alpha^* + \alpha = 2\Re(\alpha).$$

This character is the same as the character of the representation on $\mathbb{C}^2$ by matrix multiplication; in fact, these two representations are isomorphic, as the reader may show in Exercise 4.36. This is an example of the general phenomenon that will help us to classify representations: finite-dimensional representations are isomorphic if and only if their characters are equal. See Proposition 6.12. Note that while a representation is a relatively complicated object, a character is simply a function from a group to the complex numbers; it is remarkable that so much information about the complicated object is encapsulated in the simpler object.

For $n = 2$ we find that the character is

$$\chi_2\left(\begin{pmatrix} \alpha & -\beta^* \\ \beta & \alpha^* \end{pmatrix}\right) = \text{Tr}\begin{pmatrix} (\alpha^*)^2 & 2\alpha^*\beta^* & (\beta^*)^2 \\ -\alpha^*\beta & |\alpha|^2 - |\beta|^2 & \alpha\beta^* \\ \beta^2 & -2\alpha\beta & \alpha^2 \end{pmatrix} = 4\Re(\alpha)^2 - 1.$$

Like Alice, we shall never get to our goal (calculating the character of R_n for each n) at this rate! Fortunately, we can use the Spectral Theorem to find an easier way to do a more general calculation.

Proposition 4.8 *For each nonnegative integer n there is a polynomial q_n of degree n in one variable such that for each element of $SU(2)$ we have*

$$\chi_n\left(\begin{pmatrix} \alpha & -\beta^* \\ \beta & \alpha^* \end{pmatrix}\right) = q_n\left(\Re(\alpha)\right).$$

This proposition will play a crucial role in the proof of Proposition 6.14.

Proof. Loosely speaking, we can evaluate the character χ_n at an arbitrary $g \in SU(2)$ by finding a diagonal matrix with the same $\Re(\alpha)$ as g, and evaluating the character at that diagonal element of $SU(2)$.

For any diagonal element of $SU(2)$ for any basis vector $x^k y^{n-k}$ we have

$$R_n\left(\begin{pmatrix} \lambda & 0 \\ 0 & \lambda^{-1} \end{pmatrix}\right) x^k y^{n-k} = \lambda^{-k}\lambda^{n-k}x^k y^{n-k} = \lambda^{n-2k}x^k y^{n-k}.$$

So for any natural number n we have

$$\chi_n\left(\begin{pmatrix} \lambda & 0 \\ 0 & \lambda^{-1} \end{pmatrix}\right) = \sum_{k=0}^{n} \lambda^{n-2k}.$$

Next we show by induction on n that this last expression is a polynomial of degree n in $\Re(\lambda)$. First we need two base cases: for χ_0 we have

$$\sum_{k=0}^{0} \lambda^{0-2k} = 1,$$

a polynomial of degree 0, while for χ_1 we have

$$\sum_{k=0}^{1} \lambda^{1-2k} = \lambda + \lambda^{-1} = \lambda + \lambda^* = 2\Re(\lambda),$$

a polynomial of degree 1 in $\Re(\lambda)$.

For the inductive step we note that

$$\sum_{k=0}^{n} \lambda^{n-2k} = \lambda^n + \lambda^{-n} + \sum_{k=1}^{n-1} \lambda^{n-2k}$$

$$= \lambda^n + \lambda^{-n} + \sum_{k=0}^{n-2} \lambda^{n-2(k+1)}$$

$$= \lambda^n + \lambda^{-n} + \sum_{k=0}^{n-2} \lambda^{n-2-2k}.$$

By the inductive hypothesis, we know that the last term is a polynomial of degree $n-2$ in $\Re(\lambda)$. We will be done with our induction if we can show that $\lambda^{-n} + \lambda^n$ is a polynomial of degree n in $\Re(\lambda)$. Note that for $\lambda \in \mathbb{T}$ we have $\lambda^{-n} = (\lambda^n)^*$, so

$$\lambda^n + \lambda^{-n} = 2\Re(\lambda^n) = 2\Re\left(\Re(\lambda) \pm i\sqrt{1 - \Re(\lambda)^2}\right).$$

Now Exercise 1.4 implies that $\lambda^n + \lambda^{-n}$ is a polynomial of degree n in $\Re(\lambda)$. Thus we have shown that

$$\chi_n \begin{pmatrix} \lambda & 0 \\ 0 & \lambda^{-1} \end{pmatrix}$$

is a polynomial of degree n in $\Re(\lambda)$. Let q_n denote this polynomial.

Now we can verify the statement of the theorem. For any element of $SU(2)$ and any n we have

$$\chi_n \begin{pmatrix} \alpha & -\beta^* \\ \beta & \alpha^* \end{pmatrix}$$

$$= \chi_n \begin{pmatrix} \Re(\alpha) + i\sqrt{1 - \Re(\alpha)^2} & 0 \\ 0 & \Re(\alpha) - i\sqrt{1 - \Re(\alpha)^2} \end{pmatrix}$$

$$= q_n(\Re(\alpha)),$$

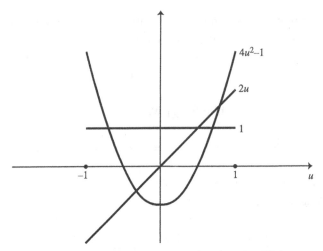

Figure 4.7. The first few character functions for $SU(2)$.

where the first equality follows from the Spectral Theorem for $SU(2)$ and the fact that χ_n is invariant under conjugation. □

Note that in the proof we have shown that $q_0(u) = 1$ and $q_1(u) = 2u$. The reader is invited to calculate more examples explicitly in Exercise 4.39. For another view of these polynomials, recall from Exercise 1.3 that

$$\sum_{k=0}^{n} \lambda^{2k-n} = \frac{\lambda^{n+1} - \lambda^{-n-1}}{\lambda - \lambda^{-1}}.$$

We will use this proposition in our classification of the representations of $SU(2)$ and $SO(3)$ (Propositions 6.14 and 6.16). Note that the converse of this proposition is false, as the reader may show in Exercise 4.23.

The first few character functions are shown in Figure 4.7.

In Section 6.5 we will use this proposition to help show that any representation of the group $SU(2)$ can be built from the R_n's. Specifically, we will make use of the fact that there is exactly one q_n for each degree n to show that the q_n's span the complex scalar product space $C[-1, 1]$.

4.8   Exercises

Exercise 4.1 *Suppose G_1 and G_2 are groups. Consider the set defined by*

$$G_1 \times G_2 := \{(g_1, g_2) : g_1 \in G_1, g_2 \in G_2\},$$

with multiplication defined by

$$(g_1, g_2)(h_1, h_2) := (g_1 h_1, g_2 h_2).$$

Show that this multiplication makes $G_1 \times G_2$ into a group. The group $G_1 \times G_2$ is called the Cartesian product *group.*

Exercise 4.2 *Show that the Cartesian product of groups defined in Exercise 4.1 is associative, i.e., for any groups G_1, G_2 and G_3 the group $(G_1 \times G_2) \times G_3$ is isomorphic to the group $G_1 \times (G_2 \times G_3)$. Conclude that for any natural number n, n-fold products of groups are well defined.*

Exercise 4.3 *Show that the set of 2×2 diagonal special unitary matrices is a group and that it is isomorphic to the group $\mathbb{T} \times \mathbb{T}$. (See Exercise 4.1 for the definition of the Cartesian product of groups.)*

Exercise 4.4 *Fix a natural number n and show that the set of $n \times n$ matrices of determinant one forms a group. Show more generally that the kernel of any group homomorphism is itself a group. Does the set of all matrices (of all finite sizes) of determinant one form a group under the usual matrix multiplication?*

Exercise 4.5 *Show that $SO(3)$, $SO(4)$, $\mathbb{T}$, the unit quaternions and $SU(2)$ are all compact.*

Exercise 4.6 *Is the group $SO(4)$ isomorphic to the Cartesian product $SO(3) \times SO(3)$?*

Exercise 4.7 *Suppose M is a matrix in $SO(3)$. Show that 1 is an eigenvalue of M. What is the geometric meaning of the associated eigenvector? Does every matrix in $SO(4)$ have 1 as an eigenvalue?*

Exercise 4.8 *Show that the subset of $T(S, S)$ of transformations T such that both T and T^{-1} are continuous is a subgroup. How many analogous constructions can you make? I.e., what other structures on sets S lead naturally to subgroups of $T(S, S)$?*

Exercise 4.9 (Used in Proposition 4.5) *Suppose G and $\tilde{G}$ are groups and $\Psi : G \to \tilde{G}$ is a surjective group homomorphism. Suppose that the kernel of Ψ contains precisely n elements. Show that Ψ is an n-to-one function, i.e., that for any $g \in \tilde{G}$ the set $\Psi^{-1}[g]$ contains precisely n elements. In particular, Ψ is injective if and only if $\Psi^{-1}[I] = I$.*

Exercise 4.10 *Consider the (topological) two-torus in $\mathbb{R}^3$ of inner radius r and outer radius R. An equation for this two-torus is*

$$\left(\sqrt{x^2 + y^2} - A\right)^2 + z^2 = B^2,$$

where $A := \frac{R+r}{2}$ and $B := \frac{R-r}{2}$. Use the rotations $\mathbf{Z}_\phi$ and $\mathbf{X}_\theta$ (and operations with constant vectors) to parameterize this torus by ϕ and θ.

Exercise 4.11 (Used in Section 6.1) Suppose $v \in \mathbb{R}^3$. Show that there is a rotation $M \in SO(3)$ such that $Mv = (r, 0, 0)^T$ for some nonnegative real number r. Is this rotation unique? Show that if v is nonzero then r is nonzero as well.

Exercise 4.12 Suppose $v, w \in \mathbb{R}^3$. Show that if there is a rotation $M \in SO(3)$ such that $Mv = (r, 0, 0)^T$ for some nonnegative real number r and $Mw = (0, s, 0)^T$ for some nonnegative real number s, then this rotation is unique. Can you find a simple criterion (a calculation in terms of v and w) for the existence of such a rotation M?

Exercise 4.13 Consider the representation $\rho : \mathbb{T} \to \mathcal{GL}(\mathbb{C}^2)$ defined by

$$e^{i\theta} \mapsto \begin{pmatrix} \cos\theta & -\sin\theta \\ \sin\theta & \cos\theta \end{pmatrix}.$$

Consider the complex scalar product defined on $\mathbb{C}^2$ by

$$\langle v, w \rangle = \lambda_1 v_1^* w_1 + \lambda_2 v_2^* w_2,$$

where λ_1 and λ_2 are both strictly positive real numbers. Show that unless $\lambda_1 = \lambda_2$, the representation is not unitary in the given scalar product.

Exercise 4.14 (Used in Proposition 5.1) Show that the degree of a polynomial in three variables is invariant under rotation. In other words, consider the natural representation ρ of $SO(3)$ on polynomials in three variables and show that the degree of a polynomial is invariant under this representation: for any polynomial q and any $g \in SO(3)$, show that the degree of q is equal to the degree of $\rho(g)q$.

Exercise 4.15 (Used in Proposition 5.1) Show that the Laplacian in three variables is invariant under rotation. In other words, consider the natural representation ρ of $SO(3)$ on twice-differentiable functions of three variables and show that for any $g \in SO(3)$ we have $\rho(g) \circ \nabla^2 = \nabla^2 \circ \rho(g)$. To put it yet another way, show that the Laplacian is a homomorphism of representations.

Exercise 4.16 Generalize Exercises 4.14 and 4.15 to n dimensions. I.e., show that the degree and the Laplacian are both invariant under rotation in $\mathbb{R}^k$. Are they both invariant under the natural action of $\mathcal{GL}(\mathbb{R}^k)$ on $\mathbb{R}^k$? Suppose T

is a linear operator on $\mathbb{R}^n$ whose determinant is 0. What can you say about the degree or the Laplacian of $p \circ T$? Here p is an arbitrary homogeneous harmonic polynomial.

Exercise 4.17 *Show that $\exp_{\mathbb{R}}: \mathbb{R} \to \mathbb{R}^{>0}$ is an isomorphism of groups. Show that $\exp_{\mathbb{C}}: \mathbb{C} \to \mathbb{C}^*$ is a homomorphism of groups but not an isomorphism of groups. Show that $\exp_{gl}: gl(n, \mathbb{R}) \to GL(n, \mathbb{R})$ is not even a homomorphism of groups. (Here $gl(n, \mathbb{R})$ denotes the additive group of all $n \times n$ matrices with real entries, and $GL(n, \mathbb{R})$ denotes the multiplicative group of all nonsingular $n \times n$ matrices with real entries. $\exp_{gl}$ is matrix exponentiation, as defined in Section 1.5.)*

Exercise 4.18 *Show that a 2×2 matrix M is a unitary transformation of determinant one on $\mathbb{C}^2$, if and only if there are complex numbers α and β such that $|\alpha|^2 + |\beta|^2 = 1$ and*

$$M = \begin{pmatrix} \alpha & -\beta^* \\ \beta & \alpha^* \end{pmatrix}.$$

Exercise 4.19 (Used in Proposition 6.4) *Suppose V_1, V_2 and V_3 are representations. Suppose $T_1: V_2 \to V_1$ and $T_2: V_3 \to V_2$ are homomorphisms of representations. Show that $T_1 \circ T_2$ is a homomorphism of representations. Suppose further that T_1 and T_2 are both isomorphisms of representations. Show that $T_1 \circ T_2$ is an isomorphism of representations.*

Exercise 4.20 *Suppose G is a group and V is a vector space. Define the trivial representation of G on V by*

$$\rho_{\text{triv}}(g) := I$$

for any $g \in G$, where I is the identity element of V. Is the trivial representation of G on $\mathbb{C}$ isomorphic to the trivial representation of G on $\mathbb{C}^2$?

Exercise 4.21 *Consider the representation $(\mathbb{T}, \mathbb{C}^2, \rho)$ defined by*

$$\rho(e^{i\theta}) := \begin{pmatrix} e^{i\theta} & 0 \\ 0 & e^{-i\theta} \end{pmatrix}$$

and the representation $(\mathbb{T}, \mathbb{C}^2, \tilde{\rho})$ defined by

$$\tilde{\rho}(e^{i\theta}) := \begin{pmatrix} e^{i\theta} & 0 \\ 0 & e^{i\theta} \end{pmatrix}.$$

Are these two representations isomorphic?

Exercise 4.22 *Consider the representation* $(\mathbb{T}, \mathbb{C}, \rho)$ *defined by*

$$\rho(e^{i\theta}) := \left(\ e^{i\theta} \ \right)$$

and the representation $(\mathbb{T}, \mathbb{C}, \tilde{\rho})$ *defined by*

$$\tilde{\rho}(e^{i\theta}) := \left(\ e^{2i\theta} \ \right).$$

Are these two representations isomorphic?

Exercise 4.23 *Show that for each integer k, the function $\rho_k \colon \mathbb{T} \to \mathcal{GL}(\mathbb{C})$ given by $\rho_k(e^{i\theta}) := e^{ik\theta}$ is a representation. Show that it is unitary. For what values of k and $\tilde{k}$ is ρ_k isomorphic to $\rho_{\tilde{k}}$?*

Exercise 4.24 (Used in Proposition 4.5.) *Show that any element of $SO(3)$ can be written in the form $Z_\phi X_\theta Z_\psi$ for some real ϕ, θ and ψ. (Hint: first express the image of $(0, 0, 1)^T$ in terms of ϕ and θ.) Recall the definition of Cartesian products of groups from Exercise 4.1. Is this map from the Lie group $\mathbb{T} \times \mathbb{T} \times \mathbb{T}$ to the Lie group $SO(3)$ a group homomorphism? Is it differentiable? One-to-one?*

Exercise 4.25 (First part used in Section 6.3) *Rewrite Equation 1.6 as a matrix multiplication of the vector $(\tilde{u}, \tilde{x}, \tilde{y}, \tilde{z})^T$ in $\mathbb{R}^4$. Write the matrix explicitly in terms of u, x, y and z. Define the group $SO(1, 3)$ to be the set of 4×4 determinant-one matrices M satisfying*

$$M^T \begin{pmatrix} 1 & 0 & 0 & 0 \\ 0 & -1 & 0 & 0 \\ 0 & 0 & -1 & 0 \\ 0 & 0 & 0 & -1 \end{pmatrix} M = \begin{pmatrix} 1 & 0 & 0 & 0 \\ 0 & -1 & 0 & 0 \\ 0 & 0 & -1 & 0 \\ 0 & 0 & 0 & -1 \end{pmatrix}.$$

What familiar condition on the quaternion $u + x\mathbf{i} + y\mathbf{j} + z\mathbf{k}$ is equivalent to requiring the corresponding matrix to be an element of $SO(1, 3)$? Use this calculation to define a group homomorphism from the set of quaternions satisfying that condition to $SO(4)$.

Exercise 4.26 *Show that there is an injective group homomorphism from $SU(2)$ to $SO(4)$. In other words, show that there is a subgroup of $SO(4)$ that is isomorphic to $SU(2)$. (Hint: use quaternions.) Is this homomorphism surjective?*

Exercise 4.27 (For topology students; used in Appendix B) *Show that the group $SU(2)$ is simply connected. (Hint: consider Exercise 4.18.)*

Exercise 4.28 (For topology students) *Show that the group $SO(3)$ is not simply connected.*

Exercise 4.29 *Show that for $G = \mathbb{T}$, $SO(2)$ or $SO(3)$, two matrices in G are similar if and only if they have the same eigenvalues (with the same multiplicities). On the other hand, find an example of two invertible matrices with the same eigenvalues (with the same multiplicities) that are not similar to one another.*

Exercise 4.30 *Show that two matrices in $SO(4)$ are similar if and only if they have the same eigenvalues (with the same multiplicities).*

Exercise 4.31 *Verify that the second and third columns of the 3×3 matrix*

$$\Phi\left(\begin{pmatrix} \alpha & -\beta^* \\ \beta & \alpha^* \end{pmatrix}\right)$$

are given correctly in Formula 4.2.

Exercise 4.32 *To show that the function Φ defined in Section 4.3 is indeed a homomorphism, it suffices to show that if*

$$\begin{pmatrix} \alpha_1 & -\beta_1^* \\ \beta_1 & \alpha_1^* \end{pmatrix}\begin{pmatrix} \alpha_2 & -\beta_2^* \\ \beta_2 & \alpha_2^* \end{pmatrix} = \begin{pmatrix} \alpha_3 & -\beta_3^* \\ \beta_3 & \alpha_3^* \end{pmatrix},$$

then the product of

$$\begin{pmatrix} |\alpha_1|^2 - |\beta_1|^2 & -2\Re(\alpha_1\beta_1) & -2\Im(\alpha_1\beta_1) \\ 2\Re(\alpha_1^*\beta_1) & \Re(\alpha_1^2 - \beta_1^2) & \Im(\alpha_1^2 - \beta_1^2) \\ 2\Im(\alpha_1^*\beta_1) & -\Im(\alpha_1^2 + \beta_1^2) & \Re(\alpha_1^2 + \beta_1^2) \end{pmatrix}$$

and

$$\begin{pmatrix} |\alpha_2|^2 - |\beta_2|^2 & -2\Re(\alpha_2\beta_2) & -2\Im(\alpha_2\beta_2) \\ 2\Re(\alpha_2^*\beta_2) & \Re(\alpha_2^2 - \beta_2^2) & \Im(\alpha_2^2 - \beta_2^2) \\ 2\Im(\alpha_2^*\beta_2) & -\Im(\alpha_2^2 + \beta_2^2) & \Re(\alpha_2^2 + \beta_2^2) \end{pmatrix}$$

is equal to

$$\begin{pmatrix} |\alpha_3|^2 - |\beta_3|^2 & -2\Re(\alpha_3\beta_3) & -2\Im(\alpha_3\beta_3) \\ 2\Re(\alpha_3^*\beta_3) & \Re(\alpha_3^2 - \beta_3^2) & \Im(\alpha_3^2 - \beta_3^2) \\ 2\Im(\alpha_3^*\beta_3) & -\Im(\alpha_3^2 + \beta_3^2) & \Re(\alpha_3^2 + \beta_3^2) \end{pmatrix}$$

Check one of the coordinates of the product. Gluttons for punishment may check more than one.

Exercise 4.33 *Is the set*

$$\{M \in SU(2) : \exists \theta \text{ such that } \Phi(M) = \mathbf{X}_\theta\}$$

a subgroup of $SO(3)$? What about

$$\{M \in SU(2) : \exists \theta \text{ such that } \Phi(M) = \mathbf{Y}_\theta\}$$

and

$$\{M \in SU(2) : \exists \theta \text{ such that } \Phi(M) = \mathbf{Z}_\theta\}?$$

Exercise 4.34 (*$SU(2)$ and the unit quaternions*) *Recall the functions f_i, f_j and f_k from Exercise 2.8. Show that the restrictions of f_i, f_j and f_k to the unit circle $\mathbb{T}$ are group homomorphisms whose range lies in the unit quaternions. Call their images $\mathbb{T}_i$, $\mathbb{T}_j$ and $\mathbb{T}_k$, respectively. Write the images of $\mathbb{T}_i$, $\mathbb{T}_j$ and $\mathbb{T}_k$ under the homomorphism Φ explicitly as 3×3 matrices.*

Exercise 4.35 (**Used in Section 10.4**) *Show that for any natural number n and any element $g \in SU(2)$ we have $R_n(-g) = (-1)^n R_n(g)$.*

Exercise 4.36 *Show that the representation of $SU(2)$ on $\mathbb{C}^2$ (matrix multiplication) is isomorphic to the $\mathcal{P}^1$ representation. Hint: Define $T : \mathbb{C}^2 \to \mathcal{P}^1$ by $T(1,0)^T := -y$ and $T(0,1)^T := x$. Is T a unitary isomorphism?*

Exercise 4.37 (**Used for an example in Section 5.6**) *Suppose $\tilde{G}$ is a nonempty subset of G and $\tilde{G}$ is closed under multiplication and inversion. Show that $\tilde{G}$ must contain the identity element I of G, and that I is the identity element of $\tilde{G}$ as well. (In this case we say that $\tilde{G}$ is a subgroup of G.*

Consider the inclusion *map i of $\tilde{G}$ into G, defined by $i(g) := g$ for each $g \in \tilde{G}$. Show that the inclusion map is a group homomorphism.*

Exercise 4.38 (**Used in proof of Proposition 4.5 and in Section 10.4**) *Calculate the three-by-three matrix*

$$\Phi\left(\begin{pmatrix} \lambda & 0 \\ 0 & \lambda^* \end{pmatrix}\right),$$

where $\lambda \in \mathbb{T}$. Calculate

$$\Phi\left(\frac{1}{\sqrt{2}}\begin{pmatrix} -i & -1 \\ 1 & i \end{pmatrix}\right).$$

Find a matrix $\begin{pmatrix} \alpha & -\beta^* \\ \beta & \alpha^* \end{pmatrix}$ *(whose entries depend on* θ *) such that*

$$\Phi\left(\begin{pmatrix} \alpha & -\beta^* \\ \beta & \alpha^* \end{pmatrix}\right) = \mathbf{X}_\theta,$$

and another matrix $\begin{pmatrix} \alpha & -\beta^* \\ \beta & \alpha^* \end{pmatrix}$ *such that*

$$\Phi\left(\begin{pmatrix} \alpha & -\beta^* \\ \beta & \alpha^* \end{pmatrix}\right) = \mathbf{Z}_\theta.$$

Exercise 4.39 (Used in Section 5.3) *Consider the characters* χ_3 *and* χ_4 *of the natural representations of* $SU(2)$ *on* $\mathcal{P}^3$ *and* $\mathcal{P}^4$. *Find the coefficients of* χ_3 *and* χ_4 *as polynomials in terms of* $\mathfrak{R}(\alpha)$.

Exercise 4.40 (Used in Proposition 6.12) *Suppose* ρ *and* $\tilde{\rho}$ *are isomorphic representations of a group* G. *Show that their characters are equal.*

Exercise 4.41 *Thought experiment: draw the graph of* $y = \sin x$ *for* x *in the interval* $[-\pi, \pi]$. *Now wrap the paper on which the graph is drawn around a cylinder so that the* $x - axis$ *forms a circle, with the point* $(\pi, 0)$ *meeting the point* $(-\pi, 0)$. *What shape does the graph of* sin *form? (Hint: consider the restrictions to the unit circle of the functions* f_1 *and* f_2 *introduced in Section 4.4.)*

Exercise 4.42 *Show that if* (G, S, σ) *is an action, then the function* $f : G \times S \to S$ *defined by*

$$f(g, s) := (\sigma(g)) s$$

satisfies:

1. *if* $g_1, g_2 \in G$ *and* $s \in S$ *then* $f(g_1 g_2, s) = f(g_1, f(g_2, s))$;

2. *if* I *denotes the identity element of* G, *then for any* $s \in S$ *we have* $f(I, s) = s$.

Conversely, show that if $f : G \times S \to S$ *satisfies the two criteria and we define* $\sigma : G \to \mathcal{GL}(S)$ *by*

$$(\sigma(g)) s = f(g, s)$$

for all $s \in S$, *then* σ *is a group homomorphism.*

Exercise 4.43 (Used in Appendix B) *Suppose S is a set, G is a group and* (S, G, σ) *is a group action. Define a relation* $\sim$ *on S by*

$$s_1 \sim s_2 \quad \text{if and only if} \quad \exists g \in G \text{ s.t. } (\sigma(g)) s_1 = s_2.$$

Show that $\sim$ *is an equivalence relation. If the action* σ *is clear from the context, then the quotient space* $S/\sim$ *is often denoted* S/G.

Exercise 4.44 *In this exercise we will show that rotation of functions is well defined on* $L^2(\mathbb{R}^3)$. *Suppose g is an element of the rotation group* $SO(3)$. *For any complex-valued function* f *on* $\mathbb{R}^3$, *let* $\tilde{f}$ *denote the function* $\mathbb{R}^3 \to \mathbb{C}$ *defined by* $\tilde{f}(x) := f(gx)$. *Show that if* f *is square-integrable, then* $\tilde{f}$ *is also square-integrable. Now suppose* f_1 *and* f_2 *are equivalent functions under the equivalence relation* $\sim$ *defined in Section 3.1. Show that* $\tilde{f}_1 \sim \tilde{f}_2$.

Exercise 4.45 *Consider the representation* ρ_2 *of* $SU(2)$ *on* $\mathcal{P}^2$. *Find the matrix of this representation in the basis* $\{x^2, xy, y^2\}$.

Exercise 4.46 *Consider the finite permutation group* S_3 *on three letters. Construct a representation* $(S_3, \mathbb{C}^3, \rho)$ *by setting* $z_1 := (1, 0, 0)^T$, $z_2 := (0, 1, 0)^T$ *and* $z_3 := (0, 0, 1)^T$ *and defining*

$$\rho(\sigma)(z_i) := z_{\sigma(i)}$$

for each $\sigma \in S_3$ *and* $i = 1, 2, 3$.

1. *Find the character of this representation.*

2. *Consider the corresponding representation* $\tilde{\rho}$ *on homogeneous polynomials of degree two in three variables. Let* σ *denote the permutation taking* $z_1 \mapsto z_2$, $z_2 \mapsto z_3$ *and* $z_3 \mapsto z_1$. *Find* $\tilde{\rho}(\sigma)p$, *where* $p(x, y, z) := x^2 + xy - 5z^2$. *Calculate the character of* $\tilde{\rho}$.

5

New Representations from Old

I cannot fix on the hour, or the spot, or the look, or the words, which laid the foundation. It is too long ago. I was in the middle before I knew that I *had* begun.

—Jane Austen, *Pride and Prejudice* [Au, Vol. III, Ch. XVIII]

In this chapter we discuss several natural ways to construct representations from other representations.

5.1   Subrepresentations

In this section we show how to construct a new representation from an old one by restricting the domain of the linear transformations. One cannot restrict the domain to any old subspace, only to *invariant* subspaces.

Definition 5.1 *An* invariant subspace W *of a representation* (G, V, ρ) *is a subspace of V such that for every $g \in G$ and every vector $w \in W$, the vector $g \cdot w$ lies in W.*

Consider, for example, the representation of the circle group $\mathbb{T}$ on the complex vector space $V := \mathbb{C}^2$ given by

$$\rho: \mathbb{T} \to \mathcal{GL}\left(\mathbb{C}^2\right)$$
$$e^{i\theta} \mapsto \begin{pmatrix} 1 & 0 \\ 0 & e^{i\theta} \end{pmatrix}. \tag{5.1}$$

In other words, for any real number θ the linear transformation $\rho(e^{i\theta})$ rotates the second entry of the complex 2-vector counterclockwise through an angle of θ radians while leaving the first entry unchanged. It is not hard to see that the (complex) one-dimensional subspace

$$\{0\} \times \mathbb{C} = \left\{ \begin{pmatrix} 0 \\ c \end{pmatrix} : c \in \mathbb{C} \right\}$$

is invariant: given any vector $(0, c)^T$ in $\{0\} \times \mathbb{C}$ and any $e^{i\theta}$ in S^1 we have $e^{i\theta} \cdot (0, c)^T = (0, e^{i\theta} c) \in \{0\} \times \mathbb{C}$. It is even easier to show that the subspace

$$\mathbb{C} \times \{0\} = \left\{ \begin{pmatrix} c \\ 0 \end{pmatrix} : c \in \mathbb{C} \right\}$$

is invariant. On the other hand, for any two nonzero complex numbers a and b, the one-dimensional subspace consisting of scalar multiples of the vector $(a, b)^T$ is not invariant, since $(-1) \cdot (a, b)^T = (a, -b)^T$, and $(a, -b)$ can only be a scalar multiple of (a, b) when either a or b is equal to zero. Thus there are precisely two one-dimensional invariant subspaces of this representation.

The zero-dimensional subspace $\{0\}$ and the two-dimensional subspace $\mathbb{C}^2$ are also invariant. In fact, we leave it to the reader to show in Exercise 5.1 that for any representation the largest and smallest subspaces are invariant.

Note that elements of an invariant subspace W are not necessarily *fixed* by the linear operators in the image of the representation. In other words, it is not necessary to have $\rho(g)w = w$ for every group element g and every $w \in W$. However, elements of W cannot be moved out of W by the representation; i.e., we do have $\rho(g)w \in W$.

For each nonnegative integer ℓ, the space $\mathcal{Y}^\ell$ of spherical harmonics of degree ℓ (see Definition 2.6) is the vector space for a representation of $SO(3)$. These representations appear explicitly in our analysis of the hydrogen atom in Chapter 7. Recall the complex scalar product space $L^2(S^2)$ from Definition 3.3.

Proposition 5.1 *Consider the natural representation of $SO(3)$ on $L^2(S^2)$. Fix any nonnegative integer ℓ. The subspace $\mathcal{Y}^\ell$ of $L^2(S^2)$ given in Definition 2.6 is an invariant subspace.*

Proof. Consider any function $y \in \mathcal{Y}^\ell$. By Definition 2.6, there is a homogeneous harmonic polynomial p of degree ℓ such that $y = p|_{S^2}$. Now rotating a polynomial preserves its degree (by Exercise 4.14), and the Laplacian is invariant under rotation (by Exercise 3.11). So for any $g \in SO(3)$ the function $g \cdot p$ is a homogeneous harmonic polynomial of degree ℓ. Hence $g \cdot y = g|_{S^2} \cdot p$ is an element of $\mathcal{Y}^\ell$. $\qquad\square$

In the proof of Proposition 6.3 we will use linear operators to identify invariant subspaces with the help of the following proposition. Recall the notion of an eigenspace of a linear operator (Exercise 2.26).

Proposition 5.2 *Suppose (G, V, ρ) is a representation and T is a linear operator on V. If T commutes with ρ, i.e., if $T\rho(g) = \rho(g)T$ for every $g \in G$, then every eigenspace of T is an invariant space for ρ.*

Proof. Suppose w is an eigenvector for T with eigenvalue λ. We must show that for any $g \in G$, the vector $\rho(g)w$ is an eigenvector for T with eigenvalue λ. We have

$$T(\rho(g)w) = \rho(g)(Tw) = \rho(g)(\lambda w) = \lambda \rho(g)w.$$

So $\rho(g)w$ is indeed an eigenvector for T with eigenvalue λ. $\qquad\square$

We can use an invariant subspace W to construct a *restriction* of the representation by restricting the linear transformation $\rho(g)$ to the subspace W for each group element g. Note that for each $g \in G$ the restriction $\rho(g)|_W$ is a function from W to W.

Definition 5.2 *Suppose (G, V, ρ) is a representation and W is an invariant subspace of V. If we define the function $\rho_W : G \to \mathcal{GL}(W)$ by*

$$g \mapsto \rho(g)\big|_W,$$

then (G, W, ρ_W) is a representation. We call this representation a subrepresentation *or, more precisely, the restriction of (G, V, ρ) to W.*

When we consider a subrepresentation ρ_W of a representation ρ it is often useful to consider the leftovers, that is, the part of ρ that is not captured by ρ_W. If the original representation ρ is unitary, then there is a particularly nice way to package those leftovers: we can put the complex scalar structure to work. Recall the notion (Definition 3.6) of the complementary subspace $W^\perp$.

Proposition 5.3 *Suppose (G, V, ρ) is a unitary representation. Suppose W is an invariant subspace. Then $W^\perp$ is also an invariant subspace. If V is finite dimensional, then the characters satisfy the relation*

$$\chi_V = \chi_W + \chi_{W^\perp},$$

where χ_V, χ_W and $\chi_{W^\perp}$ are the characters of ρ, ρ_W and $\rho_{W^\perp}$, respectively.

Proof. We must show that for each $g \in G$ and each $v \in W^\perp$ we have $\rho(g)v \in W^\perp$. Consider an arbitrary $w \in W$. Then

$$\langle \rho(g)v, w \rangle = \langle \rho(g^{-1})\rho(g)v, \rho(g^{-1})w \rangle = \langle v, \rho(g^{-1})w \rangle = 0,$$

where the first equality relies on the fact that the representation is unitary and the third uses the facts that $\rho(g^{-1})w \in W$ and $v \in W^\perp$. So $W^\perp$ is an invariant subspace.

If V is finite dimensional, then the characters are well defined. To show the additive relation of the characters, take a basis for V that is the union of a basis for W and a basis for $W^\perp$. In such a basis,

$$\operatorname{Tr} \rho(g) = \operatorname{Tr} \rho_W(g) + \operatorname{Tr} \rho_{W^\perp}(g).$$

$\square$

Note how important the unitary structure is to Proposition 5.3. If we consider a subrepresentation of a nonunitary representation, then there may not be a complementary representation. Consider, for example, the group $G = \mathbb{R}$ (with addition playing the role of the group multiplication), $V = \mathbb{C}^2$ and $\rho: G \to \mathcal{GL}(\mathbb{C}^2)$ defined by

$$\rho(r) := \begin{pmatrix} 1 & r \\ 0 & 1 \end{pmatrix}.$$

The subspace $\mathbb{C} \oplus \{0\}$ is invariant under the representation: for any $r \in \mathbb{R}$ and any $c \in \mathbb{C}$ we have

$$\begin{pmatrix} 1 & r \\ 0 & 1 \end{pmatrix} \begin{pmatrix} c \\ 0 \end{pmatrix} = \begin{pmatrix} c \\ 0 \end{pmatrix}.$$

However, there is no other subspace invariant under the representation. Every other subspace has the form

$$\mathbb{C} \begin{pmatrix} s \\ 1 \end{pmatrix} := \left\{ \begin{pmatrix} sc \\ c \end{pmatrix} : c \in \mathbb{C} \right\}.$$

Taking $r = 1$ and $c = 1$ we find

$$\rho(1) \begin{pmatrix} s \\ 1 \end{pmatrix} = \begin{pmatrix} s+1 \\ 1 \end{pmatrix},$$

which is not an element of $\mathbb{C} \begin{pmatrix} s \\ 0 \end{pmatrix}$. This example does not contradict the proposition, as the representation is not unitary: for any nonzero r we have

$$\left\| \begin{pmatrix} r \\ 1 \end{pmatrix} \right\| = \sqrt{1 + r^2} \neq 0,$$

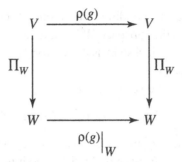

Figure 5.1. The point of Proposition 5.4 is to show that this diagram commutes.

which implies by Proposition 3.2 that $\rho(r) \notin \mathcal{U}\left(\mathbb{C}^2\right)$.

Orthogonal projection (Definition 3.11) onto an invariant subspace is a homomorphism of representations.

Proposition 5.4 *Suppose W is an invariant subspace for a unitary representation (G, V, ρ). Suppose that there is an orthogonal projection $\Pi_W \colon V \to V$ onto a subspace W. Then Π_W is a homomorphism of representations.*

Recall from Section 3.3 and Exercise 3.29 that there are infinite-dimensional W's that are not images of an orthogonal projections.

Proof. We must show that for any $g \in G$ we have

$$\Pi_W \circ \rho(g) = \rho(g) \circ \Pi_W.$$

The commutative diagram expressing this relationship is in Figure 5.1.

Let g be an arbitrary element of the group G and let v be an arbitrary element of the vector space V. Then we see that

$$\Pi_W \rho(g) v = \Pi_W \rho(g) \left(\Pi_W v + \Pi_{W^\perp} v\right)$$
$$= \Pi_W \left(\rho(g) \Pi_W v\right) + \Pi_W \left(\rho(g) \Pi_{W^\perp} v\right).$$

The subspace W is invariant under ρ by hypothesis; since ρ is a unitary representation, it follows from Proposition 5.3 that $W^\perp$ is also invariant under ρ. Thus we have $\rho(g) \Pi_W v \in W$ and $\rho(g) \Pi_{W^\perp} v \in W^\perp$. Hence

$$\Pi_W \rho(g) \Pi_W v + \Pi_W \rho(g) \Pi_{W^\perp} v = \rho(g) \Pi_W v + 0.$$

So $\Pi_W \rho(g) v = \rho(g) \Pi_W v$ for all $v \in V$ and all $g \in G$.   □

Invariant subspaces are the only physically natural subspaces. Recall from Section 4.5 that in a quantum system with symmetry, there is a natural representation (G, V, ρ). Any physically natural object must appear the same to all observers. In particular, if a subspace has physical significance, all equivalent observers must agree on the question of a particular state's membership in that subspace. So if w is an element of a physically natural subspace W, then the physical state corresponding to w in one observer's laboratory must in some sense "belong to the subspace W." But then for any $g \in G$, there is an observer who sees that physical state as the vector $\rho(g)w$. Hence $\tilde{\rho}(g)w \in W$ for any g. So W is invariant.

Consider, for example, the vector space

$$\mathcal{I} := \{f \in L^2(\mathbb{R}^3) : \forall g \in SO(3), \ g \cdot f = f\}$$

of rotation-invariant functions in $L^2(\mathbb{R}^3)$. They are also known as *radial* functions. They form a vector subspace of $L^2(\mathbb{R}^3)$, and it is easy to check that this subspace satisfies the criterion of Definition 5.1. Physically, this subspace corresponds to the s-shells of the hydrogen atom. Given any wave function in $\mathcal{I}$, the corresponding state must be a superposition of s-shell states.

In the proof of Proposition 7.7 we will need the following proposition:

Proposition 5.5 *Let $L^2(\mathbb{R}^{\geq 0})$ denote the vector space of square-integrable complex-valued functions on $\mathbb{R}^{\geq 0}$. Suppose $f \in \mathcal{I}$ and define (using spherical coordinates) $\tilde{f} : (0, \infty) \to \mathbb{C}$ by $\tilde{f}(r) := f(r, \theta, \phi)$. Then $f \in \mathcal{I}$ if and only if $r \tilde{f}(r) \in L^2(\mathbb{R}^{\geq 0})$.*

Proof. By Fubini's theorem (Theorem 3.1),

$$\int_{\mathbb{R}^3} |f|^2 = \int_0^{2\pi} \int_0^\pi \int_0^\infty |f(r, \theta, \phi)|^2 \ r^2 \sin\theta \ dr \ d\theta \ d\phi$$

$$= 4\pi \int_0^\infty \left| \tilde{f}(r) r \right|^2 \ dr.$$

The left-hand side integral is finite if and only if the right-hand side integral is finite. So $f \in \mathcal{I}$ if and only if $r \tilde{f}(r) \in L^2(\mathbb{R}^{\geq 0})$. □

Any physically natural, spherically symmetric set of states corresponds to an invariant subspace and a subrepresentation. For this reason the concepts in this section are fundamental to our analysis of the hydrogen atom. The various shells of the hydrogen atom correspond to subrepresentations of the natural representation of $SO(3)$ on $L^2(\mathbb{R}^3)$. In particular, the subspaces $\mathcal{Y}^\ell$ and $\mathcal{I}$ play a role in the analysis.

5.2   Cartesian Sums of Representations

In Definition 2.11 in Section 2.6 we introduced the Cartesian sum of vector spaces. Now fix a group G and consider a Cartesian sum in which each summand vector space has a representation of G on it. Then there is a natural way to define a representation on the Cartesian sum; such a representation is called a *Cartesian sum representation*.

Definition 5.3 *Suppose n is a natural number and $(G, V_1, \rho_1),...,(G, V_n, \rho_n)$ are n representations of the group G. Then the* Cartesian sum *of the n representations is*

$$\left(G, \bigoplus_{k=1}^{n} V_k, \bigoplus_{k=1}^{n} \rho_k \right),$$

where the function $\bigoplus_{k=1}^{n} \rho_k : G \to \mathcal{GL}\left(\bigoplus_{k=1}^{n} V_k \right)$ is defined by

$$\left(\bigoplus_{k=1}^{n} \rho_k \right)(g)(v_1, \ldots, v_n) := (\rho_1(g)v_1, \ldots, \rho_n(g)v_n)$$

for any $v_1 \in V_1, \ldots, v_n \in V_n$ and any $g \in G$. One often denotes the Cartesian sum representation simply by $\bigoplus_{k=1}^{n} \rho_k$ or, when G and $\rho_1, \ldots, \rho_n$ are known from the context, by $\bigoplus_{k=1}^{n} V_k$.

We may also write

$$\rho_1 \oplus \cdots \oplus \rho_n := \bigoplus_{k=1}^{n} \rho_k.$$

For example, consider the representations $(SU(2), \mathcal{P}^n, R_n)$ for $n = 0, 1, 2$. The vector space $\mathcal{P}^0 \oplus \mathcal{P}^1 \oplus \mathcal{P}^2$ is the set of complex-coefficient polynomials in two variables of degree two or less. Specifically, the polynomial $c_0 + c_{10}x + c_{11}y + c_{20}x^2 + c_{21}xy + c_{22}y^2$ corresponds to the element

$$\left(c_0, \ c_{10}x + c_{11}y, \ c_{20}x^2 + c_{21}xy + c_{22}y^2 \right) \in \mathcal{P}^0 \oplus \mathcal{P}^1 \oplus \mathcal{P}^2;$$

as a concrete example, $x + 2y + 10$ corresponds to $(10, x + 2y, 0)$. The representation $R_0 \oplus R_1 \oplus R_2$ (otherwise known as $\mathcal{P}^0 \oplus \mathcal{P}^1 \oplus \mathcal{P}^2$) is just the natural representation arising from the action of $SU(2)$ on the xy-plane.

Recall the projection onto the k-th summand from Definition 2.12. This projection is a homomorphism of representations.

Proposition 5.6 *Suppose (G, V_k, ρ_k) are representations for each $k = 1, \ldots, n$. Then the projection*

$$\Pi_k : V_1 \oplus \cdots \oplus V_n \to V_k$$

is a homomorphism of representations, i.e., for each $g \in G$ we have

$$\rho_k(g) \circ \Pi_k = \Pi_k \circ (\rho_1 \oplus \cdots \oplus \rho_n)(g).$$

Proof. On the one hand,

$$\rho_k(g)(\Pi_k(v_1, \ldots, v_n)) = \rho_k(g)v_k.$$

On the other hand,

$$\Pi_k(\rho_1 \oplus \cdots \oplus \rho_n)(g)(v_1, \ldots, v_n)) = \Pi_k(\rho_1(g)v_1, \ldots, \rho_n(g)v_n) = \rho_k(g)v_k.$$

Hence Π_k is a homomorphism of representations.   □

The character of a Cartesian sum of representations has a nice relationship to the characters of the summands.

Proposition 5.7 *Suppose χ is the character of a finite representation (G, V, ρ) and $\tilde{\chi}$ is the character of a finite representation $(G, \tilde{V}, \tilde{\rho})$. Then $\chi + \tilde{\chi}$ is the character of the representation $\rho \oplus \tilde{\rho}$.*

We leave the proof to the reader (Exercise 5.10).

Physically, expressing a vector space of wave functions as a Cartesian sum corresponds to decomposing any state as a superposition of states in the summand vector spaces. For instance, the fact that any bound state of hydrogen is a superposition of states in particular shells follows from the decomposition of $L^2(\mathbb{R}^3)$ as a Cartesian sum of the representations corresponding to the different shells. For another example, the state space of a spin-1/2 particle is a Cartesian sum of the pure spin-up and spin-down vector spaces. This is equivalent to the idea that any state of a spin-1/2 particle is a superposition of spin-up and spin-down states.

5.3   Tensor Products of Representations

Next we define tensor product representations. The reader may wish to recall the definition of the tensor product of two vector spaces, given in Definition 2.14.

Definition 5.4 *Suppose G is a group and (G, V, ρ) and $(G, \tilde{V}, \tilde{\rho})$ are representations. Then the* tensor product representation, *denoted $(G, V \otimes \tilde{V}, \rho \otimes \tilde{\rho})$ is defined by*

$$(\rho \otimes \tilde{\rho})(g) \colon v \otimes \tilde{v} \mapsto (\rho(g)v) \otimes (\tilde{\rho}(g)\tilde{v}).$$

Note that although not every element of $V \otimes \tilde{V}$ is a one-term product of the form $v \otimes \tilde{v}$, such one-term products span the vector space, so it suffices to define $\rho \otimes \tilde{\rho}$ on one-term products and insist that $\rho \otimes \tilde{\rho}$ be linear. We leave it to the reader to show that Proposition 2.4 applies (so that the representation is indeed a map to $\mathcal{GL}\,(V \otimes V)$) and that $\rho \otimes \tilde{\rho}$ satisfies all the criteria for a group representation: see Exercise 5.12.

Consider for example the vector space tensor product $\mathcal{P}^1 \otimes \mathcal{P}^2$ introduced in Section 2.6. Let us study the tensor product representation $R_1 \otimes R_2$, where R_1 and R_2 are the representations of $SU(2)$ on spaces of homogeneous polynomials defined in Section 4.6. As in the proof of Proposition 4.8, the character of this representation of $SU(2)$ is determined by its values on the diagonal subgroup as a consequence of the Spectral Theorem (Proposition 4.4). It is a straightforward matter to calculate the character on the diagonal subgroup: note that (using the basis given in Equation 2.7)

$$\begin{pmatrix} \alpha & 0 \\ 0 & \alpha^* \end{pmatrix} \cdot ux^2 = (\alpha^*)^3 ux^2,$$

$$\begin{pmatrix} \alpha & 0 \\ 0 & \alpha^* \end{pmatrix} \cdot uxy = (\alpha^*)^2 \alpha uxy,$$

$$\begin{pmatrix} \alpha & 0 \\ 0 & \alpha^* \end{pmatrix} \cdot uy^2 = \alpha^* \alpha^2 uy^2,$$

$$\begin{pmatrix} \alpha & 0 \\ 0 & \alpha^* \end{pmatrix} \cdot vx^2 = \alpha (\alpha^*)^2 vx^2,$$

$$\begin{pmatrix} \alpha & 0 \\ 0 & \alpha^* \end{pmatrix} \cdot vxy = \alpha^2 \alpha^* vxy,$$

$$\begin{pmatrix} \alpha & 0 \\ 0 & \alpha^* \end{pmatrix} \cdot vy^2 = \alpha^3 vy^2.$$

The character χ of the representation is the trace of the diagonal 6×6 matrix whose entries we have just calculated:

$$\chi \begin{pmatrix} \alpha & -\beta^* \\ \beta & \alpha^* \end{pmatrix} = (\alpha^*)^3 + 2\alpha(\alpha^*)^2 + 2\alpha^2\alpha^* + \alpha^3$$

$$= 8\,(\Re(\alpha))^3 - 2\Re(\alpha).$$

Using the result of Exercise 4.39, it is easy to calculate that $\chi = \chi_1\chi_2$. This is a special case of the general truth that the character of a tensor product is the product of the characters of the factors. See Proposition 5.8.

Note also that $\chi = \chi_1 + \chi_3$. From Proposition 5.7 we know that the representation $R_1 \oplus R_3$ has the same character. In fact, $R_1 \oplus R_3$ is isomorphic to $R_1 \otimes R_2$, as we are about to show. Consider the subspace V_1 spanned by

$\{uxy - vx^2, uy^2 - vxy\}$. This two-dimensional subspace is invariant, as the reader can check by a tedious but straightforward calculation:

$$\begin{pmatrix} \alpha & -\beta^* \\ \beta & \alpha^* \end{pmatrix} \cdot (uxy - vx^2) = \alpha^*(uxy - vx^2) + \alpha^*(uy^2 - vxy) \in V_1,$$

$$\begin{pmatrix} \alpha & -\beta^* \\ \beta & \alpha^* \end{pmatrix} \cdot (uy^2 - vxy) = -\alpha(uxy - vx^2) + \alpha(uy^2 - vxy) \in V_1.$$

It follows from this calculation that the restriction of the representation to V_1 is isomorphic to the representation R_1 defined in Section 4.6; the isomorphism is given by

$$uxy - vx^2 \mapsto x, \ uy^2 - vxy \mapsto y.$$

A similar (but longer) calculation shows that the four-dimensional subspace V_3 spanned by $\{ux^2, 2uxy + vx^2, vxy + 2uy^2, vy^2\}$ is invariant. (For an alternative proof, see Exercise 5.7.) The representation restricted to this subspace is isomorphic to $\mathcal{P}^3$, as the reader is invited to check in Exercise 5.8; the isomorphism is given by

$$ux^2 \mapsto x^3, (vx^2 + 2uxy) \mapsto 3x^2y, (2vxy + uy^2) \mapsto 3xy^2, vy^2 \mapsto y^3.$$

Combining these two isomorphisms, we can see that our representation is isomorphic to the representation $R_1 \oplus R_3$. This is our first example of a decomposition of a representation into its "irreducible" building blocks.

Proposition 5.8 *Suppose ρ and $\tilde{\rho}$ are two finite-dimensional representations of the same group G. Let χ denote the character of ρ and let $\tilde{\chi}$ denote the character of $\tilde{\rho}$. Then the character of the tensor product representation $\rho \otimes \tilde{\rho}$ is the function $\chi\tilde{\chi} : G \to \mathbb{C}$.*

Proof. Let $\{v_1, \ldots, v_n\}$ be a basis of the representation space V of ρ, and let $\{\tilde{v}_1, \ldots, \tilde{v}_m\}$ be a basis of the representation space $\tilde{V}$ of $\tilde{\rho}$. Then by Proposition 2.13, the set

$$\left\{v_j \otimes \tilde{v}_k : j = 1, \ldots, n; k = 1, \ldots, m\right\}$$

is a basis for $V \otimes \tilde{V}$.

Now let M denote the matrix of $\rho(g)$ in the basis $\{v_1, \ldots, v_n\}$ and let $\tilde{M}$ denote the matrix of $\tilde{\rho}(g)$ in the basis $\{\tilde{v}_1, \ldots, \tilde{v}_m\}$. Both M and $\tilde{M}$ depend tacitly on g. Then for any fixed j_0 and k_0 we have

$$(\rho \otimes \tilde{\rho})(g)(v_{j_0} \otimes \tilde{v}_{k_0}) = (\rho(g)v_{j_0}) \otimes (\tilde{\rho}(g)\tilde{v}_{k_0})$$

$$= \sum_{j=1}^{n}\sum_{k=1}^{m} M_{jj_0}\tilde{M}_{kk_0}v_j \otimes \tilde{v}_k.$$

The coefficient of $v_{j_0} \otimes \tilde{v}_{k_0}$ in this expression is $M_{j_0 j_0}\tilde{M}_{k_0 k_0}$. Hence the character of the representation is

$$\sum_{j_0=1}^{n}\sum_{k_0=1}^{m} M_{j_0 j_0}\tilde{M}_{k_0 k_0} = \left(\sum_{j_0=1}^{n} M_{j_0 j_0}\right)\left(\sum_{k_0=1}^{m}\tilde{M}_{k_0 k_0}\right) = \chi(g)\tilde{\chi}(g).$$

$\square$

If both factors are unitary representations, then so is the tensor product. If both V and $\tilde{V}$ have complex scalar products defined on them, then there is a natural complex scalar product on the tensor product $V \otimes \tilde{V}$ of vector spaces. Specifically, we define

$$\langle v \otimes \tilde{v}, w \otimes \tilde{w}\rangle := \langle v, w\rangle \langle \tilde{v}, \tilde{w}\rangle \tag{5.2}$$

for one-term products. The reader should check that this bracket is well defined and satisfies all the requirements for a complex scalar product (Exercise 5.16).

Proposition 5.9 *Suppose* (G, V, ρ) *and* $(G, \tilde{V}, \tilde{\rho})$ *are unitary representations. Then the tensor product representation* $(G, V \otimes \tilde{V}, \rho \otimes \tilde{\rho})$ *is unitary also.*

Proof. Fix any $g \in G$ and consider the effect of $\rho \otimes \tilde{\rho}(g)$ on a bracket of one-term products. For simplicity we write $g \cdot v \otimes \tilde{v}$ as a shorthand for $((\rho \otimes \tilde{\rho})(g)) v \otimes \tilde{v}$. The bracket of arbitrary one-term tensor products $v \otimes \tilde{v}$ and $w \otimes \tilde{w}$ is

$$\langle g \cdot v \otimes \tilde{v}, g \cdot w \otimes \tilde{w}\rangle = \langle (\rho(g)v) \otimes (\tilde{\rho}(g)\tilde{v}), (\rho(g)w) \otimes (\tilde{\rho}(g)\tilde{w})\rangle$$

$$= \langle \rho(g)v, \rho(g)w\rangle \langle \tilde{\rho}(g)\tilde{v}, \tilde{\rho}(g)\tilde{w}\rangle = \langle v, w\rangle \langle \tilde{v}, \tilde{w}\rangle$$

$$= \langle v \otimes \tilde{v}, w \otimes \tilde{w}\rangle.$$

By the linearity of the bracket, because $\rho \otimes \tilde{\rho}$ preserves brackets of all one-term products, it preserves all brackets. In other words, the representation $\rho \otimes \tilde{\rho}$ is unitary.

$\square$

Tensor products of representations arise naturally in physics. To obtain the space of states of two particles, take the tensor product of the two spaces of states. Thus the state space for two mobile particles in $\mathbb{R}^3$ is $L^2(\mathbb{R}^3) \otimes L^2(\mathbb{R}^3)$. Also, if one wants to study two qualities of one particle (say, its motion in $\mathbb{R}^3$ and its spin-1 spin state), one takes a tensor product $(L^2(\mathbb{R}^3) \otimes \mathbb{C}^3)$. (Some readers may already know that spin-1 spin states are described by vectors in $\mathbb{C}^3$; others might see Section 10.4.) We will use tensor products in Proposition 7.7, our mathematical description of the elementary states of the hydrogen atom.

5.4   Dual Representations

In Section 4.4 we saw how to build a representation from the action of a group on a set; the new representation space is a space of functions. In this section, we apply this idea to linear functions on a vector space of a representation to define the *dual representation*.

To define the dual representation we first must define dual vector spaces.

Definition 5.5 *Suppose V is a complex vector space. The* dual vector space *of V, denoted V^* and pronounced "V-dual," is the complex vector space of linear transformations from V to $\mathbb{C}$.*

Recall from Exercise 2.14 that V^* is indeed a complex vector space.

For example, if we think of $\mathbb{C}^3$ as the set of column three-vectors with complex entries, then with the help of matrix multiplication we can think of $(\mathbb{C}^3)^*$ as the set of row three-vectors with complex entries. In other words, given any row vector of the form (T_1, T_2, T_3), where $T_1, T_2, T_3 \in \mathbb{C}$, we can define a linear transformation $\alpha : \mathbb{C}^3 \to \mathbb{C}$ by

$$
\alpha \begin{pmatrix} v_1 \\ v_2 \\ v_3 \end{pmatrix} := \begin{pmatrix} T_1 & T_2 & T_3 \end{pmatrix} \begin{pmatrix} v_1 \\ v_2 \\ v_3 \end{pmatrix} = T_1 v_1 + T_2 v_2 + T_3 v_3,
$$

and every linear transformation from $\mathbb{C}^3$ to $\mathbb{C}$ is of this form: explicitly, set $T_j := \alpha(e_j)$ for $j = 1, 2, 3$, where e_j denotes the jth standard basis vector.

Another way to make the dual space $(\mathbb{C}^3)^*$ concrete is to use the complex scalar product and think of elements of the dual space as column vectors. Recall the * notation for the conjugate transpose of a vector. In this interpretation the linear transformation

$$\alpha : \begin{pmatrix} v_1 \\ v_2 \\ v_3 \end{pmatrix} \mapsto T_1 v_1 + T_2 v_2 + T_3 v_3$$

corresponds to the column vector

$$\begin{pmatrix} T_1^* \\ T_2^* \\ T_3^* \end{pmatrix} = \begin{pmatrix} T_1 & T_2 & T_3 \end{pmatrix}^*,$$

via the calculation

$$\alpha \begin{pmatrix} v_1 \\ v_2 \\ v_3 \end{pmatrix} = \left\langle \begin{pmatrix} T_1^* \\ T_2^* \\ T_3^* \end{pmatrix}, \begin{pmatrix} v_1 \\ v_2 \\ v_3 \end{pmatrix} \right\rangle.$$

This is a special case of a more general construction. If there is a complex scalar product on V, then there is a natural linear transformation $\tau : V \to V^*$ defined by

$$\tau(v) := \langle v, \cdot \rangle. \tag{5.3}$$

We leave it to the reader to show that τ is injective and, if V is finite dimensional, also surjective (Exercise 5.20).[1] It follows that $\dim V = \dim V^*$ for any Hilbert space or finite-dimensional complex scalar product space V.

In fact, $\tau(v)$ is the adjoint of v in the sense of Definition 3.9. If we think of v as a function from $\mathbb{C}$ into V, then for any $c \in \mathbb{C}$ and any $w \in V$ we have $\langle w, vc \rangle_V = c(\tau(v)w)^* = \langle \tau(v)w, c \rangle_{\mathbb{C}}$, so $v^* = \tau(v)$.

We can use τ to define a complex scalar product on V^*.

Proposition 5.10 *For any complex scalar product* $\langle \cdot, \cdot \rangle$ *on a finite-dimensional vector space* V, *there is an associated complex scalar product* $\langle \cdot, \cdot \rangle_*$ *on* V^*, *given by*

$$\langle \alpha, \alpha \rangle_* := \langle \tau^{-1}\alpha, \tau^{-1}\alpha \rangle.$$

With this complex scalar product on the dual space V^* in hand, we can make the relationship between the dual and the adjoint clear. The definition

---

[1] If V is a *bona fide* Hilbert space, in the strict mathematical sense, then τ is surjective even if V is infinite dimensional. This fact is known as the *Riesz Representation Theorem* or the *Riesz Lemma*. See, e.g., [RS, Theorem II.4].

of the dual space (in Exercise 2.14) makes no reference to any complex scalar product. In other words, there is no need to specify a complex scalar product before defining V^* from V, and even if there are different possible complex scalar products on V, the dual space V^* will be the same. However, once we have specified a complex scalar product $\langle \cdot, \cdot \rangle$ on V, then there is a natural complex scalar product on V^* given by Proposition 5.10. Furthermore, for any $v \in V$ the adjoint $v^* := \tau(v)$ of v is an element of the dual space V^*. Finally, if V is finite dimensional, then we can identify $(V^*)^*$ with V (as in Exercise 2.15). For any v we have $(v^*)^* = v$, since for any $w \in V$ and $c \in \mathbb{C}$ we have

$$\langle (v^*)^* c, w \rangle_V = \langle c, (v^*) w \rangle_{\mathbb{C}} = c^* \langle v, w \rangle = \langle cv, w \rangle.$$

So when there is one fixed complex scalar product on a vector space V, it is consistent to use the notation v^* for both dual and adjoint. In a unitary basis, the asterisk means coordinate transpose.

Next we define the dual representation.

Definition 5.6 *Suppose* (G, V, ρ) *is a group representation. The* dual repre- *sentation of* (G, V, ρ) *is the representation* (G, V^*, ρ^*), *where*

$$\rho^*(g)T := T \circ \rho(g)^{-1}$$

for every $T \in V^*$.

The character of a dual representation is the complex conjugate of the character of the original.

Proposition 5.11 *Suppose* (G, V, ρ) *is a finite-dimensional unitary repre- sentation with character* χ. *Then the character of the dual representation* (G, V^*, ρ^*) *is* χ^*. *(Recall that* χ^* *denotes the complex conjugate of the* $\mathbb{C}$- *valued function* χ.) *Furthermore,* (G, V^*, ρ^*) *is a unitary representation with respect to the natural complex scalar product on* V^*.

Proof. Suppose $g \in G$. Recall the function τ defined in Equation (5.3). For the purpose of this proof we let $[\rho(g)^{-1}]$ denote the matrix of the linear op- erator $\rho(g)^{-1}$ in an orthonormal basis $\{v_1, \ldots, v_n\}$, and let $[\rho^*(g)]$ denote the matrix of $\rho^*(g)$ in the basis $\{\gamma_1, \ldots, \gamma_n\}$, where $\gamma_j := \tau(v_j)$ for each $j = 1, \ldots, n$. We can calculate the coefficient of γ_j in the expansion of $\rho^*(g)\gamma_j$ by applying $\rho^*(g)\gamma$ to the vector v_j:

$$[\rho^*(g)]_{jj} = \left(\rho^*(g)\gamma_j \right) v_j = \gamma_j \left(\rho(g)^{-1} v_j \right) = [\rho(g)^{-1}]_{jj} = [\rho(g)]_{jj}^*,$$

where the final equality holds because ρ is a unitary representation. We conclude that the character of V^* is

$$\operatorname{Tr} \rho^*(g) = \sum_{j=1}^{n} [\rho^*(g)]_{jj} = \sum_{j=1}^{n} [\rho(g)]_{jj}^* = (\chi(g))^*.$$

Next we show that the dual representation is unitary. By Exercise 5.20, for any $\gamma, U \in V^*$ there are elements $v, w \in V$ such that $\gamma = \tau(v)$ and $U = \tau(w)$. Then, for any for any $g \in G$ we have

$$\begin{aligned}
\langle \rho^*(g)\gamma, \rho^*(g)U \rangle_* &= \langle \rho^*(g)\tau(v), \rho^*(g)\tau(w) \rangle_* \\
&= \langle \tau(v) \circ \rho(g)^{-1}, \tau(w) \circ \rho(g)^{-1} \rangle_* \\
&= \langle \rho(g)v, \rho(g)w \rangle \\
&= \langle v, w \rangle = \langle \gamma, U \rangle_*,
\end{aligned}$$

where the second equality follows from the definition of the dual representation and the third equality follows from the fact that for any $u \in V$ we have

$$\tau(v)(\rho(g)^{-1}w) = \langle v, \rho(g)^{-1}w \rangle = \langle \rho(g)v, w \rangle$$

because ρ is a unitary representation.   □

For example, consider the representation of $SU(2)$ on $\mathbb{C}^3$ defined by

$$\rho \begin{pmatrix} \alpha & -\beta^* \\ \beta & \alpha^* \end{pmatrix} := \begin{pmatrix} 1 & 0 & 0 \\ 0 & \alpha & -\beta^* \\ 0 & \beta & \alpha^* \end{pmatrix}.$$

The dual representation ρ^* is a representation of $SU(2)$ on $(\mathbb{C}^3)^*$. To calculate ρ^* explicitly, we fix an element

$$g = \begin{pmatrix} \alpha & -\beta^* \\ \beta & \alpha^* \end{pmatrix} \in SU(2)$$

and an element $\gamma \in (\mathbb{C}^3)^*$. If we use the first (matrix multiplication) interpretation, we think of γ as a row vector (T_1, T_2, T_3), and for any column vector $v \in \mathbb{C}^*$ we have

$$(\rho^*(g)\gamma)(v) = \gamma(\rho(g^{-1})v)$$

$$= (T_1 \ \ T_2 \ \ T_3) \begin{pmatrix} 1 & 0 & 0 \\ 0 & \alpha^* & \beta^* \\ 0 & -\beta & \alpha \end{pmatrix} \begin{pmatrix} v_1 \\ v_2 \\ v_3 \end{pmatrix}.$$

So $\rho^*(g)$ is right multiplication of the row vector (T_1, T_2, T_3) by the matrix

$$\begin{pmatrix} 1 & 0 & 0 \\ 0 & \alpha^* & \beta^* \\ 0 & -\beta & \alpha \end{pmatrix}.$$

Readers who find right multiplication unfamiliar or mysterious should take the time to convince themselves that this correspondence between group elements and right multiplication is indeed a group homomorphism. The point is that the order of operations must be preserved. Multiplying on the right toggles the order, as does taking inverses. Hence the order is preserved by multiplication on the right by the inverse.

In order to calculate ρ^* in terms of left multiplication, we can use the interpretation of $(\mathbb{C}^3)^*$ as column vectors via the complex scalar product. Here we think of α as the column vector $(T_1, T_2, T_3)^*$ and we have

$$\left(\rho^* \begin{pmatrix} \alpha & -\beta^* \\ \beta & \alpha^* \end{pmatrix} \gamma \right) (v) = \gamma \left(\rho \begin{pmatrix} \alpha^* & \beta^* \\ -\beta & \alpha \end{pmatrix} v \right)$$

$$= \left\langle \begin{pmatrix} T_1^* \\ T_2^* \\ T_3^* \end{pmatrix}, \begin{pmatrix} 1 & 0 & 0 \\ 0 & \alpha^* & \beta^* \\ 0 & -\beta & \alpha \end{pmatrix} \begin{pmatrix} v_1 \\ v_2 \\ v_3 \end{pmatrix} \right\rangle$$

$$= \left\langle \begin{pmatrix} 1 & 0 & 0 \\ 0 & \alpha^* & \beta^* \\ 0 & -\beta & \alpha \end{pmatrix}^* \begin{pmatrix} T_1^* \\ T_2^* \\ T_3^* \end{pmatrix}, \begin{pmatrix} v_1 \\ v_2 \\ v_3 \end{pmatrix} \right\rangle.$$

Hence

$$\rho^* \begin{pmatrix} \alpha & -\beta^* \\ \beta & \alpha^* \end{pmatrix} (\gamma) = \begin{pmatrix} 1 & 0 & 0 \\ 0 & \alpha^* & \beta^* \\ 0 & -\beta & \alpha \end{pmatrix} \begin{pmatrix} T_1 \\ T_2 \\ T_3 \end{pmatrix}.$$

In this section we have shown how a representation on a vector space determines a representation on the dual of the vector space. We will find the dual representation useful in Section 5.5. More generally, duality is an important theoretical concept in many mathematical settings. Physically, momentum space is dual to position space, so the name "momentum space" in the physics literature often connotes duality.

5.5   The Representation Hom

Recall from Section 5.3 that one can interpret $(\mathbb{C}^3)^* \otimes \mathbb{C}^2$ as a vector space of matrices, which in turn can be interpreted as linear transformations from $\mathbb{C}^3$

to $\mathbb{C}^2$. This suggests that there may be a relationship between spaces of linear transformations and tensor products involving a dual space. In this section we show how to create a new representation $\text{Hom}(V, W)$ out of any two representations V and W of the same group G. Finally, we express $\text{Hom}(V, W)$ as a tensor product of representations.

The set of all linear transformations (not necessarily homomorphisms of representations) from a representation V to a representation W forms a vector space too. This vector space is denoted $\text{Hom}(V, W)$. (Here "Hom" refers to the fact that a linear transformation can be considered a "homomorphism" of vector spaces.) There is a natural representation of G on this vector space.

Proposition 5.12 *Suppose* (G, V, ρ) *and* $(G, W, \tilde{\rho})$ *are representations of the same group* G. *Let* $\text{Hom}(V, W)$ *denote the vector space of linear transformations from* V *to* W. *Define a function*

$$\sigma : G \to \mathcal{GL}\left(\text{Hom}(V, W)\right)$$

by setting, for each $g \in G$ *and each* $T \in \text{Hom}(V, W)$,

$$\sigma(g)T := \tilde{\rho}(g)T \left(\rho(g)\right)^{-1}.$$

Then $(G, \text{Hom}(V, W), \sigma)$ *is a representation.*

This representation is often denoted simply $\text{Hom}(V, W)$.

Proof. We must show that σ is a group homomorphism. So suppose $g_1, g_2 \in G$. Then for any $T \in \text{Hom}(V, W)$ we have

$$\sigma(g_1)\sigma(g_2)T = \tilde{\rho}(g_1)\tilde{\rho}(g_2)T \left(\rho(g_2)\right)^{-1} \left(\rho(g_1)\right)^{-1}$$
$$= \tilde{\rho}(g_1 g_2)T \left(\rho(g_1 g_2)\right)^{-1} = \sigma(g_1 g_2)T.$$

So σ is indeed a group homomorphism. $\qquad\qquad\Box$

There is special notation for the set of linear transformations that are homomorphisms of representations.

Definition 5.7 *Suppose* (G, V, ρ) *and* $(G, W, \tilde{\rho})$ *are representations of the same group* G. *We define*

$$\text{Hom}_G(V, W) := \{T \in \text{Hom}(V, W) : \tilde{\rho}(g) \circ T = T \circ \rho(g) \text{ for all } g \in G\}.$$

Note that $\text{Hom}_G(V, W)$ is a vector subspace of $\text{Hom}(V, W)$. Also, its elements are precisely the fixed points of the representation σ defined in Proposition 5.12.

Proposition 5.13

$$\text{Hom}_G(V, W) = \{T \in \text{Hom}(V, W) : \sigma(g)T = T \text{ for all } g \in G\}.$$

Proof. For each $g \in G$ we have $\tilde{\rho}(g) \circ T = T \circ \tilde{\rho}(g)$ if and only if $T = \tilde{\rho}(g)^{-1} \circ T \circ \rho(g)$. Hence the linear transformation T lies in $\text{Hom}_G(V, W)$ if and only if $\sigma(g)T = T$ for every $g \in G$. $\qquad\square$

In other words, the natural representation of G on $\text{Hom}_G(V, W)$ is trivial. Still, $\text{Hom}_G(V, W)$ does carry important information. In Section 6.4 we will find the vector space dimension of $\text{Hom}_G(V, W)$ to be useful.

Even when there is no unitary structure (i.e., no complex scalar product) on vector spaces V and W, there is a natural complex scalar product on the vector space $\text{Hom}(V, W)$, given by

$$\langle T, U \rangle := \text{Tr}(T^*U),$$

where $T^* \in \text{Hom}(W, V)$ denotes the adjoint operator of T (Definition 3.9). If (G, V, ρ) and $(G, V, \tilde{\rho})$ are unitary representations, then the representation $(G, \text{Hom}(V, W), \sigma)$ defined in Proposition 5.12 is unitary, since

$$
\begin{aligned}
\langle \sigma(g)T, \sigma(g)U \rangle &= \left\langle \tilde{\rho}(g)T\rho(g)^{-1}, \tilde{\rho}(g)U\rho(g)^{-1} \right\rangle \\
&= \text{Tr}\left((\rho(g)^{-1})^* T^* \tilde{\rho}(g)^* \tilde{\rho}(g)U\rho(g)^{-1} \right) \\
&= \text{Tr}\left(\rho(g)T^*U\rho(g)^{-1} \right) \\
&= \text{Tr}(T^*U) = \langle T, U \rangle,
\end{aligned}
$$

where the second-to-last equality is a consequence of Proposition 2.10.

The next proposition expresses $\text{Hom}(V, W)$ as a tensor product of representations and shows how to calculate the character of $\text{Hom}(V, W)$ in terms of the characters of V and W.

Proposition 5.14 *Suppose G is a group. Suppose (G, V, ρ) and $(G, V, \tilde{\rho})$ are group representations. Then the representation $\text{Hom}(V, W)$ is isomorphic to the representation $V^* \otimes W$ of G. Furthermore, if V and W are both finite dimensional, then the dimension of $\text{Hom}(V, W)$ equals the product of the dimensions of V and W and the character of the representation $\text{Hom}(V, W)$ is the product of the characters of V^* and W.*

We will use this proposition in Propositions 6.8 and 11.1.

Proof. Let us define an isomorphism $\mu : V^* \otimes W \to \text{Hom}(V, W)$. First we define μ on elementary tensors, i.e., products of the form $\alpha \otimes w$, where $w \in W$

and $\alpha \in V^*$. Recall that α is a linear transformation from V to $\mathbb{C}$. We let $\mu(\alpha \otimes w)$ equal the linear transformation

$$A_{\alpha,w} : V \to W$$
$$v \mapsto \alpha(v)w.$$

Because every element of $V^* \otimes W$ can be written as a finite sum of elementary tensors, we can define μ on all of $V^* \otimes W$ by linearity. Note that μ is one-to-one and onto, i.e., $\mu : V^* \otimes W \to \text{Hom}(V, W)$ is an isomorphism of vector spaces.

Next we show that μ is an isomorphism of representations. Let $\sigma : G \to \mathcal{GL}(\text{Hom}(V, W))$ denote the representation on Hom from Proposition 5.12. Recall that for any linear transformation $A \in \text{Hom}(V, W)$ and any $g \in G$ we have

$$\sigma(g)(A) = \tilde{\rho}(g)A\rho(g)^{-1}.$$

Hence

$$\left(\mu \circ (\rho^* \otimes \tilde{\rho})(g)\right)(\alpha \otimes w) = \mu\left((\rho^*(g)\alpha) \otimes (\tilde{\rho}(g)w)\right)$$
$$= A_{\rho^*(g)\alpha, \tilde{\rho}(g)w}.$$

Note that for any $v \in V$ we have

$$A_{\rho^*(g)\alpha, \tilde{\rho}(g)w}v = \rho^*(g)\alpha(v)\tilde{\rho}(g)w = \alpha(\rho(g)^{-1}v)\tilde{\rho}(g)w,$$

and hence

$$A_{\rho^*(g)\alpha, \tilde{\rho}(g)w} = \tilde{\rho}(g)A_{\alpha,w}\rho(g)^{-1} = \sigma(g)\left(A_{\alpha,w}\right)$$
$$= \sigma \circ \mu(\alpha \otimes w).$$

Putting it all together we have $\mu \circ (\rho^* \otimes \tilde{\rho}) = \sigma \circ \mu$, so μ is a homomorphism of representations. Because μ is a vector space isomorphism, it follows that μ is an isomorphism of representations.

To verify the last statement of the proposition note that by Propositions 2.13 and 5.11,

$$\dim(\text{Hom}(V, W)) = \dim(V^* \otimes W) = \dim(V^*)\dim(W)$$
$$= \dim(V)\dim(W),$$

while letting χ_τ denote the characters of a representation τ, we have

$$\chi_\sigma = \chi_{\rho^* \otimes \tilde{\rho}} = \chi_{\rho^*}\chi_{\tilde{\rho}},$$

where the second equality follows from Proposition 5.8.   □

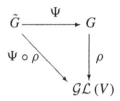

Figure 5.2. Pulling a representation back.

In this section we have seen how representations on two spaces V and W determine a representation on the set of homomorphisms of representations from V to W. Familiarity with this kinds of categorical construction is often the key to finding simple, direct proofs of interesting results such as Proposition 6.8.

5.6   Pullback and Pushforward Representations

In this section we show how to use group homomorphisms to construct a representation of one group from a representation of another group.

Proposition 5.15 *Suppose (G, V, ρ) is a representation, $\tilde{G}$ is a group and $\Psi: \tilde{G} \to G$ is a group homomorphism. Then $(\tilde{G}, V, \rho \circ \Psi)$ is a representation. If ρ is unitary, then so is $\rho \circ \Psi$.*

The representation $\rho \circ \Psi$ is called the *pullback representation*; see Figure 5.2.

Proof. First we show that $(\tilde{G}, V, \rho \circ \Psi)$ is a representation by checking the criteria given in Definition 4.7. We know by hypothesis that $\tilde{G}$ is a group and V is a vector space. Because both ρ and Ψ are group homomorphisms, it follows from Proposition 4.3 that $\rho \circ \Psi$ is a group homomorphism from $\tilde{G}$ to $\mathcal{GL}(V)$. Hence $\rho \circ \Psi$ is a representation.

If ρ is a unitary representation then $\rho: G \to \mathcal{U}(V)$. Hence $\rho \circ \Psi: \tilde{G} \to \mathcal{U}(V)$, and so $\rho \circ \Psi$ is also unitary. $\square$

Consider for example the inclusion map i of a subgroup $\tilde{G}$ of a group G, defined in Exercise 4.37. By that exercise, the inclusion map is a group homomorphism. Note that for any representation ρ of G, the pullback representation $\rho \circ i$ is just the restriction of ρ to the subgroup $\tilde{G}$.

We saw in Section 4.5 how to build a representation from the symmetries of the physical space of a quantum system. Some quantum systems have even

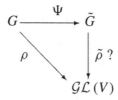

Figure 5.3. Pushing a representation forward.

more symmetry. These additional symmetries are called *hidden symmetries*. Mathematically, hidden symmetries correspond to a representation of a group G which contains as a subgroup the group $\tilde{G}$ of symmetries of physical space; restricting the representation to $\tilde{G}$ should yield the natural physical representation. One of the most beautiful examples is the hydrogen atom itself, whose Hilbert space of states has a representation of $SO(4) \supset SO(3)$. One can exploit this hidden symmetry to make more precise predictions about the structure of the shells of the hydrogen atom; for details, see Chapters 8 and 9.

It is not usually possible to *push* a representation *forward*, i.e., to use a representation on the domain of a group homomorphism to obtain a representation on the image. See Exercise 5.4. However, in certain circumstances a pushforward representation can be defined. See Figure 5.3.

Proposition 5.16 *Suppose (G, V, ρ) is a representation, $\tilde{G}$ is a group and $\Psi: G \to \tilde{G}$ is a group homomorphism. Suppose further that Ψ is surjective and for all $g \in G$ satisfying $\Psi(g) = I \in \tilde{G}$ we have $\rho(g) = I \in \mathcal{GL}(V)$. Then the function $\tilde{\rho}: \tilde{G} \to \mathcal{GL}(V)$ is well defined by the formula*

$$\tilde{\rho}(\tilde{g}) = \rho(g) \text{ whenever } \tilde{g} = \Psi(g)$$

and $(\tilde{G}, V, \tilde{\rho})$ is a representation. Furthermore, if ρ is unitary then so is $\tilde{\rho}$.

Proof. First we prove that $\tilde{\rho}$ is well defined. Fix any $\tilde{g} \in \tilde{G}$. Because Ψ is surjective, there is at least one g in the set $\Psi^{-1}(\tilde{g})$ we can use to define $\tilde{\rho}(\tilde{g})$. It remains to show that the value of $\tilde{\rho}(\tilde{g})$ does not depend on the choice of $g \in \Psi^{-1}(\tilde{g})$. Suppose $g_1, g_2 \in G$ and $\Psi(g_1) = \Psi(g_2) = \tilde{g}$. We must show that $\rho(g_1) = \rho(g_2)$. Note that

$$\Psi(g_1^{-1}g_2) = \Psi(g_1)^{-1}\Psi(g_2) = \Psi(g_1)^{-1}\Psi(g_1) = I \in \tilde{G},$$

so by hypothesis

$$\rho(g_1)^{-1}\rho(g_2) = \rho(g_1^{-1}g_2) = I \in \mathcal{GL}(V)$$

and hence $\rho(g_1) = \rho(g_2)$ by the uniqueness of the inverse (Proposition 4.1). So $\tilde{\rho}$ is well defined.

Second, we must show that $(\tilde{G}, V, \tilde{\rho})$ is a representation. We use the fact that Ψ and ρ are group homomorphisms to check that $\tilde{\rho}$ preserves multiplication. If $\Psi(g_1) = \tilde{g}_1$ and $\Psi(g_2) = \tilde{g}_2$, then $\Psi(g_1 g_2) = \tilde{g}_1 \tilde{g}_2$ and hence

$$\tilde{\rho}(\tilde{g}_1)\tilde{\rho}(\tilde{g}_2) = \rho(g_1)\rho(g_2) = \rho(g_1 g_2) = \tilde{\rho}(\tilde{g}_1 \tilde{g}_2).$$

Third, we must show that if ρ is unitary, then so is $\tilde{\rho}$. If ρ is unitary, then $\rho: G \to \mathcal{U}(V)$, and hence for each g we have $\tilde{\rho}(g) \in \mathcal{U}(V)$ as well. So $\tilde{\rho}$ is unitary. $\qquad\square$

In fact, $\tilde{\rho}$ is the unique representation of $\tilde{G}$ satisfying $\tilde{\rho} \circ \Psi = \rho$. In other words, $\tilde{\rho}$ is the only representation whose pullback under Ψ is ρ; see Exercise 5.3.

In Section 6.6 we will both push representations forward and pull them back along the two-to-one group homomorphism $\Phi: SU(2) \to SO(3)$ introduced in Section 4.2.

5.7   Exercises

Exercise 5.1 *Suppose (G, V, ρ) is a representation. Show that both the trivial subspace $\{0\}$ and the entire subspace V are invariant subspaces for the representation.*

Exercise 5.2 *Show that the intersection of any two invariant subspaces is an invariant subspace.*

Exercise 5.3 (Used in Proposition 6.15) *Under the hypotheses of Proposition 5.16, show that $\tilde{\rho}$ is the unique representation satisfying $\tilde{\rho} \circ \Psi = \rho$.*

Exercise 5.4 *Find a representation ρ and a group homomorphism Ψ such that ρ cannot be pushed forward via Ψ.*

Exercise 5.5 *Suppose $V_1, \ldots, V_n$ are linearly independent subspaces of W. Suppose further that $\bigcup_{k=1}^{n} V_n$ spans W. Let $B_1, \ldots, B_n$ denote bases of the subspaces $V_1, \ldots, V_n$. Then $\bigcup_{k=1}^{n} B_n$ is a basis of W.*

Exercise 5.6 *Recall the representations R_n of $SU(2)$ on homogeneous polynomials introduced in Section 4.6. Show that the representation $R_1 \otimes R_2$ introduced in the beginning of Section 5.3 has dimension six, while the representation $R_1 \oplus R_2$ has dimension five. Then prove that these two representations are not isomorphic.*

Exercise 5.7 *Recall the representations R_n of $SU(2)$ on homogeneous polynomials introduced in Section 4.6. Find a complex scalar product on the vector space of the representation $R_1 \otimes R_2$ such that the representation is unitary. Consider the subspace V_1 spanned by $\{uxy - vx^2, uy^2 - vxy\}$ and the subspace V_3 spanned by $\{ux^2, 2uxy + vx^2, 2vxy + uy^2, vy^2\}$. Use this complex scalar product to find $V_1^\perp$. Is your answer isomorphic to V_3? Is it equal to V_3?*

Exercise 5.8 *Recall the representations R_n of $SU(2)$ on homogeneous polynomials introduced in Section 4.6. Check that the representation on the subspace V_3 (defined in Exercise 5.7) is isomorphic to the representation R_3 on $\mathcal{P}^3$. Use the suggested isomorphism, and check that it satisfies Definition 4.9. It suffices to check that $R_3 \circ T(p) = T \circ \rho(p)$ for each of the four basis vectors p. (Here ρ is the representation on V_3 and T is the alleged isomorphism.)*

Exercise 5.9 *Recall the representations R_n of $SU(2)$ on homogeneous polynomials introduced in Section 4.6. For any natural number n, consider the character χ of the representation $R_1 \otimes R_n$. Show that $\chi = \chi_{n-1} + \chi_{n+1}$.*

Exercise 5.10 *Prove Proposition 5.7. (Hint: pick a basis of V and a basis of $\tilde{V}$.)*

Exercise 5.11 (Used in Proposition 7.3) *Suppose k is a natural number. Suppose (G, V_i, ρ_i) and (G, W_i, σ_i) are representations for $i = 1, \ldots, k$. Suppose further that $T_i : V_i \to W_i$ is a homomorphism of representations for $i = 1, \ldots, k$. Show that the function*

$$\bigoplus_{i=1}^k T_i : \bigoplus_{i=1}^k V_i \to \bigoplus_{i=1}^k W_i$$
$$(v_1, \ldots, v_k) \mapsto (T_1(v_1), \ldots, T_k(v_k))$$

is a homomorphism of representations.

Next, replace every instance of the word "homomorphism" in the paragraph above by the word "isomorphism" and show that the resulting paragraph is true.

Exercise 5.12 *Show that the function $\rho \otimes \tilde{\rho}$ of Definition 5.4 is in fact a representation. First check that it is a well-defined linear function — note that we defined it on all one-term products (not just on a basis). Then check that it is a group homomorphism from G into $\mathcal{GL}\left(V \otimes \tilde{V}\right)$.*

Exercise 5.13 *Can you use tensor products to construct a group operation on finite-sized square matrices of determinant one?*

Exercise 5.14 (Used in Proposition 11.1) *Consider the isomorphism between $V^* \otimes W$ and $\mathrm{Hom}(V, W)$ given in Proposition 5.14. Show that $x \in V^* \otimes W$ is elementary if and only if the corresponding linear transformation $X \in \mathrm{Hom}(V, W)$ has rank one.*

Exercise 5.15 *Suppose (G, V, ρ) and $(G, \tilde{V}, \tilde{\rho})$ are representations. Suppose $T : V \to \tilde{V}$ is a homomorphism of representations. Suppose W is an invariant subspace of V. Then $T|_W$ is a homomorphism of representations.*

Exercise 5.16 *Show that the bracket operation defined in Equation 5.2 is well defined on $V \otimes \tilde{V}$ and that it is a complex scalar product.*

Exercise 5.17 *Consider the finite permutation group S_3 on three letters. Construct a representation $(S_3, \mathbb{C}^3, \rho)$ by setting $z_1 := (1, 0, 0)^T$, $z_2 := (0, 1, 0)^T$ and $z_3 := (0, 0, 1)^T$ and defining*

$$\rho(\sigma)(z_i) := z_{\sigma(i)}$$

for each $\sigma \in S_3$ and $i = 1, 2, 3$.

1. *Find a one-dimensional invariant subspace W_1 of $\mathbb{C}^3$. Show that the representation (S_3, W_1, ρ) is trivial.*

2. *Find a complementary two-dimensional invariant subspace W_2 of $\mathbb{C}^3$ such that $\mathbb{C}^3 = W_1 \oplus W_2$.*

3. *Find the character χ of the restriction of ρ to W_2. You can do this by hand (find vectors that span W_2, write down the matrix in that basis and calculate the trace) or, more easily, by applying one of the results of Chapter 5.*

Exercise 5.18 *Define a representation $(\mathbb{T}, \mathbb{C}, \rho)$ by $\rho(e^{i\theta}) := (e^{i\theta})$. Define another representation $(\mathbb{T}, \mathbb{C}, \tilde{\rho})$ by $\tilde{\rho}(e^{i\theta}) := (e^{2i\theta})$. Write the tensor representation $\rho \otimes \tilde{\rho}$ explicitly as a matrix depending on $e^{i\theta}$. What is the character of $\rho \otimes \tilde{\rho}$?*

Exercise 5.19 *Define a representation $(\mathbb{T}, \mathbb{C}, \rho)$ by $\rho(e^{i\theta}) := (e^{7i\theta})$. Define another representation $(\mathbb{T}, \mathbb{C}^2, \tilde{\rho})$ by*

$$\tilde{\rho}(e^{i\theta}) := \begin{pmatrix} e^{i\theta} & 0 \\ 0 & e^{2i\theta} \end{pmatrix}.$$

Write the tensor representation $\rho \otimes \bar{\rho}$ explicitly as a matrix depending on $e^{i\theta}$. Find two invariant subspaces U and V of $\mathbb{C} \otimes \mathbb{C}^2$ such that $\mathbb{C} \otimes \mathbb{C}^2 = U \oplus V$ as representations.

Exercise 5.20 (Used in Proposition 5.10) *Suppose V is a vector space with a scalar product $\langle \cdot, \cdot \rangle$. Then the function $\tau : V \to V^*$ defined in Equation 5.3 is injective. If V is finite dimensional, then τ is also surjective onto V^*.*

Exercise 5.21 *Suppose V is a complex vector space. Show that $V^* = \mathrm{Hom}(V, \mathbb{C})$. Now suppose that V has a complex scalar product. Do the natural complex scalar products induced on V^* (defined in Exercise 3.19) and $\mathrm{Hom}(V, \mathbb{C})$ (defined in Exercise 3.20) agree?*

Exercise 5.22 (Used in Proposition 11.1) *Suppose V and W are complex scalar product spaces. Recall (from Exercise 3.19, Exercise 3.20 and Equation 5.2) the natural complex scalar products $\langle \cdot, \cdot \rangle_{V^* \otimes W}$ and $\langle \cdot, \cdot \rangle_{\mathrm{Hom}(V,W)}$ induced on $V^* \otimes W$ and $\mathrm{Hom}(V, W)$, respectively. Consider the isomorphism $\mu : V^* \otimes W \to \mathrm{Hom}(V, W)$ defined in the proof of Proposition 5.14. Show that μ is unitary, i.e., that for any $x, y \in V^* \otimes W$ we have*

$$\langle x, y \rangle_{V^* \otimes W} = \langle \mu(x), \mu(y) \rangle_{\mathrm{Hom}(V,W)} \, .$$

Exercise 5.23 *Consider the representation $(\mathbb{T}, \mathbb{C}, \rho)$ defined by $\rho(e^{i\theta}) := e^{16i\theta}$. Check that the function $\Psi : \mathbb{T} \to \mathbb{T}$ defined by $\Psi(e^{i\theta}) := e^{8i\theta}$ is a group homomorphism. Show that ρ can be pushed forward via Ψ and find the character of the pushforward representation.*

6

Irreducible Representations and Invariant Integration

With me it's all er nuthin';
Is it all er nuthin' with you?
It cain't be in between, it cain't be now and then,
No half-and-half romance will do!
I'm a one-woman man, home-lovin' type,
All complete with slippers and pipe.
Take me like I am er leave me be!
If you cain't give me all, give me nuthin'
And nuthin's whut you'll git from me.

— Oscar Hammerstein II, "All er Nuthin'," from the musical *Oklahoma* [Ham]

Irreducible representations are the building blocks of all other representations. Just as each molecule is made up of particular atoms, each representation is made up of particular irreducible representations. Unlike a molecule, whose properties are determined not only by which atoms it is made of, but also by their configuration, a representation is merely the sum of its irreducible parts. Mathematically, irreducible representations are useful because one can often reduce an idea or a calculation involving representations to an easier one involving only irreducible representations. Physically, irreducible representations correspond to fundamental physical entities.

In Section 6.1 we define irreducible representations. Then we state, prove and illustrate Schur's lemma. Schur's lemma is the statement of the all-or-nothing personality of irreducible representations.[1] In the Section 6.2 we discuss the physical importance of irreducible representations. In Section 6.3 we introduce invariant integration and apply it to show that characters of irreducible representations form an orthonormal set. In the optional Section 6.4 we use the technology we have developed to show that finite-dimensional unitary representations are no more than the sum of their irreducible parts. The remainder of the chapter is devoted to classifying the irreducible representations of $SU(2)$ and $SO(3)$.

6.1   Definitions and Schur's Lemma

In this section we will use the idea of invariant subspaces of a representation (see Definition 5.1) to define irreducible representations. Then we will prove Schur's lemma, which tells us that irreducible representations are indeed good building blocks.

For some representations, the largest and smallest subspaces are the only invariant ones. Consider, for example, the natural representation of the group $G = SO(3)$ on the three-dimensional vector space $\mathbb{C}^3$. Suppose W is an invariant subspace with at least one nonzero element. We will show that $W = \mathbb{C}^3$. In other words, we will show that only $\mathbb{C}^3$ itself (all) and the trivial subspace $\{0\}$ (nothing) are invariant subspaces of this representation. It will suffice to show that the vector $(1, 0, 0)^T$ lies in W, since W would then have to contain both

$$\begin{pmatrix} 0 \\ 1 \\ 0 \end{pmatrix} = \begin{pmatrix} 0 & -1 & 0 \\ 1 & 0 & 0 \\ 0 & 0 & 1 \end{pmatrix} \begin{pmatrix} 1 \\ 0 \\ 0 \end{pmatrix}$$

and

$$\begin{pmatrix} 0 \\ 0 \\ 1 \end{pmatrix} = \begin{pmatrix} 0 & 0 & -1 \\ 0 & 1 & 0 \\ 1 & 0 & 0 \end{pmatrix} \begin{pmatrix} 1 \\ 0 \\ 0 \end{pmatrix},$$

and the set $\{(1, 0, 0)^T, (1, 0, 0)^T, (1, 0, 0)^T\}$ spans the complex vector space $\mathbb{C}^3$. Note that both square matrices are elements of $SO(3)$.

To show that $(1, 0, 0)^T$ lies in W, consider any nonzero vector $w \in W$. Note that $w \in \mathbb{C}^3$, and it might not be pure real or pure imaginary. Define

---

[1] We would like to use the word "character" here, but it has a previous commitment.

real vectors u and v by $u + iv := w$. On the one hand, if $v = 0$, then u is nonzero (because w is nonzero) and, by Exercise 4.11, we can choose a rotation $M \in SO(3)$ such that $Mw = Mu = (r, 0, 0)^T$ and r is nonzero. Hence, since W is an invariant subspace, it must contain $r^{-1}Mw = (1, 0, 0)^T$. So if $v = 0$ then, by the argument above, we have $W = \mathbb{C}^3$. On the other hand, if $v \neq 0$, then again by Exercise 4.11 we can choose a rotation $M \in SO(3)$ such that $Mv = (r, 0, 0)^T$ for some nonzero real number r. Thus the invariant subspace W contains the vector $Mw = (a + ir, b, c)^T$ for some real numbers a, b and c. It follows that the subspace W also contains the vector

$$(2a + 2ir)^{-1} \left(Mw + \begin{pmatrix} 1 & 0 & 0 \\ 0 & -1 & 0 \\ 0 & 0 & -1 \end{pmatrix} Mw \right) = \begin{pmatrix} 1 \\ 0 \\ 0 \end{pmatrix}.$$

Note that because r is nonzero, so is $2a + 2ir$. So in this case as well we have $(1, 0, 0)^T \in W$ and hence, as argued above, $W = \mathbb{C}^3$. This shows that the only nonzero invariant subspace of the representation is the whole space $\mathbb{C}^3$.

Such a representation is called *irreducible*.

Definition 6.1 *A representation (G, V, ρ) is* irreducible *if its only invariant subspaces are V itself and the trivial subspace $\{0\}$. Representations that are not irreducible are called* reducible.

We can summarize our work above by writing that the natural representation of $SO(3)$ on $\mathbb{C}^3$ is irreducible. In contrast, we have seen in Section 5.1 that the representation of the circle group defined by Formula 5.1 is not irreducible.

We sometimes speak of *irreducible vector spaces* as well, especially if the group G and group homomorphism ρ are clear from the context. Recall the definition of a subrepresentation (Definition 5.2).

Definition 6.2 *Suppose (G, V, ρ) is a representation and (G, W, ρ_W) is a subrepresentation. Suppose that (G, W, ρ_W) is an irreducible representation. Then we call W an* irreducible subspace *or an* irreducible invariant subspace *of (G, V, ρ).*

Now we come to the key technical propositions, which tell us that irreducible representations cannot be mixed up in any clever ways. As with Will Parker in the musical comedy *Oklahoma*, with irreducible representations it's all or nothing. The following proposition will be useful in Chapter 7.

Proposition 6.1 *Suppose (G, V_1, ρ_1) and (G, V_2, ρ_2) are representations. Suppose $T: V_1 \rightarrow V_2$ is a homomorphism of representations. If V_1 is an irreducible representation, then either the kernel of T is trivial or the image* *of T is trivial. If V_2 is irreducible, then either T is surjective or T is the trivial homomorphism.*

Proof. Consider the kernel K_T of T. This subspace of V_1 is an invariant space for the representation ρ_1, since for any $v \in V_1$ such that $Tv = 0 \in V_2$ and for any $g \in G$ we have

$$T\rho_1(g)v = \rho_2(g)Tv = 0,$$

so $\rho_1(g)v \in K_T$. Since ρ_1 is irreducible, we conclude that either $K_T = V_1$ or $K_T = \{0\}$.

To prove the second statement, consider the image $T[V_1]$ of T. Because T is linear, $T[V_1]$ must be a subspace of V_2. In fact, $T[V]$ is an invariant subspace: for any $w \in T[V_1]$ there is a $v \in V_1$ such that $Tv = w$ and hence for any $g \in G$ we have

$$\rho_2(g)w = \rho_2(g)Tv = T\rho_1(g)v \in T[V_1].$$

Because V_2 is irreducible, either $T[V_1] = V_2$ or $T[V_1] = \{0\}$. $\square$

The next proposition is a workhorse of representation theory.

Proposition 6.2 (Schur's lemma) *Suppose (G, V_1, ρ_1) and (G, V_2, ρ_2) are irreducible representations of the same group G. Suppose that $T : V_1 \to V_2$ is a homomorphism of representations. Then there are only two possible cases:*

- *The function T is the zero function, i.e., $Tv = 0$ for all $v \in V_1$.*

- *The representations (G, V_1, ρ_1) and (G, V_2, ρ_2) are isomorphic (and T is an isomorphism).*

Proof. Let K_T denote the kernel of T. If $K_T = V_1$, then the function T is the zero function, and the conclusion of the theorem is satisfied. So suppose $K_T \neq V$; then $K_T = \{0\}$ by the first part of Proposition 6.1. By the second part of Proposition 6.1, it follows that T is surjective onto V_2. Hence T must be an isomorphism between (G, V_1, ρ_1) and (G, V_2, ρ_2). $\square$

The next proposition says that there are no interesting homomorphisms from an irreducible representation to itself. We will use this consequence of Schur's lemma in our first prediction for the hydrogen atom, Proposition 7.7. For the statement of the proposition, some terminology is convenient.

Definition 6.3 *Suppose (G, V, ρ) is a representation. A linear operator $T : V \to V$ commutes with ρ if and only if, for each $g \in G$ we have*

$$T \rho(g) = \rho(g)T.$$

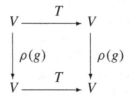

Figure 6.1. Commutative diagram for $T\rho(g) = \rho(g)T$.

In other words, $T : V \to V$ commutes with ρ if and only if T is a homomorphism of representations.

For example, each linear operator on $\mathbb{C}^2$ that is diagonal in the standard basis of $\mathbb{C}^2$ commutes with the representation of the circle group $\mathbb{T}$ defined by Formula 5.1. One often expresses commutation with a diagram. For the diagram version of Definition 6.3, see Figure 6.1.

Proposition 6.3 *Suppose (G, V, ρ) is a finite-dimensional irreducible representation. Then every linear operator $T : V \to V$ that commutes with ρ is a scalar multiple of the identity. In other words, if $T : V \to V$ is a homomorphism of representations, then T is a scalar multiple of the identity.*

Proof. Suppose that (G, V, ρ) is irreducible and the linear transformation $T : V \to V$ commutes with ρ. We must show that T is a scalar multiple of the identity. Because V is finite dimensional there must be at least one eigenvalue λ of T (by Proposition 2.11). By Proposition 5.2, the eigenspace corresponding to λ must be an invariant space for ρ. This space is not trivial, so because ρ is irreducible it must be all of V. In other words, $T = \lambda I$. So T is a scalar multiple of the identity.   □

There is an elegant summary of our results so far involving the concept of the vector space $\text{Hom}_G(V_1, V_2)$ from Section 5.5. We will use this proposition in the proof of Proposition 6.8.

Proposition 6.4 *Suppose (G, V_1, ρ_1) and (G, V_2, ρ_2) are irreducible representations of the same group G. Then there are two possible cases:*

- dim $\text{Hom}_G(V_1, V_2) = 0$.

- dim $\text{Hom}_G(V_1, V_2) = 1$.

Proof. Either the representations V_1 and V_2 are isomorphic, or they are not. If they are not isomorphic, then by Schur's lemma the only element of $\text{Hom}_G(V_1, V_2)$ is the zero function. In this case dim $\text{Hom}_G(V_1, V_2) = 0$.

Now suppose that the representations V_1 and V_2 are indeed isomorphic. Let T and $\tilde{T}$ denote isomorphisms (of representations) from V_1 to V_2. It suffices to show that $\tilde{T}$ must be a scalar multiple of T. Consider the linear transformation $\tilde{T} \circ T^{-1} \colon V_2 \to V_2$. By Exercise 4.19, this linear transformation is an isomorphism of representations. Hence by Proposition 6.3, there must be a complex number λ such that $\tilde{T} \circ T^{-1} = \lambda I$, and hence $\tilde{T} = \lambda T$. Note that because T is an isomorphism, $\lambda \neq 0$. $\qquad\square$

The following consequence of Schur's lemma will be useful in the proof that every polynomial restricted to the two-sphere is equal to a harmonic polynomial restricted to the two-sphere (Proposition 7.3). The idea is that once we decompose a representation into a Cartesian sum of irreducibles, every irreducible subrepresentation appears as a term in the sum.

Proposition 6.5 *Suppose G is a group and $(G, V_0, \rho_0), \ldots, (G, V_n, \rho_n)$ are finite-dimensional irreducible representations of G. Suppose that for all $j = 1, \ldots, n$, ρ_0 is not isomorphic to ρ_j. Then*

$$\dim \operatorname{Hom}_G(V_0, V_1 \oplus \cdots \oplus V_n) = 0.$$

Proof. Suppose $T \colon V_0 \to V_1 \oplus \cdots \oplus V_n$ is a homomorphism of representations. We must show that T is trivial. Fix any $j = 1, \ldots, n$. Consider the projection Π_j onto V_j introduced in Definition 2.12. By Proposition 5.6, this projection is a homomorphism of representations. Hence $\Pi_j \circ T$ is a homomorphism of representations. Its domain V_0 and its range V_j are both irreducible representations. By hypothesis these representations are not isomorphic; hence Schur's lemma implies that $\Pi_j \circ T$ is trivial. Because $\Pi_j \circ T$ is trivial for each $j = 1, \ldots, n$, the homomorphism T must be trivial. $\qquad\square$

For unitary representations we have a converse to Proposition 6.3. Unitary irreducible representations are sometimes called *unirreps* for short.

Proposition 6.6 *Suppose V is a finite-dimensional complex vector space with a complex scalar product. Suppose (G, V, ρ) is a unitary representation. Suppose that every linear operator $T \colon V \to V$ that commutes with ρ is a scalar multiple of the identity. Then (G, V, ρ) is irreducible.*

Proof. Suppose every linear transformation $T \colon V \to V$ that commutes with ρ is a scalar multiple of the identity. Suppose also that W is an invariant subspace for (G, V, ρ). We must show that $W = V$. By Proposition 3.5, because V is finite dimensional there is an orthogonal projection $\Pi_W \colon V \to V$ whose image is W. Since ρ is unitary, we can apply Proposition 5.4 to show that the linear transformation Π_W is a homomorphism of representations. So, by

Definition 4.9 we know that Π_W commutes with every $\rho(g)$. Hence by hypothesis P_W must be a scalar multiple of the identity. If the scalar is nonzero, then $W = V$. If the scalar is zero, then $W = \{0\}$.

We have shown that V (all) and $\{0\}$ (nothing) are the only invariant subspaces of V. So (G, V, ρ) is irreducible.   □

The following technical proposition will be useful in Proposition 7.6.

Proposition 6.7 *Suppose (G, V_1, ρ_1) and (G, V_2, ρ_2) are subrepresentations of a unitary representation (G, V, ρ). Suppose V_1 is irreducible, and suppose that V_2 is finite dimensional. Suppose that ρ_1 not isomorphic to any subrepresentation of (G, V_2, ρ_2). Then V_1 is perpendicular to V_2; that is, for any $v_1 \in V_1$ and any $v_2 \in V_2$ we have $\langle v_1, v_2 \rangle = 0$.*

Proof. By Proposition 3.5, since V_2 is finite dimensional we know that there is an orthogonal projection Π_2 with range V_2. Because ρ is unitary, the linear transformation Π_2 is a homomorphism of representations by Proposition 5.4. Thus by Exercise 5.15 the restriction of Π_2 to V_1 is a homomorphism of representations. By hypothesis, this homomorphism cannot be injective. Hence Schur's lemma (Proposition 6.2) implies that since V_1 is irreducible, $\Pi_2[V_1]$ is the trivial subspace. In other words, V_1 is perpendicular to V_2.   □

Schur's lemma is both elementary and far-reaching. By showing that besides the trivial homomorphism and the identity homomorphism there are no homomorphisms between irreducible representations, Schur's lemma ensures that irreducible representations make good building blocks, solid and incorruptible. It allows us to generalize the notion of eigenspaces: just as a vector space can be seen as a Cartesian sum of eigenspaces of a single linear operator, a vector space can also be seen as a Cartesian sum of invariant spaces for whole representation's worth of linear operators. We suggest that the reader keep an eye out for the crucial use of Schur's lemma and its consequences in the remainder of the text.

6.2   Elementary States of Quantum Mechanical Systems

We saw in Section 4.5 that a quantum mechanical system with symmetry determines a unitary representation of the symmetry group. It is natural then to ask about the physical meaning of representation-theoretic concepts. In this section, we consider the meaning of invariant subspaces and irreducible representations.

Consider a complex scalar product space V that models the states of a quantum system. Suppose G is the symmetry group and (G, V, ρ) is the natural representation. By the argument in Section 5.1, the only physically natural subspaces are invariant subspaces. Suppose there are invariant subspaces $U_1, U_2, W \subset V$ such that $W = U_1 \oplus U_2$. Now consider a state w of the quantum system such that $w \in W$, but $w \notin U_1$ and $w \notin U_2$. Then there is a nonzero $u_1 \in U_1$ and a nonzero $u_2 \in U_2$ such that $w = u_1 + u_2$. This means that the state w is a superposition of states u_1 and u_2. It follows that w is not an elementary state of the system — by the principle of superposition, anything we want to know about w we can deduce by studying u_1 and u_2.

We know from numerous experiments that every quantum system has *elementary states*. An elementary state of a quantum system should be observer-independent. In other words, any observer should be able (in theory) to recognize that state experimentally, and the observations should all agree. Second, an elementary state should be indivisible. That is, one should not be able to think of the elementary state as a superposition of two or more "more elementary" states. If we accept the model that every recognizable state corresponds to a vector subspace of the state space of the system, then we can conclude that elementary states correspond to irreducible representations. The independence of the choice of observer compels the subspace to be invariant under the representation. The indivisible nature of the subspace requires the subspace to be irreducible. So elementary states correspond to irreducible representations. More specifically, if a vector w represents an elementary state, then w should lie in an *irreducible* invariant subspace W, that is, a subspace whose only invariant subspaces are itself and 0. In fact, every vector in W represents a state "indistinguishable" from w, as a consequence of Exercise 6.6.

The reader should consider the argument in this section carefully: it is the core philosophy of this book. It implies that every elementary state of a quantum system with symmetry corresponds to an irreducible representation of the symmetry group (namely, the restriction of ρ to the irreducible invariant subspace containing the state). Thus, classifying the irreducible representations of the symmetry group makes concrete predictions about the quantum system. We will see in Chapter 7 that we can think a representation as the sum of its irreducible parts; physically, this means that *if we know enough mathematics to find what the irreducible parts of a given quantum mechanical representation are*, then we can predict what the elementary building blocks of that system should be. We apply this idea to the hydrogen atom in Section 7.3 and again in Chapter 8. In Section 10.4 we will apply it to the spin of elementary particles.

6.3   Invariant Integration and Characters of Irreducible Representations

A fundamental tool in the study of compact[2] groups (such as $SU(2)$, tori and $SO(n)$ for any n) is *invariant integration*. An integral on a group G allows us to define a complex vector space $L^2(G)$. An integral invariant under multiplication gives particularly nice results when applied to characters of representations. In this section we define invariant integrals on the circle and, more importantly for our purpose, on $SU(2)$. Then we use invariant integration to prove a proposition about the orthogonality (more precisely, the orthonormality) of characters of irreducible representations.

As a simple example of the ideas we will develop in this section, consider integrating functions on the circle group $\mathbb{T} = \{\lambda \in \mathbb{C}: |\lambda| = 1\}$. One way to define an integral is to introduce a coordinate θ by parameterizing the circle as $\mathbb{T} = \{e^{i\theta} : \theta \in [0, 2\pi]\}$. Then we can integrate functions over the circle by thinking of them as functions on the interval $[0, 2\pi]$ and using techniques of integration from calculus. Notice that this parameterization of the circle group $\mathbb{T}$ is not unique: for example, we could have used $e^{i\theta^3}$ on the interval $[0, (2\pi)^{1/3}]$ instead. However, the standard parameterization is undoubtedly nicer than many others. Here is one particularly nice feature: if we "rotate" a function, its standard integral does not change. To put it more rigorously, given any integrable function $f : \mathbb{T} \to \mathbb{C}$ and any fixed $\lambda_0 \in \mathbb{T}$, we have

$$\int_0^{2\pi} f(\lambda_0^{-1} e^{i\theta})d\theta = \int_0^{2\pi} f(e^{i\theta})d\theta,$$

as the reader is invited to check in Exercise 6.7. To say the same thing in yet another way, note that there is an action of $\mathbb{T}$ on itself by left multiplication; this action induces a representation ρ of $\mathbb{T}$ on the vector space of complex-valued functions on $\mathbb{T}$ (as we saw in Section 4.4). For any $\lambda_0 \in \mathbb{T}$ and any integrable function $f : \mathbb{T} \to \mathbb{C}$ we have $(\rho(\lambda_0)f)(e^{i\theta}) = f(\lambda_0^{-1} e^{i\theta})$ and hence, writing $\lambda_0 = e^{i\theta_0}$,

$$\int_0^{2\pi} (\rho(\lambda_0)f)(e^{i\theta})d\theta = \int_0^{2\pi} f(e^{i\theta-\theta_0})d\theta = \int_0^{2\pi} f(e^{i\theta})d\theta.$$

---

[2]Compactness for matrix groups is no different from compactness for subsets of Euclidean space (see Definition 3.16). In fact, every matrix group is a subset of Euclidean space, since an $n \times n$ matrix can be construed as a point in $\mathbb{R}^{n^2}$ or $\mathbb{C}^{n^2} = \mathbb{R}^{2n^2}$. Furthermore, students of topology will appreciate that if a group has a topological structure (as any manifold, and hence any Lie group has), then the more general topological definition in terms of open covers can be applied to that group to determine whether it is compact.

To put it more succinctly, this integral is unchanged by the action of the group on itself by left multiplication. A similar argument shows that the integral is invariant under right multiplication as well. In summary, the integral on the group defined by the standard parameterization is invariant under multiplication; it is an *invariant integral*.

The existence of an invariant integral on the circle is no accident. Every compact Lie group has an invariant integral, usually written $\int_G dg$. For a proof of the existence of the invariant integral on an arbitrary compact group, see Bröcker and tom Dieck [BtD, Proposition 5.5]. One can normalize the invariant integral by insisting that the value of the integral of the constant function 1 be 1. Intuitively, this means that the "volume" according to this integral should be 1. This choice of invariant integral allows us to interpret integrals over the groups as averages. Our standard parameterization of the circle fails the volume-one criterion, as

$$\int_0^{2\pi} 1 d\theta = 2\pi.$$

However, a slight modification will bring the circle in line with the customary invariant integration. Parametrizing the circle by

$$\mathbb{T} = \left\{ e^{2\pi i t} : t \in [0, 1] \right\},$$

we get the volume-one invariant integral taking a function f on the circle to

$$\int_{\mathbb{T}} f dg := \int_0^1 f(e^{2\pi i t}) dt.$$

Let us double check that the integral is invariant under left multiplication. Any element of the group can be written $e^{2\pi i t_0}$ for some $t_0 \in \mathbb{R}$, so we have

$$\int_0^1 f(e^{2\pi i t_0} e^{2\pi i t}) dt = \int_0^1 f(e^{2\pi i (t+t_0)}) dt = \int_0^1 f(e^{2\pi i (t)}) dt.$$

Note that the integral is invariant under right multiplication as well:

$$\int_0^1 f(e^{2\pi i t} e^{2\pi i t_0}) dt = \int_0^1 f(e^{2\pi i (t-t_0)}) dt = \int_0^1 f(e^{2\pi i (t)}) dt.$$

Right invariance follows from left invariance for all compact groups. The general theorem and its proof are in [BtD, Theorem 5.12]. We will prove the special case of $SU(2)$ below. The invariant, volume-one integral on $SU(2)$

plays an important role in our story. We will use it in Section 6.5 to prove that the list of irreducible representations of $SU(2)$ found in Section 4.6 is comprehensive. We will find an integral on $SU(2)$ by identifying $SU(2)$ with the three-sphere S^3 in $\mathbb{R}^4$ and pulling the natural volume element on S^3 back to $SU(2)$. This integral turns out to be invariant under multiplication (on left or right) by elements of $SU(2)$. From Section 4.2 we know that there is a group isomorphism from the unit quaternions (i.e., the three-sphere in $\mathbb{R}^4$) to $SU(2)$. In spherical coordinates this group isomorphism takes the form

$$
\begin{pmatrix}
\cos\psi + i\sin\psi\sin\theta\cos\phi & -\sin\psi\sin\theta\sin\phi + i\sin\psi\cos\theta \\
\sin\psi\sin\theta\sin\phi + i\sin\psi\cos\theta & \cos\psi + i\sin\psi\sin\theta\cos\phi
\end{pmatrix}.
$$

Note that the transformation $\psi \mapsto -\psi$ corresponds to complex conjugation.

Consider the natural integral on the unit three-sphere S^3 (the Euclidean integral inherited from $\mathbb{R}^4$, in which S^3 sits). We pull this back to get an integral on the group $SU(2)$. In spherical coordinates (up to a constant factor) we have

$$
\int_{SU(2)} f = \frac{1}{2\pi^2} \int_0^{2\pi} \int_0^\pi \int_0^\pi f(\phi, \theta, \psi)\sin^2\psi\sin\theta\, d\psi\, d\theta\, d\phi. \tag{6.1}
$$

for any function f on S^3. See Exercise 1.11. Since surface area on S^3 inside $\mathbb{R}^4$ is spherically symmetric, this integral is invariant under the action of $SO(4)$ on S^3 by matrix multiplication of column vectors. The constant $\frac{1}{2\pi^2}$ ensures that we have a volume-one integral since

$$
\int_{SU(2)} 1 = \frac{1}{2\pi^2} \int_0^{2\pi} \int_0^\pi \int_0^\pi \sin^2\psi\sin\theta\, d\psi\, d\theta\, d\phi = 1.
$$

Note also the effect of complex conjugation:

$$
\int_{SU(2)} f(g^*)dg = \frac{1}{2\pi^2} \int_0^{2\pi} \int_0^\pi \int_0^\pi f(\phi, \theta, -\psi)\sin^2\psi\sin\theta\, d\psi\, d\theta\, d\phi
$$

$$
= \frac{1}{2\pi^2} \int_0^{2\pi} \int_0^\pi \int_{-\pi}^0 f(\phi, \theta, \psi)\sin^2\psi\sin\theta\, d\psi\, d\theta\, d\phi
$$

$$
= \frac{1}{2\pi^2} \int_0^{2\pi} \int_0^\pi \int_0^\pi f(\phi, \theta, \psi)\sin^2\psi\sin\theta\, d\psi\, d\theta\, d\phi = \int_{SU(2)} f(g)dg,
$$

where the second-to-last equality holds by substituting $\psi + \pi$ for ψ and noting that $\sin^2$ is a function with period π. See Figure 6.2. So the integral is invariant under complex conjugation.

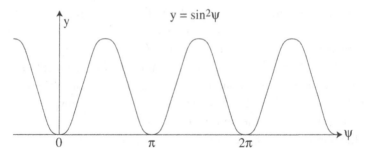

Figure 6.2. The period of sin^2 is π.

Next we show that this integral is invariant under group multiplication on the left. Recall from Section 4.2 that $SU(2)$ is isomorphic to the unit quaternions. From Exercise 4.25 we know that multiplication of a unit quaternion $\mathbf{q}$ on the left by a unit quaternion $\mathbf{q}_0$ corresponds to the product of a matrix in $SO(4)$ (corresponding to $\mathbf{q}_0$) and a vector in $S^3 \subset \mathbb{R}^4$ (corresponding to $\mathbf{q}$). See Figure 6.3.

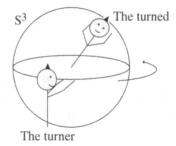

Figure 6.3. A unit quaternion rotating his fellow.

But the integral is invariant under such a change in coordinates because the volume element on the three-sphere is unchanged by rotations. So for any $g_0 \in SU(2)$ we have

$$\int_{SU(2)} f(g_0 g)dg = \int_{SU(2)} f(g)dg.$$

Finally, we must show that the integral is invariant under group multiplication on the right. Let f be any integrable function on $SU(2)$, and let

$\bar{f} : SU(2) \to \mathbb{C}$ denote the function defined by $\bar{f}(g) := f(g^*)$. Then

$$\int_{SU(2)} f(gg_0)\,dg = \int_S U(2)\bar{f}(g_0^* g^*)\,dg = \int_{SU(2)} \bar{f}(g_0^* g)\,dg$$

$$= \int_{SU(2)} \bar{f}(g)\,dg = \int_{SU(2)} f(g^*)\,dg$$

$$= \int_{SU(2)} f(g)\,dg.$$

Here we have used the invariance of the integral under conjugation and left multiplication. So the integral is invariant under group multiplication on the right as well on the left.

We are most interested in integrating products of characters of representations. In this case, we can use the Spectral Theorem (Proposition 4.4) to simplify the expression of the integral. The proposition implies that for any function f invariant under conjugation, we have

$$f\left(\begin{pmatrix} \alpha & -\beta^* \\ \beta & \alpha^* \end{pmatrix}\right) = f\left(\begin{pmatrix} \cos\psi + i\sin\psi & 0 \\ 0 & \cos\psi - i\sin\psi \end{pmatrix}\right),$$

where we have used the fact that $\Re(\alpha) = \cos\psi$ in spherical coordinates. Setting

$$\tilde{f}(\cos\psi) := f\left(\begin{pmatrix} \alpha & -\beta^* \\ \beta & \alpha^* \end{pmatrix}\right)$$

we have

$$\int_{SU(2)} f(g)\,dg = \frac{1}{2\pi^2}\int_0^{2\pi}\int_0^{\pi}\int_0^{\pi} \tilde{f}(\cos\psi)\sin^2\psi\sin\theta\,d\psi d\theta d\phi$$

$$= \frac{2}{\pi}\int_0^{\pi} \tilde{f}(\cos\psi)\sin^2\psi\sin\theta\,d\psi.$$

Changing variables ($x = \cos\psi$) we find

$$\int_{SU(2)} f(g)\,dg = \frac{2}{\pi}\int_{-1}^{1} \tilde{f}(x)\sqrt{1 - x^2}\,dx. \tag{6.2}$$

We can use the invariant integral on a compact group G to define a complex scalar product on the vector space of complex-valued functions on G:

$$\langle f, \tilde{f} \rangle := \int_G f^*(g)\tilde{f}(g)\,dg.$$

The next proposition shows that characters of unitary irreducible representations form a Hermitian orthonormal subset of the vector space of complex-valued functions on the group. This theoretical result will help us to ascertain that we have found *all* irreducible representations when the characters of the irreducible representations we know span the set of functions invariant under conjugation.

Proposition 6.8 *Suppose G is a group with a volume-one invariant integral $\int_G dg$. Suppose that two finite-dimensional representations (G, V_1, ρ_1) and (G, V_2, ρ_2) are both unitary and irreducible. Let χ_1 and χ_2 be the characters of the representations. Then if V_1 and V_2 are not isomorphic we have $\langle \chi_1, \chi_2 \rangle = 0$, while if $V_1 \cong V_2$ we have $\langle \chi_1, \chi_2 \rangle = 1$.*

For the proof, it is helpful to recall the vector space $\mathrm{Hom}(V_1, V_2)$, the vector space of linear transformations from V_1 to V_2, as well as the subspace $\mathrm{Hom}_G(V_1, V_2)$ of homomorphisms of representations from V_1 to V_2. These were introduced in Section 5.5.

Proof. We calculate the scalar product by constructing a linear operator P whose trace is equal to the scalar product. Consider the representation $(G, \mathrm{Hom}(V_1, V_2), \sigma)$ defined in Proposition 5.12. Let χ denote the character of this representation. By Proposition 5.14 we know that $\chi = \chi_1^* \chi_2$. Consider the linear operator

$$P := \int_G \sigma(g)\, dg.$$

In other words, $P: \mathrm{Hom}(V_1, V_2) \to \mathrm{Hom}(V_1, V_2)$ is defined by

$$PT := \int_G \sigma(g) T\, dg$$

for each $T \in \mathrm{Hom}(V_1, V_2)$.

Next we will show that P is a projection with image $\mathrm{Hom}_G(V_1, V_2)$. To show that P is a projection it suffices to show that $P^2 = P$. But since the integral is invariant under right multiplication we have

$$P^2 = \int_G \sigma(g)\, dg \int_G \sigma(\tilde{g})\, d\tilde{g} = \int_G \int_G \sigma(g\tilde{g})\, dg d\tilde{g}$$

$$= \int_G \sigma(g)\, dg \int_G d\tilde{g} = \int_G \sigma(g)\, dg = P.$$

Now suppose that $T \in \mathrm{Hom}_G(V_1, V_2)$. Then

$$PT = \int_G \sigma(g) T\, dg = \int_G T\, dg = T,$$

so T lies in the image of P. On the other hand, if T lies in the image of P, then there is a U such that $T = PU$, and hence $PT = P^2U = PU = T$, so for any $\tilde{g} \in G$ we have

$$\sigma(\tilde{g})T = \sigma(\tilde{g})PT = \int_G \sigma(\tilde{g}g)T\,dg = \int_G \sigma(g)T\,dg = PT = T.$$

In this case T is fixed by the representation σ and hence T is a homomorphism of representations, i.e., $T \in \mathrm{Hom}_G(V_1, V_2)$. We conclude that the image of P is precisely $\mathrm{Hom}_G(V_1, V_2)$.

By Proposition 2.9, the trace of P must be equal to the dimension of $\mathrm{Hom}_G(V_1, V_2)$. By Proposition 6.4, the dimension of $\mathrm{Hom}_G(V_1, V_2)$ is 0 if V_1 and V_2 are not isomorphic, and 1 if they are isomorphic. Hence

$$\langle \chi_1, \chi_2 \rangle = \int_G \chi_1^*(g)\chi_2(g)\,dg = \int_G \chi(g)\,dg = \int_G \mathrm{Tr}\,\sigma(g)\,dg$$

$$= \mathrm{Tr}\,P = \begin{cases} 0 & V_1 \cong V_2 \\ 1 & \text{otherwise.} \end{cases} \qquad \square$$

As an example of Proposition 6.8, consider the characters χ_0 and χ_1 of the representations of $SU(2)$ on the spaces of constant and degree-one (respectively) homogeneous polynomials of two variables. The proposition implies that $\langle \chi_0, \chi_1 \rangle = 0$. We can check this result by direct calculation: using the formulas from Section 4.6 and Equation 6.2 we have

$$\langle \chi_0, \chi_1 \rangle = \int_{SU(2)} \chi_0^* \chi_1\,dg = \frac{2}{\pi} \int_{-1}^1 (1)(2u)\sqrt{1-u^2}\,du = 0,$$

since the integrand is odd. See Exercise 6.11.

One can also use invariant integration to show that every finite-dimensional representation of a finite group is unitary in some complex scalar product space; see Exercise 6.13. More important for our purposes, invariant integration and Proposition 6.8 are indispensable in our proof in Proposition 6.14, that the representations $(SU(2), \mathcal{P}^n, R_n)$ introduced in Section 4.6 are (up to an isomorphism) the only irreducible unitary representations of $SU(2)$.

6.4   Isotypic Decompositions (Optional)

Just as any natural number can be written uniquely as a product of primes, any representation of a compact group can be written uniquely as a sum of irreducible representations. Such a sum is called an *isotypic decomposition*. The goal of this section is to prove the existence and uniqueness of this decomposition. The results in this section are so widely useful and important that we could not resist including them. However, we will not have occasion to use them in the rest of the text.

We start with a convenient definition. Just as prime powers play a particular role in number theory, Cartesian sums of copies of one irreducible representation play a particular role in representation theory.

Definition 6.4 *Suppose (G, W, ρ) is a group representation and k is a natural number. Set*

$$W^k := \bigoplus_{j=1}^{k} W, \qquad \rho^k := \bigoplus_{j=1}^{k} \rho.$$

Note that (G, W^k, ρ^k) is a group representation.

This definition makes the isotypic decomposition given in Proposition 6.11 easier to write down.

Our first proposition in this section establishes a useful isomorphism.

Proposition 6.9 *Suppose (G, W, ρ) and (G, V, σ) are finite-dimensional representations of the same group G. Set*

$$k := \dim \mathrm{Hom}_G(W, V) \in \mathbb{N}.$$

Then there is an isomorphism of representations

$$W^k \cong \mathrm{Hom}_G(W, V) \otimes W.$$

Proof. Choose a basis $\{T_1, \ldots, T_k\}$ of $\mathrm{Hom}_G(W, V)$. Consider the linear transformation

$$W^k \to \mathrm{Hom}_G(W, V) \otimes W$$
$$(w_1, \ldots, w_k) \mapsto T_1 \otimes w_1 + \cdots + T_k \otimes w_k.$$

Since the T_j's are linearly independent, this linear transformation is injective. Since the T_j's span $\mathrm{Hom}_G(W, V)$, the linear transformation is surjective. Finally, let us show that the linear transformation is a homomorphism of representations. For any $g \in G$ we have

$$\rho^k(g)(w_1, \ldots, w_k) = (\rho(g)w_1, \ldots, \rho(g)w_k)$$
$$\mapsto T_1 \otimes (\rho(g)w_1) + \cdots + T_k \otimes (\rho(g)w_k)$$
$$= (I \otimes \rho)(g)(T_1 \otimes w_1 + \cdots + T_k \otimes w_k),$$

where I denotes the identity operator on $\text{Hom}_G(W, V)$. Recall from Proposition 5.13 that the natural representation of G on $\text{Hom}_G(W, V)$ is trivial. So our injective, surjective linear transformation is a homomorphism of representations. Hence it is an isomorphism of representations.   □

Our second proposition relates the dimension of $\text{Hom}_G(W, V)$ to the size of the largest power of the irreducible representation W appearing inside V.

Proposition 6.10 *Suppose* (G, W, ρ) *is an irreducible representation. Suppose* (G, V, σ) *is a representation of the same group* G *and*

$$k := \dim \text{Hom}_G(W, V) \in \mathbb{N}.$$

Then W^k *is isomorphic to a subrepresentation of* V. *However, for any natural number* k' *such that* $k' > k$, *the representation* $W^{k'}$ *is not isomorphic to a subrepresentation of* V.

In other words, if $k = \dim \text{Hom}_G(W, V)$, then W^k is the largest power of W that occurs as a subrepresentation of V. This result will help with Proposition 6.11. Schur's lemma plays an important role in the proof.

Proof. We show that $\text{Hom}_G(W, V) \otimes W$ is isomorphic to a subrepresentation of V. We define a linear transformation

$$\text{Hom}_G(W, V) \otimes W \to V$$
$$T \otimes w \mapsto Tw.$$

We first show that this linear transformation is a homomorphism of representations. The crucial calculation is, for any $g \in G$,

$$T \otimes \rho(g)w \mapsto T\rho(g)w = \sigma(g)(Tw),$$

where the equality follows from the fact that T is a homomorphism of group representations. Next we show that the homomorphism is injective. Suppose $Tw = 0$. Then $w \in \ker T$. Because W is irreducible we can apply Schur's lemma to conclude that either $T = 0$ or $w = 0$. In either case we conclude that $T \otimes w = 0$. Hence $\text{Hom}_G(W, V) \otimes W$ is isomorphic to a subrepresentation of V.

Next we apply Proposition 6.9, which says that the representation W^k is isomorphic to the representation $\text{Hom}_G(W, V) \otimes W$. Hence W^k is isomorphic to a subrepresentation of V.

In the proof of the final statement, it helps to know the dimension of $\dim \text{Hom}_G(W, W^{k'})$. For $j = 1, \ldots, k'$, we define $T_j : W \to W^k$ by

$$T_j(w) := (0, \ldots, w, \ldots, 0),$$

where only the jth entry can be nonzero. It follows that $\{T_1, \ldots, T_{k'}\}$ is a basis for $\mathrm{Hom}_G(W, W^{k'})$ and hence

$$\dim \mathrm{Hom}_G(W, W^{k'}) = k'.$$

(All we really need to know here is that the T_j's are linearly independent.)

Finally, we must show that if $k' > k$, then $W^{k'}$ is not isomorphic to a subrepresentation of V. We prove the contrapositive. Suppose that $k' \in \mathbb{N}$ and there is an isomorphism of representations from $W_{k'}$ to a subrepresentation of V. Then

$$k = \dim \mathrm{Hom}_G(W, V) \geq \dim \mathrm{Hom}_G(W, W^{k'}) = k'.$$

$$\square$$

Now we prove the existence of the isotypic decomposition for finite-dimensional representations. Just as any natural number has a prime factorization, every finite-dimensional representation of a compact group has an isotypic decomposition. This decomposition tells us what irreducible representations appear as subrepresentations and what their multiplicities are. Note that Proposition 6.11 guarantees uniqueness as well, since the selection of irreducible representations and exponents are uniquely determined.

Proposition 6.11 *Suppose (G, V, ρ) is a finite-dimensional representation of a compact group G. Then there are a finite number of distinct (i.e., not isomorphic) irreducible representations (G, W_j, ρ_j) such that*

$$c_j := \mathrm{Hom}_G(W_j, V) \neq 0.$$

Moreover, the representation (G, V, ρ) is isomorphic to the representation

$$\bigoplus_j W_j^{c_j}. \tag{6.3}$$

This Cartesian sum representation is called the isotypic decomposition *of V. The list of representations W_j and their* multiplicities c_j *is called the* isotype *of V.*

Schur's lemma plays an important role in the proof.

Proof. We proceed by induction on the dimension of V. If V is one dimensional, then it is irreducible. By Schur's lemma (in the guise of Proposition 6.4) we know that $\dim \mathrm{Hom}_G(V, V) = 1$ while $\dim \mathrm{Hom}_G(V, W) = 0$ for every irreducible representation W that is not isomorphic to V. Moreover, the representation (6.3) reduces to V, which is trivially isomorphic to V. This proves the base case of the induction.

Next, fix a natural number n and suppose that the result is known for all natural numbers $k < n$. Because every finite-dimensional representation contains at least one irreducible representation, we can choose one and call it W_0. Set $c_0 := \dim \operatorname{Hom}_G(W_0, V)$. Then by Proposition 6.10 we know that $W_0^{c_0}$ is isomorphic to a subrepresentation U of V. Since the representation V is unitary, we can consider the complementary unitary representation $U^\perp$, whose dimension is strictly less than n.

If $U^\perp = 0$, then $V = U \cong W_0^{c_0}$. By Proposition 6.5, any irreducible representation isomorphic to a subrepresentation of $W_0^{c_0}$ must be isomorphic to W_0. Thus the conclusion holds in this special case.

If $U^\perp \neq 0$, then, by the inductive hypothesis, we have a finite list $W_1, \ldots, W_k$ of distinct irreducible representation such that

$$c_j := \dim \operatorname{Hom}_G(W_j, U^\perp) \neq 0$$

and

$$U^\perp \cong \bigoplus_{j=1}^{k} W_j^{c_j}.$$

We have

$$V \cong U \oplus U^\perp \cong W_0^{c_0} \oplus \left(\bigoplus_{j=1}^{k} W_j^{c_j} \right) \cong \bigoplus_{j=0}^{k} W_j^{c_j}.$$

By Proposition 6.5 we conclude that any irreducible representation W with $\dim \operatorname{Hom}_G(W, V) > 0$ must be one of the W_j's (where j can now take the value 0).

Finally, we must show that the W_j's are distinct. Since $W_1, \ldots, W_k$ arise in the isotypic decomposition of $U^\perp$, they must be distinct. It remains to show that for any $j = 1, \ldots, k$, the representations W_0 and W_j are not isomorphic. Since

$$\dim \operatorname{Hom}_G(W_0, V) = k,$$

Proposition 6.10 implies that there is no injective homomorphisms from $W_0^{k'}$ into V for any natural number $k' > k$. Thus there can be no subrepresentation (other than W_0^k) isomorphic to a power of W_0 in the isotypic decomposition of $U^\perp$. This completes the inductive step. $\square$

Proposition 6.11 has many applications. One is the fact that a character completely determines a representation. Compared to representations, characters are relatively simple objects — complex-valued functions on the group. Yet they carry all the information about the representation T.

Proposition 6.12 *Suppose G is a group with a volume-one invariant integral $\int_G dg$. Suppose that (G, V, ρ) and $(G, \tilde{V}, \tilde{\rho})$ are both finite-dimensional unitary representations. Let χ and $\tilde{\chi}$ be the characters of the representations. Then V is isomorphic to $\tilde{V}$ if and only if $\chi = \tilde{\chi}$.*

Proof. One direction is easy and is left to the reader in Exercise 4.40.

For the other direction, let us suppose that $\chi = \tilde{\chi}$ and show that $\rho \cong \tilde{\rho}$. By Proposition 6.11, we can write the isotypic decomposition of V:

$$V \cong \bigoplus_{j=1}^{k} W_j^{c_j}$$

and apply Proposition 5.7 several times to see that

$$\tilde{\chi} = \chi = \sum_{j=1}^{k} k\chi_j, \tag{6.4}$$

where for each $j = 1, \ldots, k$, the function χ_j is the character of the irreducible representation W_j. But Proposition 6.8 tells us that characters of finite-dimensional unitary representations are linearly independent. Hence the sum in Equation 6.4 is the unique way to express $\tilde{\chi}$ as a sum of characters of unitary irreducible representations. Therefore the isotypic decomposition of $\tilde{V}$ must be the same as the isotypic decomposition of V, i.e.,

$$\tilde{V} \cong \bigoplus_{j=1}^{k} W_j^{c_j} \cong V.$$

$\square$

Proposition 6.11 implies that irreducible representations are the identifiable basic building blocks of all finite-dimensional representations of compact groups. These results can be generalized to infinite-dimensional representations of compact groups. The main difficulty is not with the representation theory, but rather with linear operators on infinite-dimensional vector spaces. Readers interested in the mathematical details ("dense subspaces" and so on) should consult a book on functional analysis, such as Reed and Simon [RS].

6.5 Classification of the Irreducible Representations of $SU(2)$

In this section we classify the finite-dimensional irreducible representations of the Lie group $SU(2)$. First we show that each of the representations R_n defined in Section 4.6 is irreducible. Then we show that there are essentially no other finite-dimensional irreducible representations.

First we show that each representation R_n is irreducible.

Proposition 6.13 *Fix any nonnegative integer n. Then the representation $\left(SU(2), \mathcal{P}^n, R_n\right)$ is irreducible.*

The definition of these representations is given in Section 4.6. For an alternative proof using characters, see Exercise 6.12.

Proof. By Proposition 6.6 it suffices to show that if a linear transformation from $\mathcal{P}^n$ to $\mathcal{P}^n$ commutes with R_n, then that linear transformation is a scalar multiple of the identity. So suppose T is such a linear transformation. Consider the basis $\{x^n, x^{n-1}y, \ldots, xy^{n-1}, y^n\}$ of $\mathcal{P}^n$. We can think of the linear transformation T as a $(n+1) \times (n+1)$ matrix in this basis. Likewise, for any $g \in SU(2)$, we can consider $R_n(g)$ as a $(n+1) \times (n+1)$ matrix. For example, because for each integer $k \in [0, n]$ and each real number θ we have

$$R_n\left(\begin{pmatrix} e^{-i\theta} & 0 \\ 0 & e^{i\theta} \end{pmatrix}\right)(x^{n-k}y^k) = (e^{i\theta}x)^{n-k}(e^{-i\theta}y)^k = e^{i(n-2k)\theta}x^{n-k}y^k,$$

it follows that in our chosen basis the matrix of $R_n\left(\begin{pmatrix} e^{i\theta} & 0 \\ 0 & e^{-i\theta} \end{pmatrix}\right)$ is

$$\begin{pmatrix} e^{-in\theta} & 0 & \cdots & 0 & 0 \\ 0 & e^{i(2-n)\theta} & & 0 & 0 \\ & & \ddots & & \\ 0 & 0 & & e^{i(n-2)\theta} & 0 \\ 0 & 0 & \cdots & 0 & e^{-in\theta} \end{pmatrix}.$$

Notice that we can choose a value of θ such that the entries of this diagonal matrix are distinct. It follows (applying Proposition 2.7) that the matrix of T must commute with this diagonal matrix and hence must be diagonal in the chosen basis. We can now write the matrix of T explicitly:

$$\begin{pmatrix} a_0 & & 0 \\ & \ddots & \\ 0 & & a_n \end{pmatrix};$$

equivalently, note that there are numbers $a_0, \ldots, a_n$ such that for each integer $k \in [0, n]$ we have $T(x^{n-k} y^k) = a_k x^{n-k} y^k$.

To show that all the diagonal entries of the matrix of T are equal, we will consider one particular element of $SU(2)$, namely $g := \frac{\sqrt{2}}{2} \begin{pmatrix} 1 & -1 \\ 1 & 1 \end{pmatrix}$. Note that $R_n(g)x^n = \left(\frac{\sqrt{2}}{2} \right)^n (x+y)^n$ and hence

$$\left(\frac{\sqrt{2}}{2} \right)^n \sum_{k=0}^{n} a_k \begin{pmatrix} n \\ k \end{pmatrix} x^{n-k} y^k = \left(\frac{\sqrt{2}}{2} \right)^n T(x+y)^n$$

$$= T R_n(g)(x^n) = R_n(g) T(x^n) = R_n(g)(a_0 x^n)$$

$$= a_0 \left(\frac{\sqrt{2}}{2} \right)^n (x+y)^n = \left(\frac{\sqrt{2}}{2} \right)^n \sum_{k=0}^{n} a_0 \begin{pmatrix} n \\ k \end{pmatrix} x^{n-k} y^k,$$

where the third equality depends on the hypothesis that T commutes with R_n. We conclude that for all integers $k \in [0, n]$ we have $a_k = a_0$. Hence the matrix of T is diagonal with all diagonal entries equal; i.e., T is a scalar multiple of the identity. Because T was an arbitrary linear transformation commuting with R_n, Proposition 6.6 tells us that the representation $(SU(2), \mathcal{P}^n, R_n)$ is irreducible. □

Our remaining task in this section is to show that our family contains all of the finite-dimensional unitary irreducible representations, without repeats.

Proposition 6.14 *Every finite-dimensional unitary irreducible Lie group representation of $SU(2)$ is isomorphic to $(SU(2), \mathcal{P}^n, R_n)$ for some n. In addition, $(SU(2), \mathcal{P}^n, R_n)$ is isomorphic to $(SU(2), \mathcal{P}^{n'}, R_{n'})$ if and only if $n = n'$.*

In other words, the representations $(SU(2), \mathcal{P}^n, R_n)$, for nonnegative integers n, form a complete list of the finite-dimensional unitary irreducible representations of $SU(2)$, without repeats. Complete lists without repeats are called *classifications*.

Proof. Suppose $(SU(2), V, \rho)$ is a finite-dimensional unitary irreducible Lie group representation. Let χ denote its character. Define the function $f_\chi : [-1, 1] \to \mathbb{C}$ by

$$f_\chi(u) := \chi \left(\begin{pmatrix} u + i\sqrt{1-u^2} & 0 \\ 0 & u - i\sqrt{1-u^2} \end{pmatrix} \right).$$

Since ρ is a Lie group homomorphism, $\chi = \text{Tr} \circ \rho$ is continuous and hence f_χ is continuous. Since $f_\chi(1) = \chi(I) = \dim V$, continuity implies that there is a nontrivial open interval $(a, 1)$ on which $f_\chi \neq 0$. Hence $f_\chi(u)\sqrt{1 - u^2} \neq 0$ for $u \in (a, 1)$, so $f_\chi(u)\sqrt{1 - u^2} \neq 0$ as an element of $C[-1, 1]$. By Proposition 3.8 we conclude that there exists a polynomial p such that

$$\int_{-1}^{1} p^*(u) f_\chi(u)\sqrt{1 - u^2} du \neq 0.$$

Suppose p has the minimum degree of all such polynomials and set $n :=$ deg p. Then for any $k < n$ we have

$$\int_{-1}^{1} u^k f_\chi(u)\sqrt{1 - u^2}\, du = 0.$$

Consider the character χ_n of the representation R_n. Recall from Proposition 4.8 that there is a polynomial q_n such that

$$\chi_n \begin{pmatrix} \alpha & -\beta^* \\ \beta & \alpha^* \end{pmatrix} = q_n(\Re(\alpha)).$$

Because p and q_n are polynomials of the same degree, there is a nonzero complex scalar c such that

$$q_n(u) = cp(u) + \text{lower order terms}.$$

Then we have

$$\int_{SU(2)} \chi_n^* \chi \, dg = \int_{-1}^{1} q_n^*(u) f_\chi(u)\sqrt{1 - u^2} du$$

$$= \int_{-1}^{1} (cp(u) + \text{lower order terms})^* f_\chi(u)\sqrt{1 - u^2} du$$

$$= c \int_{-1}^{1} p^*(u) f_\chi(u)\sqrt{1 - u^2} du$$

$$\neq 0.$$

Note that by Proposition 4.7 we know that R_n is unitary, while ρ is unitary by hypothesis. Hence we can apply Proposition 6.8 to find that the representation $(SU(2), V, \rho)$ is isomorphic to the representation $(SU(2), \mathcal{P}^n, R_n)$.

Since dim $\mathcal{P}^n = n + 1$, as calculated in Section 2.2, we know from Proposition 4.6 that R_n is isomorphic to $R_{n'}$ if and only if $n = n'$. $\qquad\square$

In this section we have shown that the representations on homogeneous polynomials of fixed degree form a complete list of the finite-dimensional unitary irreducible representations of $SU(2)$, with no repeats. In other words, we have classified the finite-dimensional unitary irreducible representations of the group $SU(2)$. In fact, all irreducible representations of $SU(2)$ on complex scalar product spaces are finite dimensional because G is compact. We will not prove this fact; the interested reader might consult Bröcker and tom Dieck [BtD, Chapter III, Corollary 5.8]. In addition, any representation on a complex scalar product space is unitary with respect to some complex scalar product space by Exercise 6.13. So Proposition 6.14 classifies all irreducible representations of $SU(2)$ on complex scalar product spaces. This classification in this section will help us classify the irreducible representations of the group $SO(3)$ in Section 6.6.

6.6 Classification of the Irreducible Representations of $SO(3)$

In this section we classify the finite-dimensional irreducible representations of $SO(3)$. Compared to the work we did classifying the irreducible representations of $SU(2)$ in Section 6.5, the calculation in this section is a piece of cake. However, the reader should note that we use the $SU(2)$ classification in this section. So our classification for $SO(3)$ is not inherently easier. Our trick is to use the group homomorphism $\Phi: SU(2) \rightarrow SO(3)$ (defined in Section 4.3) to show that any representation of $SO(3)$ is just a representation of $SU(2)$ in disguise. At the end of the section we show how to use "weights" to identify irreducible representations.

We can push an irreducible representation $\mathcal{P}^k$ of $SU(2)$ forward to a representation of $SO(3)$ if and only if k is even. If k is odd, then the representation takes different values at I and $-I \in SU(2)$, so there is no good way to define the pushforward of $R_k(I)$ under the group homomorphism Φ. On the other hand, for even k we can push the representation forward:

Proposition 6.15 *Suppose n is a nonnegative even integer. Then we can push the representation $(SU(2), \mathcal{P}^n, R_n)$ forward under the group homomorphism Φ. Let $(SO(3), \mathcal{P}^n, Q_n)$ denote the pushforward representation. Then Q_n is unitary and irreducible.*

Proof. We must first check that Φ and R_n satisfy the hypotheses of Proposition 5.16. By Proposition 4.5, Φ is surjective and its kernel is $\{I, -I\}$. So we must check that $I, -I \in SU(2)$ are both in the kernel of R, i.e., that $R_n(I) = R_n(-I) = I \in \mathcal{GL}(\mathcal{P}^n)$. But for any basis vector $x^k y^{n-k}$ in $\mathcal{P}^n$, we have

$$(R_n(\pm I))\,x^k y^{n-k} = (\pm 1)^k x^k (\pm 1)^{n-k} y_{n-k} = (\pm 1)^n x^k y^{n-k}$$
$$= x^k y^{n-k}.$$

So $R_n(\pm I) = I \in \mathcal{GL}(\mathcal{P}^n)$. Hence we can apply Proposition 5.16 to find that the pushforward representation Q_n is well defined and unitary.

Next we must check that Q_n is irreducible. Suppose W is a subspace of $\mathcal{P}^n$ invariant under Q_n. Then W must be invariant under R_n, since for any $g \in SU(2)$ and $w \in W$ we have, by the definition of the pushforward representation,

$$R_n(g)w = Q_n(\Phi(g))w \in W.$$

Since R_n is irreducible, it follows that W is either the zero subspace or is all of $\mathcal{P}^n$. Hence Q_n is irreducible. $\qquad\square$

The Q_n's are essentially the only finite-dimensional irreducible representations of $SO(3)$.

Proposition 6.16 *Every finite-dimensional, unitary, irreducible representation of $SO(3)$ is isomorphic to Q_n for some even n. In addition, Q_n is isomorphic to $Q_{n'}$ if and only if $n = n'$.*

In this proposition, as in Proposition 6.14, it is possible to drop the hypothesis that the representation be unitary. See Exercise 6.13. We will apply this classification of irreducibles of $SO(3)$ in Section 7.1 to show that for each nonnegative integer n the set of homogeneous harmonic polynomials of degree n forms an irreducible representation of $SO(3)$.

Proof. Suppose $(SO(3), V, \rho)$ is an irreducible unitary representation. By Proposition 5.15, $(SU(2), V, \rho \circ \Phi)$ is also a unitary representation; in addition, $(SU(2), V, \rho \circ \Phi)$ is irreducible. To prove irreducibility of $\rho \circ \Phi$, suppose W is an invariant subspace for $\rho \circ \Phi$. By Proposition 4.5, we know that Φ is surjective onto $SO(3)$; hence the invariance of W under $\rho \circ \Phi$ implies the invariance under ρ: for any $g \in SO(3)$, there is a $\tilde{g} \in SU(2)$ such that $\Phi(\tilde{g}) = g$ and hence for any $w \in W$ we have

$$\rho(g)w = \rho(\Phi(\tilde{g}))w = \rho \circ \Phi(\tilde{g})w \in W,$$

where the inclusion follows from the invariance of W under $\rho \circ \Phi$. Hence by the irreducibility of ρ we conclude that W is either all of V or is the trivial subspace. Since W was an arbitrary invariant subspace, this proves irreducibility of the representation $\rho \circ \Phi$.

It then follows from Proposition 6.14 that there must be a nonnegative integer n such that $\rho \circ \Phi$ is isomorphic to R_n. Since $\Phi(-I) = I \in SO(3)$, we know that $R_n(-I) = I \in \mathcal{GL}(\mathcal{P}^n)$. Hence in particular

$$x^n = I x^n = R_n(-I) x^n = (-1)^n x^n,$$

and so n must be even. Hence there is a nonnegative even integer n such that $\rho \circ \Phi$ is isomorphic to R_n. By the uniqueness of the pushforward (see Exercise 5.3), this implies that $(SO(3), V, \rho)$ is isomorphic to $(SO(3), \mathcal{P}^{2n}, Q_{2n})$.

To prove the final statement of the proposition, note that the dimension of $\mathcal{P}^n$ is $n + 1$. Hence if $n \neq n'$, then Q_n cannot be isomorphic to $Q_{n'}$. $\quad\square$

Finally, we will need a way (other than counting dimensions) to distinguish between the various irreducible representations of $SO(3)$. To this end we define *weights* and *weight vectors*. Weight vectors are certain eigenvectors, and weights give eigenvalues as a function of a parameter. Recall the subgroup $\{X_\theta : \theta \in \mathbb{R}\}$ of $SO(3)$ defined in Section 4.2.

Definition 6.5 *Suppose $(SO(3), V, \rho)$ is a representation and n is an integer. Suppose a nonzero vector $v \in V$ satisfies*

$$\rho(X_\theta)v = e^{in\theta} v$$

for all real θ. Then n is a weight *of the representation ρ, and v is a* weight *vector (of weight n) of the representation.*

For example, consider the representation ρ of $SO(3)$ on $\mathbb{C}^3$ by matrix multiplication. Then the vector $(1, 0, 0)^T$ is a weight vector of weight 0:

$$\rho(X_\theta) \begin{pmatrix} 1 \\ 0 \\ 0 \end{pmatrix} = \begin{pmatrix} 1 & 0 & 0 \\ 0 & \cos\theta & -\sin\theta \\ 0 & \sin\theta & \cos\theta \end{pmatrix} \begin{pmatrix} 1 \\ 0 \\ 0 \end{pmatrix} = \begin{pmatrix} 1 \\ 0 \\ 0 \end{pmatrix}.$$

The vector $(0, 1, -i)^T$ is a weight vector of weight 1:

$$\rho(X_\theta) \begin{pmatrix} 0 \\ 1 \\ -i \end{pmatrix} = \begin{pmatrix} 1 & 0 & 0 \\ 0 & \cos\theta & -\sin\theta \\ 0 & \sin\theta & \cos\theta \end{pmatrix} \begin{pmatrix} 0 \\ 1 \\ -i \end{pmatrix} = e^{i\theta} \begin{pmatrix} 0 \\ 1 \\ -i \end{pmatrix}.$$

Finally, the vector $(0, 1, i)^T$ is a weight vector of weight -1:

$$\rho(X_\theta) \begin{pmatrix} 0 \\ 1 \\ i \end{pmatrix} = \begin{pmatrix} 1 & 0 & 0 \\ 0 & \cos\theta & -\sin\theta \\ 0 & \sin\theta & \cos\theta \end{pmatrix} \begin{pmatrix} 0 \\ 1 \\ i \end{pmatrix} = e^{-i\theta} \begin{pmatrix} 0 \\ 1 \\ -i \end{pmatrix}.$$

Given an even nonnegative even integer n, it is not hard to find the weights and weight vectors of the representation Q_n. Note that

$$\Phi \begin{pmatrix} e^{i\theta/2} & 0 \\ 0 & e^{-i\theta/2} \end{pmatrix} = \mathbf{X}_\theta.$$

Hence for $k = 0, \ldots, n$ we have

$$Q_n\,(\mathbf{X}_\theta)\,x^k y^{n-k} = e^{-i\frac{k}{2}\theta} e^{i(\frac{n-k}{2})\theta} x^k y^{n-k} = e^{i(\frac{n}{2}-k)\theta} x^k y^{n-k},$$

so $x^k y^{n-k}$ is a weight vector of weight $\frac{n}{2} - k$. Because these weight vectors span the vector space $\mathcal{P}^n$, the weights we have found are the only weights for the representation.

Proposition 6.17 *Suppose $(SO(3), V, \rho)$ is a finite-dimensional unitary representation of the group $SO(3)$. Suppose this representation has a vector of weight n. Then $\dim V \geq 2n + 1$.*

We will use this proposition in the proof of Proposition 7.2.

Proof. Let w denote a weight vector of weight n. Let W denote the smallest invariant subspace containing w. Since $w \neq 0$ by the definition of a weight vector, we have $W \neq \{0\}$. Let $\tilde{W}$ be a nontrivial irreducible invariant subspace of W and note that $\tilde{w} := \Pi_{\tilde{W}} w \neq 0$, because otherwise $\tilde{W}^\perp$ would contain w and would be smaller than W, contrary to the definition of W.

Recall from Proposition 5.4 that orthogonal projection onto an invariant subspace of a unitary representation is a homomorphism of representations. Hence for any $\mathbf{X}_\theta$ we have

$$\rho(\mathbf{X}_\theta)\tilde{w} = \rho(\mathbf{X}_\theta)\Pi_{\tilde{W}} w = \Pi_{\tilde{W}}\rho(\mathbf{X}_\theta)w = \Pi_{\tilde{W}} e^{in\theta} w = e^{in\theta} \tilde{w}.$$

So $\tilde{w}$ is a weight vector of weight n.

Now $\tilde{W}$ is a finite-dimensional, unitary, irreducible representation, so by Proposition 6.16 there must be a nonnegative even integer $\tilde{n}$ and an isomorphism $T: \mathcal{P}^{\tilde{n}} \to \tilde{W}$ of representations. Because T is an isomorphism, the list of weights for $\mathcal{P}^{\tilde{n}}$ must be the same as the list of weights for $\tilde{W}$. Hence $-\frac{\tilde{n}}{2} \leq n \leq \frac{\tilde{n}}{2}$. So

$$\dim V \geq \dim W = \tilde{n} + 1 \geq 2n + 1.$$

$\square$

In this section we have classified the finite-dimensional irreducible Lie group representations of $SO(3)$. What about infinite-dimensional irreducible representations? It turns out that there are no infinite-dimensional unitary irreducible representations of any compact Lie group, including $SO(3)$. A proof of this fact can be found in the book of Bröcker and tom Dieck [BtD, Section III.5]. While the discussion there does not completely rule out the existence of nonunitary infinite-dimensional irreducible representations, it makes clear that any infinite-dimensional representation would have to be on a fairly ugly vector space.

6.7   Exercises

Exercise 6.1 *Show that any one-dimensional representation is irreducible.*

Exercise 6.2 *Consider the representation of the circle group* $\mathbb{T}$ *on the complex vector space* $V = \mathbb{C}^3$ *with*

$$\rho: \mathbb{T} \to \mathcal{GL}\left(\mathbb{C}^3\right)$$

$$\lambda \mapsto \begin{pmatrix} 1 & 0 & 0 \\ 0 & \Re(\lambda) & -\Im(\lambda) \\ 0 & \Im(\lambda) & \Re(\lambda) \end{pmatrix}.$$

Find all invariant subspaces of this representation.

Exercise 6.3 *Consider the representation of* $SU(2)$ *on* $\mathbb{C}^2$ *defined by matrix multiplication. Consider the group homomorphism* $\Psi: \mathbb{T} \to SU(2)$ *defined by*

$$\Psi(e^{i\theta}) := \begin{pmatrix} e^{i\theta} & 0 \\ 0 & e^{-i\theta} \end{pmatrix}.$$

Calculate the pullback representation of $\mathbb{T}$ *on* $\mathbb{C}^2$*. Is it irreducible?*

Exercise 6.4 *Suppose* (G, V, ρ) *is a representation and* $w \in V$*. Let* W *denote the span of the set* $\{g \cdot w : g \in G\}$*. Show that* W *is the smallest invariant subspace containing* w*. Give an example to show that* $\{g \cdot w : g \in G\}$ *is not necessarily a subspace. Can you find an example where* $\{g \cdot w : g \in G\}$ *is indeed a subspace?*

Exercise 6.5 *Use Proposition 6.3 to prove that every irreducible representation of the circle group* $\mathbb{T}$ *is one dimensional. Then generalize this result to prove that every irreducible representation of an n-fold product of circles* $\mathbb{T} \times \cdots \times \mathbb{T}$ *(otherwise known as an n-torus) is one dimensional. (As always in this text, representations are complex vector spaces, so "one dimensional" refers to one complex dimension.)*

Exercise 6.6 *Suppose* (G, V, ρ) *is an irreducible representation. Show that for any* $v_0 \in V$ *we have*

$$V = \{\lambda\rho(g)v_0 : \lambda \in \mathbb{C}, g \in G\}.$$

In other words, an irreducible representation is the closure of an orbit under scalar multiplication. Is the converse true? I.e., if a representation satisfies the condition above for every v_0, *is the representation necessarily irreducible? What if the condition is satisfied for one particular* v_0?

Exercise 6.7 *Show that for any* $\lambda \in \mathbb{T}$ *and any integrable function* f *on* $\mathbb{T}$ *we have*

$$\int_0^{2\pi} f(\lambda_0^{-1}e^{i\theta})d\theta = \int_0^{2\pi} f(e^{i\theta})d\theta.$$

(Hint: for any λ_0, *there is a real* θ_0 *such that* $\lambda_0^{-1}e^{i\theta} = e^{i(\theta-\theta_0)}$.) *On the other hand, find a* $\lambda \in \mathbb{T}$ *and an integrable function* f *on* $\mathbb{T}$ *such that*

$$\int_0^{2\pi} f(\lambda_0^{-1}e^{i\theta^3})d\theta \neq \int_0^{2\pi} f(e^{i\theta^3})d\theta.$$

Exercise 6.8 *Recall the Lie group homomorphism* $\Psi : \mathbf{Q} \to SU(2)$ *defined in Section 4.2 and show that for any* $\mathbf{q} \in \mathbf{Q}$ *we have*

$$\Psi(\mathbf{q}^*) = \Psi(\mathbf{q})^*,$$

where the asterisk on the left denotes conjugation of quaternions and the asterisk on the right denotes conjugate transposition on $SU(2)$.

Exercise 6.9 *Use Euler angles to write an explicit formula for invariant integration on* $SO(3)$.

Exercise 6.10 *Show that the invariant integral on* $SU(2)$ *given in Equation 6.1 is invariant under the group action.*

Exercise 6.11 *Calculate a few integrals of products of characters of the representations* R_n *defined in Section 4.6 to confirm (by techniques of just plain calculus) that these characters are mutually orthogonal.*

Exercise 6.12 *Suppose that* G *is a Lie group with a volume-one invariant integral. Suppose that* (G, V, ρ) *is a representation with character* χ. *Show that* ρ *is irreducible if and only if* $\int_G |\chi(g)|^2 \, dg = 1$.

Use this result to give an alternative proof of Proposition 6.13. That is, show that for each nonnegative number n, *the character* χ_n *corresponding to the natural representation of* $SU(2)$ *on* $\mathcal{P}^n$ *satisfies* $\int_G |\chi_n(g)|^2 \, dg = 1$.

Exercise 6.13 *Suppose (G, V, ρ) is a Lie group representation where G is a Lie group with a volume-one invariant integral and V is a complex scalar product space $\langle \cdot, \cdot \rangle$. Then there is a complex scalar product $\langle \cdot, \cdot \rangle_\rho$ on V such that ρ is a unitary representation on V with respect to $\langle \cdot, \cdot \rangle_\rho$. (Hint: define*

$$\langle v, w \rangle_\rho := \int_G \langle \rho(g)v, \rho(g)w \rangle \, dg$$

and check that it is a complex scalar product.)

Exercise 6.14 *Use the Gram–Schmidt technique of orthogonalization to find a recursive formula for an orthogonal basis of $C[-1, 1]$ with the property that the kth basis vector is a polynomial of degree n (for $n = 0, 1, 2, \ldots$). Show (from general principles) that the nth basis element is precisely the character χ_n of the representation of $SU(2)$ on $\mathcal{P}^n$. Use the recursive formula to calculate χ_3 and χ_4.*

7

Representations and the Hydrogen Atom

I was only a child, but I was already aware of it, — *Qfwfq narrated* — I was acquainted with all the hydrogen atoms, one by one, and when a new atom cropped up, I noticed it right away. When I was a kid, the only playthings we had in the whole universe were the hydrogen atoms, and we played with them all the time, I and another youngster my age whose name was Pfwfp.
— Italo Calvino, *Cosmicomics* [Cal, p. 63]

The goal of this chapter is to apply the technology developed in the previous chapters to the study of the hydrogen atom. We have fixed a model of the hydrogen atom: a single particle (the electron) moving in a spherically symmetric space. What experimental predictions does this model make? We will give an answer in Section 7.3. Our answer depends on the fact that the spherical harmonics of any given degree form an irreducible representation of the rotation group $SO(3)$, as shown in Section 7.2. This fact depends in turn on the content of Section 7.1, namely, that homogeneous harmonic polynomials of any fixed degree form an irreducible representation.

7.1 Homogeneous Harmonic Polynomials of Three Variables

In this section we consider the vector spaces $\mathbb{H}^\ell$ of homogeneous harmonic polynomials of degree ℓ in three variables, where ℓ ranges over the nonnegative integers. We show that for every nonnegative integer ℓ, the dimension of $\mathbb{H}^\ell$ is $2\ell + 1$. From Exercises 4.14 and 4.15 we know that every $\mathbb{H}^\ell$ has a natural representation of $SO(3)$; we will show that every $\mathbb{H}^\ell$ is an irreducible subspace for this natural representation of $SO(3)$. In other words, the natural representation of $SO(3)$ on $\mathbb{H}^\ell$ is irreducible.

To calculate the dimension of the vector space $\mathbb{H}^\ell$ for every nonnegative integer ℓ we will use the Fundamental Theorem of Linear Algebra (Proposition 2.5), which we repeat here: if T is a linear transformation from a finite-dimensional vector space V to a finite-dimensional vector space W, then we have

$$\dim V = \dim(\text{kernel } T) + \dim(\text{image } T).$$

Proposition 7.1 *Suppose ℓ is a nonnegative integer. Then the dimension of the vector space $\mathbb{H}^\ell$ of homogeneous harmonic polynomials of degree ℓ in three variables is $2\ell + 1$.*

Proof. Consider the vector spaces $\mathcal{P}_3^\ell$ of homogeneous polynomials of degree ℓ and $\mathcal{P}_3^{\ell-2}$ of homogeneous polynomials of degree $\ell - 2$ in three variables. (Sticklers for rigor should define $\mathcal{P}_3^{-1} := \mathcal{P}_3^{-2} := \{0\}$.) Let ∇_ℓ^2 denote the restriction of the Laplacian $\nabla^2 := \partial_x^2 + \partial_y^2 + \partial_z^2$ to $\mathcal{P}_3^\ell$. By Exercise 2.21 we know that the image of the linear transformation ∇_ℓ^2 lies in $\mathcal{P}_3^{\ell-2}$.

Our goal is to calculate the dimension of the kernel of ∇_ℓ^2, since this kernel consists precisely of $\mathbb{H}^\ell$, the harmonic functions in $\mathcal{P}_3^\ell$. From Section 2.2 we know that the dimension of $\mathcal{P}_3^\ell$ is $\frac{1}{2}(\ell + 1)(\ell + 2)$. So, by the Fundamental Theorem of Linear Algebra (Proposition 2.5) it suffices to calculate the dimension of the image of the the the linear transformation ∇_ℓ^2.

We already know from Exercise 2.21 that this image is contained in $\mathcal{P}_3^{\ell-2}$; we will now show that this image is all of $\mathcal{P}_3^{\ell-2}$. In other words, we will show that the dimension of the image is $\frac{1}{2}(\ell - 1)\ell$ by showing that the restricted Laplacian ∇_ℓ^2 is surjective. Our (slightly informal) argument is based on a triangular arrangement of the monomial bases of the domain and range. Sticklers should see Exercise 7.1.

Consider Figure 7.1. The reader should imagine the corresponding two-dimensional figure for an arbitrary ℓ. The lighter monomials (such as x^4 and xy^2z) form a basis for the domain $\mathcal{P}_3^\ell$ of the restricted Laplacian. The darker monomials (such as x^2 and xy) form a basis for the range $\mathcal{P}_3^{\ell-2}$. The arrows encode some information about the restricted Laplacian. For instance, the two arrows emanating from x^2y^2 encode the fact that $\nabla^2(x^2y^2)$ is a linear combination of y^2 and x^2. The precise recipe for the arrows is as follows. For any monomial $x^i y^j z^k$ of $\mathcal{P}_3^\ell$, expand $\nabla^2(x^i y^j z^k)$ as a linear combination of monomials in $\mathcal{P}_3^{\ell-2}$. No more than three of the monomials in this expansion

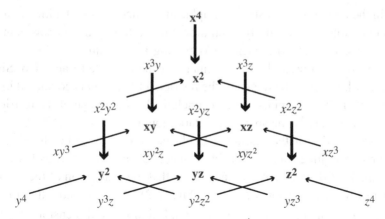

Figure 7.1. The image of basis vectors of $\mathcal{P}_3^4$ under the Laplacian.

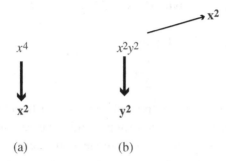

(a)   (b)

Figure 7.2. (a) Arrows emanating from x^4. (b) Arrows emanating from $x^2 y^2$.

will have nonzero coefficients. We draw an arrow from the monomial in $\mathcal{P}_3^\ell$ to each monomial in $\mathcal{P}_3^{\ell-2}$ with a nonzero coefficient. The reader should verify that the pattern of arrows is correct.

We will show surjectivity by showing that every monomial in the range (i.e., every dark monomial; see Figure 7.1) is in the image of the restricted Laplacian. We argue by induction on the rows of the triangular array. In the interest of clarity, we refer to the specific case of $\ell = 4$, but an analogous proof works for any ℓ. We start by considering the single light monomial in the top row of the diagram (x^4). See Figure 7.2(a). The single arrow emanating from x^4 tells us that $\nabla^2(x^4)$ is a nonzero scalar multiple of x^2. So x^2, the top dark monomial is in the image of ∇^2. Similarly, by applying ∇^2 to the second row of light monomials we see that each of the two dark monomials in the second dark row is in the image of ∇^2. Now consider the monomials in the third light row, such as $x^2 y^2$. See Figure 7.2(b). The arrows emanating from $x^2 y^2$ tell us that $\nabla^2(x^2 y^2)$ can be written as the sum of a nonzero scalar multiple of y^2 and a linear combination of dark monomials we already know to be in the image of the restricted Laplacian. Hence y^2 is in the image of the restricted Laplacian. Similarly, each monomial in the third dark row is in the image of the restricted Laplacian. Continuing in the same vein, we can see that every dark monomial is in the image of the restricted Laplacian. Since we already know that the image of the restricted Laplacian is contained in the span of the dark monomials, we can conclude that the image of the restricted Laplacian is precisely the span of the dark monomials, namely, $\mathcal{P}_3^{\ell-2}$.

The surjectivity of the restricted Laplacian allows us to finish our computation of the dimension of the vector space $\mathbb{H}^\ell$ of homogeneous harmonic polynomials of degree ℓ. We already knew that the dimension of the domain of the restricted Laplacian was $\frac{1}{2}(\ell+1)(\ell+2)$. We now know that the dimension of the image of the restricted Laplacian is the dimension of $\mathcal{P}_3^{\ell-2}$, that is, $\frac{1}{2}(\ell-1)\ell$. Hence by Proposition 2.5 the dimension of the space $\mathbb{H}^\ell$ of harmonic homogeneous polynomials of degree ℓ is

$$\dim(\ker \nabla_\ell^2) = \dim \mathcal{P}_3^\ell - \dim \operatorname{Image} \nabla_\ell^2$$
$$= \frac{1}{2}(\ell+1)(\ell+2) - \frac{1}{2}(\ell-1)\ell = 2\ell+1. \quad \square$$

Combining this last result with our knowledge of the classification of the irreducible representations of the group $SO(3)$, we can show that the representation of the rotation group on homogeneous harmonic polynomials of any fixed degree is irreducible.

Proposition 7.2 *Suppose ℓ is a nonnegative integer. Then the natural representation of $SO(3)$ on $\mathbb{H}^\ell$ is irreducible.*

Proof. We will show that there is a polynomial $p \in \mathbb{H}^\ell$ of weight ℓ with respect to the circle group $\mathbb{T}$. Using Proposition 6.17 from Section 6.6 we will conclude that the dimension of one of the irreducible components of $\mathbb{H}^\ell$ must be at least $2\ell+1$. Because the total dimension of $\mathbb{H}^\ell$ is $2\ell+1$, we will conclude that $\mathbb{H}^\ell$ is itself irreducible.

Consider the polynomial $(y-iz)^\ell$, a harmonic polynomial of degree ℓ. Note that the polynomial is indeed harmonic, since all polynomials of degree zero or one are harmonic, while for $\ell \geq 2$,

$$\nabla^2(y-iz)^\ell = \ell(\ell-1)(y-iz)^{\ell-2} + i\ell i(\ell-1)(y-iz)^{\ell-2} = 0.$$

For every real number θ, the corresponding group element $\mathbf{X}_\theta$ acts on polynomials on $\mathbb{R}^3$ by taking

$$x \mapsto x$$
$$y \mapsto (\cos\theta)y + (\sin\theta)z$$
$$z \mapsto (\cos\theta)z - (\sin\theta)y.$$

It follows easily that this action takes the linear polynomial $y - iz$ to $e^{i\theta}(y - iz)$. Hence the action takes the polynomial $(y - iz)^\ell$ to $e^{i\ell\theta}(y - iz)^\ell$:

$$(y - iz)^\ell \mapsto e^{i\ell\theta}(y - iz)^\ell.$$

Thus $(y - iz)^\ell$ has weight ℓ with respect to the action of the given circle subgroup of $SO(3)$. By Proposition 6.17 the dimension of one of the irreducible components of the representation on $\mathbb{H}^\ell$ must be at least $2\ell + 1$. However, the dimension of $\mathbb{H}^\ell$ itself is $2\ell + 1$, so the $(2\ell + 1)$-dimensional irreducible component must be all of $\mathbb{H}^\ell$. Hence $\mathbb{H}^\ell$ is irreducible.   □

Proposition 7.2 is crucial to our proof in Section 7.2 that the spherical harmonics span the complex scalar product space $L^2(S^2)$ of square-integrable functions on the two-sphere.

7.2   Spherical Harmonics

In this section we use the results of Section 7.1 and our knowledge of irreducible representations to show that the spherical harmonic functions span the space $L^2(S^2)$ of square-integrable functions on the two-sphere. In other words, the set of spherical harmonics is a complete set for expansions: we can write any function on the sphere as a (possibly infinite) sum of spherical harmonic functions. This justifies the physicists' practice of using spherical harmonic functions (which have nice properties and are relatively easy to calculate with) to draw conclusions about arbitrary functions of angular variables.

Our first proposition is a useful technical tool for the proof of Proposition 7.4.

Proposition 7.3 *Suppose ℓ is a nonnegative integer. The natural representation of $SO(3)$ on the vector space $\mathcal{P}_3^\ell$ of homogeneous complex-coefficient polynomials in three variables is isomorphic to the Cartesian sum*

$$\mathbb{H}^\ell \oplus \mathbb{H}^{\ell-2} \oplus \cdots \oplus \mathbb{H}^\epsilon$$

of representations, where $\epsilon = (1 - (-1)^{\ell})/2$ (i.e., ϵ is 0 if ℓ is even and 1 if ℓ is odd) and each summand carries the natural representation of $SO(3)$. The isomorphism is given explicitly by

$$\mathbb{H}^{\ell} \oplus \mathbb{H}^{\ell-2} \oplus \cdots \oplus \mathbb{H}^{\epsilon} \quad \rightarrow \quad \mathcal{P}_3^{\ell}$$
$$(p_{\ell}, \ldots, p_{\epsilon}) \quad \mapsto \quad p_{\ell} + r^2 p_{\ell-2} + \cdots + r^{(\ell-\epsilon)} p_{\epsilon},$$

where r^2 denotes multiplication by the sum of the squares of the three variables. (That is, if the variables are x, y and z, then r^2 is multiplication by $x^2 + y^2 + z^2$.)

The fact that multiplication by r^2 is a linear operator follows from Exercise 2.9. Readers familiar with the Fourier transform should note that the linear transformation r^2 is essentially the Fourier transform of the Laplacian ∇^2 (Exercise 7.5). In the proof, we will find it useful to have a natural name for the isomorphism given in the statement of the proposition; we will call this isomorphism

$$1 \oplus r^2 \oplus r^4 \oplus \cdots \oplus r^{(\ell-\epsilon)}.$$

Proof. We use induction on ℓ. For $\ell = 0$ and $\ell = 1$ the result is trivial: $\mathcal{P}_3^0 = \mathbb{H}^0$ and $\mathcal{P}_3^1 = \mathbb{H}^1$; the isomorphism is the identity. Fix any natural number $\ell \geq 2$ and suppose that

$$1 \oplus r^2 \oplus r^4 \oplus \cdots \oplus r^{(\ell-\epsilon)-2}: \mathbb{H}^{\ell-2} \oplus \mathbb{H}^{\ell-4} \oplus \cdots \oplus \mathbb{H}^{\epsilon} \rightarrow \mathcal{P}_3^{\ell-2}$$

is an isomorphism of representations. We wish to show that

$$1 \oplus r^2 \oplus r^4 \oplus \cdots \oplus r^{(\ell-\epsilon)}: \mathbb{H}^{\ell} \oplus \mathbb{H}^{\ell-2} \oplus \cdots \oplus \mathbb{H}^{\epsilon} \rightarrow \mathcal{P}_3^{\ell}$$

is an isomorphism of representations. Consider the linear transformation

$$1 \oplus r^2: \mathbb{H}^{\ell} \oplus \mathcal{P}_3^{\ell-2} \rightarrow \mathcal{P}_3^{\ell}$$
$$(h(x, y, z), p(x, y, z)) \mapsto h(x, y, z) + (x^2 + y^2 + z^2)p(x, y, z).$$

We have

$$1 \oplus r^2 \oplus r^4 \oplus \cdots \oplus r^{(\ell-\epsilon)}$$
$$= (1 \oplus r^2) \circ \left(1 \oplus \left(1 \oplus r^2 \oplus r^4 \oplus \cdots \oplus r^{(\ell-\epsilon)-2}\right)\right).$$

Note that $\left(1 \oplus r^2 \oplus r^4 \oplus \cdots \oplus r^{(\ell-\epsilon)-2}\right)$ is an isomorphism from $\mathbb{H}^{\ell-2} \oplus \cdots \oplus \mathbb{H}^{\epsilon}$ to $\mathcal{P}_3^{\ell-2}$ by the inductive hypothesis. Furthermore, the middle 1 on the right-hand side of the equation above (the 1 to the right of the composition sign ∘) is an isomorphism from $\mathbb{H}^\ell$ to itself. Finally, by Exercise 5.11, the Cartesian sum of isomorphisms is an isomorphism, and by Exercise 4.19 the composition of isomorphisms is an isomorphism. Hence we will be done if we can show that $1 \oplus r^2$ is an isomorphism of representations. It is easy to check that $1 \oplus r^2$ is a homomorphism of representations. We must show that it is injective and surjective.

Let us show injectivity. Note that r^2 is an injective homomorphism of representations. Hence the subspace $r^2\left[\mathcal{P}_3^{\ell-2}\right]$ of $\mathcal{P}_3^\ell$ is an invariant subspace and has dimension equal to the dimension of $\mathcal{P}_3^{\ell-2}$, namely

$$\dim r^2\left[\mathcal{P}_3^{\ell-2}\right] = \frac{1}{2}(\ell-1)\ell.$$

We know that $\mathbb{H}^\ell$ is an irreducible invariant subspace of $\mathcal{P}_3^\ell$ by Proposition 7.2. By Proposition 6.5 and Proposition 7.1 we know that $\mathbb{H}^\ell$ is not isomorphic to any subrepresentation of the Cartesian sum

$$\mathbb{H}^{\ell-2} \oplus \mathbb{H}^{\ell-4} \oplus \cdots \oplus \mathbb{H}^\epsilon.$$

By induction we know that $\mathcal{P}_3^{\ell-2}$ is isomorphic to this Cartesian sum; also, we know that r^2 is injective. Hence $\mathbb{H}^\ell$ is not isomorphic to any subrepresentation of $r^2\left[\mathcal{P}_3^{\ell-2}\right]$. Hence by Proposition 6.7, the subspace $\mathbb{H}^\ell$ must be perpendicular to the subspace $r^2\left[\mathcal{P}_3^{\ell-2}\right]$. We conclude that $1 \oplus r^2(h, p) = 0$ if and only if $h = 0$ and $r^2 p = 0$, if and only if $(h, p) = 0$. So $1 \oplus r^2$ is injective.

Now we will prove surjectivity: we will show that the subspace

$$\mathbb{H}^\ell \oplus r^2\left[\mathcal{P}_3^{\ell-2}\right]$$

of $\mathcal{P}^\ell$ is actually equal to $\mathcal{P}^\ell$. By Proposition 7.1, the dimension of $\mathbb{H}^\ell$ is $2\ell + 1$. Since $\mathbb{H}^\ell$ is perpendicular to $r^2\left[\mathcal{P}_3^{\ell-2}\right]$, the dimension of the Cartesian sum is the sum of the dimensions of the two summands:

$$(2\ell + 1) + \frac{1}{2}(\ell-1)\ell = \frac{1}{2}(\ell+1)(\ell+2) = \dim \mathcal{P}_3^\ell.$$

Hence

$$\mathbb{H}^\ell \oplus r^2\left[\mathcal{P}_3^{\ell-2}\right] = \mathcal{P}_3^\ell.$$

It follows that $1 \oplus r^2 \oplus \cdots \oplus r^{\ell-\epsilon}$ is an isomorphism from $\mathbb{H}^\ell \oplus \cdots \oplus \mathbb{H}^\epsilon$ to $\mathcal{P}_3^\ell$. This completes the inductive step.   □

Proposition 7.3 implies that any polynomial on the two-sphere S^2 in $\mathbb{R}^3$ can be written as a sum of harmonic polynomials. See Exercise 7.3. This fact is important to the proof of Proposition 7.4. The point is that we cannot apply the Stone–Weierstrass theorem directly to harmonic functions (see Exercise 3.22). However, we can apply the Stone–Weierstrass theorem to polynomials. Proposition 7.3 is the link we need.

Recall the vector space $\mathcal{Y}$ of spherical harmonics from Definition 2.6: $\mathcal{Y}$ is the set of restrictions to S^2 of homogeneous harmonic polynomials. Recall also the definition of spanning (Definition 3.7). The set $\mathcal{Y}$ of spherical harmonics spans $L^2(S^2)$:

Proposition 7.4 *In the complex scalar product space $L^2(S^2)$ we have*

$$\mathcal{Y}^\perp = 0.$$

Proof. Suppose $f \in L^2(S^2)$ and $\langle y, f \rangle = 0$ for every $y \in \mathcal{Y}$. We must show that $f = 0$. We show first that f is perpendicular to any homogeneous polynomial. Consider any homogeneous polynomial p, and let ℓ denote the degree of p. By Proposition 7.3, there are harmonic polynomials $h_\ell, h_{\ell-2}, \ldots, h_\epsilon$ (where $\epsilon = 0$ if ℓ is even and $\epsilon = 1$ if ℓ is odd) such that

$$p(x, y, z) = h_\ell(x, y, z) + (x^2 + y^2 + z^2)h_{\ell-2}$$
$$+ \cdots + (x^2 + y^2 + z^2)^{(\ell-\epsilon)/2}h_\epsilon(x, y, z).$$

On the unit sphere S^2 we have $x^2 + y^2 + z^2 = 1$. Because the complex scalar product in $L^2(S^2)$ depends only on the values of the functions on the sphere itself, we have
$$\langle p, f \rangle = \langle h_\ell + \cdots + h_\epsilon, f \rangle = 0,$$

because $h_\ell + \cdots + h_\epsilon \in \mathcal{Y}$.

Finally, since any polynomial is a finite sum of homogeneous polynomials, we conclude that $\langle q, f \rangle = 0$ for any polynomial q. Hence by Proposition 3.9 we have $f = 0$.

We have shown that if f is perpendicular in $L^2(S^2)$ to any $y \in \mathcal{Y}$, then $f = 0$. In other words, $\mathcal{Y}^\perp = 0$ in $L^2(S^2)$.   □

One consequence of this proposition is that any function in $L^2(S^2)$ can be approximated by finite sums of spherical harmonics. Heuristically this means that to prove something about all functions in $L^2(S^2)$, it often suffices to prove it for finite sums of spherical harmonics. This is an enormous simplification, often exploited by physicists. Newcomers to the physics literature might sometimes be confused by the restriction of a problem to spherical harmonic (or other special functions); the point is that solutions in this special case can be used to construct solutions in an enormous class of more complicated cases.

Analogously, the next proposition shows that any function in $L^2(\mathbb{R}^3)$ can be approximated by finite sums of terms of the form $f(r)y(\theta, \phi)$. This justifies the physicists' practice of using separation of variables to solve partial differential equations. (See Section 1.6 for an example of this technique.) Recall the subspace $\mathcal{I}$ of rotation-invariant functions in $L^2(\mathbb{R}^3)$ defined in Section 5.1. Note that there is a natural correspondence between $\mathcal{I} \otimes L^2(S^2)$ and a subset of $L^2(\mathbb{R}^3)$ given by

$$f \otimes y \mapsto fy.$$

In other words, the tensor product of a function f of radius alone and a function y of spherical angles alone is the function $(r, \theta, \phi) \mapsto f(r)y(\theta, \phi)$. Thus the tensor product of $\mathcal{I}$ and $L^2(S^2)$ is a subspace of $L^2(\mathbb{R}^3)$. In fact, it spans $L^2(\mathbb{R}^3)$.

Proposition 7.5 *In the complex scalar product space $L^2(\mathbb{R}^3)$ we have*

$$(\mathcal{I} \otimes \mathcal{Y})^{\perp} = 0.$$

We will use this proposition in the proof of Proposition 7.7 and again in the proof of Proposition A.3.

Proof. Recall the ball B_R of radius R around 0 in $\mathbb{R}^3$, where R is a strictly positive real number. We consider the set

$$\mathcal{I}_R := \left\{ f\big|_{B_R} : f \in \mathcal{I} \right\}.$$

The elements of $\mathcal{I}_R$ are restrictions of rotation-invariant functions to the ball of radius R. We will apply the Stone–Weierstrass Theorem (Theorem 3.2) and Proposition 3.7 to show that $\mathcal{I}_R \otimes \mathcal{Y}$ spans $L^2(B_R)$, which will imply that $\mathcal{I} \otimes \mathcal{Y}$ spans $L^2(\mathbb{R}^3)$.

The set $\mathcal{I}_R \otimes \mathcal{Y}$ satisfies the hypotheses of the Stone–Weierstrass theorem. The set B_R is compact by Exercise 3.30. The set $\mathcal{I}_R \otimes \mathcal{Y}$ is a complex vector space because it is a tensor product of vector spaces. To see that $\mathcal{I}_R \otimes \mathcal{Y}$ is closed under multiplication, it suffices to consider products of elements of the form $f \otimes y$: we have

$$(f_1 \otimes y_1)(f_2 \otimes y_2) = (f_1 f_2) \otimes (y_1 y_2).$$

Since both $\mathcal{I}$ and $\mathcal{Y}$ are complex vector spaces of functions, they are closed under complex conjugation, and hence so is their tensor product. The tensor product separates points, since any two points of different radius can be separated by $\mathcal{I}_R$ and any two points of different spherical angle can be separated by $\mathcal{Y}$. Finally, the function

$$1_R : (r, \theta, \phi) \mapsto \begin{cases} 1, & 0 \le r \le R \\ 0, & R < r \end{cases}$$

is rotation-invariant and square-integrable, so it lies in $\mathcal{I}_R$, while the spherical harmonic function $Y_{0,0}$ is a nonzero constant. Hence for any point (r, θ, ϕ) we have $(1_R \otimes Y_{0,0})(r, \theta, \phi) \neq 0$. Thus $\mathcal{I}_R \otimes \mathcal{Y}$ satisfies all criteria of the Stone–Weierstrass Theorem.

It follows that the conclusion of the Stone–Weierstrass Theorem holds: any continuous function in $L^2(B_R)$ can be uniformly approximated by elements of $\mathcal{I}_R \otimes \mathcal{Y}$. Hence by Proposition 3.7, any element of $L^2(B_R)$ can be approximated in the norm by an element of $\mathcal{I}_R \otimes \mathcal{Y}$.

Now we are ready to show that $(\mathcal{I} \otimes \mathcal{Y})^{\perp} = 0$. For any function $q \in L^2(\mathbb{R}^3)$, let q_R denote the restriction

$$q_R := q\big|_{B_R} \in L^2(B_R).$$

Note that if $q \in L^2(\mathbb{R}^3)$, then $q_R \in L^2(B_R)$. (Sticklers for rigor should see Exercise 7.7.) Similarly, for any $p \in L^2(B_R)$, let $\tilde{p}$ denote the element of $L^2(\mathbb{R}^3)$ defined by

$$\tilde{p}(r, \theta, \phi) := \begin{cases} p(r, \theta, \phi), & 0 \le r \le R \\ 0, & R < r. \end{cases}$$

Now suppose the function $h \in L^2(\mathbb{R}^3)$ satisfies $\langle h, q \rangle = 0$ for every $q \in \mathcal{I} \otimes \mathcal{Y}$. Let R be an arbitrary strictly positive real number. Then h_R is square-integrable, and for any $p \in \mathcal{I}_R \otimes \mathcal{Y}$ we have $\tilde{p} \in \mathcal{I} \otimes \mathcal{Y}$.

$$\langle h_R, p \rangle = \langle h, \tilde{p} \rangle = 0,$$

where the first complex scalar product is taken in $L^2(B_R)$ and the second is taken in $L^2(\mathbb{R}^3)$. Hence the restriction $h_R = 0$. But R was an arbitrary real number, so it follows that $h = 0$. $\qquad \square$

Next we show that the spaces of spherical harmonics of various degrees are the only irreducible subspaces of the natural representation of $SO(3)$ on square-integrable functions on S^2. While *a priori* it seems possible that V might be infinite dimensional, the proposition implies that V is finite dimensional.

Proposition 7.6 *Suppose that V is a nontrivial irreducible invariant subspace of the natural representation of $SO(3)$ on $L^2(S^2)$. Then there is a nonnegative integer ℓ such that $V \cong \mathcal{Y}^\ell$.*

Proof. Since each $\mathcal{Y}^\ell$ is finite-dimensional, we can use Proposition 3.5 to define the orthogonal projection $\Pi_\ell : V \to L^2(S^2)$ onto the subspace $\mathcal{Y}^\ell$. Since V is not trivial, Proposition 7.4 implies that V is not orthogonal to all of the spherical harmonics. Hence there must be at least one ℓ such that the orthogonal projection $\Pi_\ell[V]$ is not trivial.

From Proposition 5.1 we know that $\mathcal{Y}^\ell$ is an invariant subspace. Since the natural representation of $SO(3)$ on $L^2(\mathbb{R}^3)$ is unitary, Proposition 5.4 implies that Π_ℓ is a homomorphism of representations. Since V and $\mathcal{Y}^\ell$ are irreducible, it follows from Schur's Lemma and the nontriviality of $\Pi_\ell[V]$ that Π_ℓ gives an isomorphism of representations from V to $\mathcal{Y}^\ell$.

Proposition 3.5 also guarantees the existence of $\Pi_{(\mathcal{Y}^\ell)^\perp} : V \to L^2(S^2)$, the orthogonal projection of V onto the subspace of $L^2(S^2)$ perpendicular to $\mathcal{Y}^\ell$. Set $W := \Pi_{(\mathcal{Y}^\ell)^\perp}[V]$. By Proposition 6.1, either W is trivial or W is isomorphic to V and hence to $\mathcal{Y}^\ell$. In either case (either by triviality or by Proposition 6.7) W must be perpendicular to $\mathcal{Y}^{\tilde{\ell}}$ for any $\tilde{\ell} \neq \ell$. Furthermore, since W is defined as an image under projection onto the subspace perpendicular to $\mathcal{Y}^\ell$, we know that W is perpendicular to $\mathcal{Y}^\ell$. So W is perpendicular to all the spherical harmonics. In other words, $W \subset \mathcal{Y}^\perp$, so by Proposition 7.4 we know that W must be trivial. Hence $V \subset \mathcal{Y}^\ell$. But V is isomorphic to $\mathcal{Y}^\ell$, so the subspace V has the same dimension as $\mathcal{Y}^\ell$. Hence $V = \mathcal{Y}^\ell$. $\square$

In this section we have verified mathematically what physicists have tested with long use. In spherically symmetric problems in $L^2(\mathbb{R}^3)$, the spherical harmonics of various degrees are the sensible building blocks: they leave nothing out (Proposition 7.5) and they have no substitutes (Proposition 7.6).

7.3 The Hydrogen Atom

In this section we discuss the scientific consequences of our work so far. To better appreciate the results, the reader may wish to review the experimental facts in Section 1.3.

The first statement of Proposition 7.7 below contains the minimum needed to make the strongest experimentally verifiable prediction we can make so far. The second statement, while not verifiable, could be contradicted by experiment. We discuss the physical implications after the proof. In this proposition invariance refers to the natural representation of $SO(3)$ on $L^2(\mathbb{R}^3)$ and its subspaces.

Proposition 7.7 *Suppose $f \in \mathcal{I}$ is nonzero and ℓ is a nonnegative integer. Let F denote the one-dimensional subspace of $\mathcal{I}$ spanned by f. Then $F \otimes \mathcal{Y}^\ell$ is an invariant, irreducible, nontrivial subspace of $L^2(\mathbb{R}^3)$. Furthermore, every invariant, irreducible, nontrivial subspace of $L^2(\mathbb{R}^3)$ has this form.*

Proof. First we show that $F \otimes \mathcal{Y}^\ell$ is invariant, irreducible and nontrivial. Invariance follows because both F and $\mathcal{Y}^\ell$ are invariant. More explicitly, note that because F is one-dimensional, every element of $F \otimes \mathcal{Y}^\ell$ is of the form $f \otimes y$ for some $y \in \mathcal{Y}^\ell$. But we have

$$g \cdot (f \otimes y) = (g \cdot f) \otimes (g \cdot y) = f \otimes \tilde{y},$$

where $\tilde{y} := g \cdot y$ is in $\mathcal{Y}^\ell$ because $\mathcal{Y}^\ell$ is invariant. Hence $F \otimes \mathcal{Y}^\ell$ is invariant. Irreducibility follows from the irreducibility of $\mathcal{Y}^\ell$: suppose W is a nontrivial subspace of $F \otimes \mathcal{Y}^\ell$. Then there is a nontrivial element $y_W \in \mathcal{Y}^\ell$ such that $f \otimes y_W \in W$. But $\mathcal{Y}^\ell$ is irreducible, so for any $y \in \mathcal{Y}^\ell$ we have $f \otimes y \in W$. Hence $W = F \otimes \mathcal{Y}^\ell$. Finally, because $\mathcal{Y}^\ell$ and F are both nontrivial, $F \otimes \mathcal{Y}^\ell$ is nontrivial.

To prove the final statement of the proposition, we define a family of linear transformations from $L^2(\mathbb{R}^3)$ to $L^2(S^2)$. For any function $\alpha \in \mathcal{I}$ we can apply Fubini's Theorem (Theorem 3.1) to define a linear transformation $T_\alpha : L^2(\mathbb{R}^3) \to L^2(S^2)$ by

$$(T_\alpha h)(\theta, \phi) := \int_0^\infty \alpha^*(r) h(r, \theta, \phi) r^2 \, dr.$$

Fubini's Theorem guarantees that for each $h \in L^2(\mathbb{R}^3)$, the function $T_\alpha h$ is well defined as a measurable function (i.e., up to Lebesgue equivalence).

Since $h \in L^2(\mathbb{R}^3)$, we have

$$\int_{\mathbb{R}^3} |h(r, \theta, \phi)|^2 r^2 \sin\theta \, d\theta d\phi dr < \infty,$$

and hence it follows from Fubini's Theorem that the function

$$F : S^2 \to \mathbb{R}$$

$$(\theta, \phi) \mapsto \int_0^\infty |h(r, \theta, \phi)|^2 r^2 dr$$

is well defined as a measurable function on S^2 and satisfies

$$\int_{S^2} F(\theta, \phi) \sin \theta \, d\theta \, d\phi < \infty. \tag{7.1}$$

By the Schwarz inequality (Proposition 3.6) on $L^2(\mathbb{R}^{\geq 0})$ we have

$$|T_\alpha h|^2 (\theta, \phi) = \left| \int_0^\infty \alpha^*(r) h(r, \theta, \phi) r^2 dr \right|^2$$
$$\leq \left(\int_0^\infty |\alpha(r)|^2 r^2 dr \right) F(\theta, \phi).$$

Note that the integral in the parentheses is constant in θ, ϕ, so this inequality, along with Inequality 7.1, implies that $T_\alpha h \in L^2(S^2)$. It is easy to check that T_α is a homomorphism of representations for the natural representations of $SO(3)$ on domain and range.

Now consider the restrictions of the T_α's to any invariant, irreducible, non-trivial subspace $W \subset L^2(\mathbb{R}^3)$. Because W is nontrivial, it contains a function $w \neq 0$. By Proposition 7.5, there must be a function $\alpha_w \in \mathcal{I}$ and a function $y \in \mathcal{Y}$ such that

$$0 \neq \langle \alpha_w \otimes y, w \rangle = \int_{S^2} \int_0^\infty \alpha_w^*(r) y^*(\theta, \phi) w(r, \theta, \phi) r^2 \, dr \, \sin \theta \, d\theta \, d\phi$$
$$= \langle y, T_{\alpha w} w \rangle.$$

Hence $T_{\alpha w}|_W$ is not trivial. Because W is an irreducible invariant subspace, Proposition 7.6 implies that there is a nonnegative integer ℓ such that $T_{\alpha w}[W] = \mathcal{Y}^\ell$.

We can apply Schur's lemma (Proposition 6.2) to see that for any function $\alpha \in \mathcal{I}$ the linear transformation T_α is a constant multiple of $T_{\alpha w}$. Consider the linear transformation $T_\alpha \circ (T_{\alpha w})^{-1} : \mathcal{Y}^\ell \to L^2(S^2)$. By Proposition 6.1, this linear transformation must be either trivial or an isomorphism onto its image. In the first case, $T_\alpha = 0$; in the second case, the image must be $\mathcal{Y}^\ell$ and hence by Proposition 6.3, the linear transformation $T_\alpha \circ (T_{\alpha w})^{-1} : \mathcal{Y}^\ell \to \mathcal{Y}^\ell$ must be a constant multiple of the identity. In either case we have $T_\alpha = cT_{\alpha w}$ for some $c \in \mathbb{C}$.

It remains to find a function $f \in \mathcal{I}$ such that $V = F \otimes \mathcal{Y}^\ell$, where F is the one-dimensional vector space spanned by f. To this end, we choose any element $y \in \mathcal{Y}^\ell$ such that $\|y\| = 1$, define $h_V := (T_{\alpha v})^{-1} y$, and set

$$f(r) := \langle y, h_V(r, \cdot, \cdot) \rangle = \int_{S^2} y^*(\theta, \phi) h_V(r, \theta, \phi) \sin \theta \, d\theta \, d\phi$$

for any nonnegative real number r. We will see that this function f satisfies the conclusion of the theorem.

Consider the linear transformation $U: V \rightarrow L^2(\mathbb{R}^3)$ defined by

$$h \mapsto f T_{\alpha v} h - h.$$

The image of this linear transformation is a subspace of $L^2(\mathbb{R}^3)$; we would like to show that the image is the trivial subspace. Note that the linear transformation U is a homomorphism of representations and its domain is irreducible. Therefore, by Proposition 6.1, either the image of U is trivial or the kernel of U is trivial. Thus to show that the image of U is trivial it suffices to find one nonzero element h of V such that $Uh = 0$.

We will show that $Uh_V = 0$ by showing that for any α we have $T_\alpha(Uh_V) = 0$. By the argument above we know that there is a complex number c such that $T_\alpha h_V = c T_{\alpha v} h_V$ and hence

$$T_\alpha \left(f T_{\alpha v} h_V \right) = T_\alpha \left(f y \right) = \left(\int_0^\infty \alpha(r) f(r) r^2 \, dr \right) y$$

$$= \left(\int_0^\infty \alpha(r) \int_{S^2} y^*(\theta, \phi) h_V(r, \theta, \phi) r^2 \sin\theta \, d\theta \, d\phi \, dr \right) y$$

$$= \left(\int_{S^2} y^*(\theta, \phi) \int_0^\infty \alpha(r) h_V(r, \theta, \phi) r^2 \, dr \sin\theta \, d\theta \, d\phi \right) y$$

$$= \left(\int_{S^2} y^*(\theta, \phi) T_\alpha h_V \sin\theta \, d\theta \, d\phi \right) y$$

$$= c \left(\int_{S^2} y^*(\theta, \phi) T_{\alpha v} h_V \sin\theta \, d\theta \, d\phi \right) y$$

$$= c \left(\int_{S^2} y^*(\theta, \phi) y(\theta, \phi) \sin\theta \, d\theta \, d\phi \right) y$$

$$= c T_{\alpha v} h_V = T_\alpha h_V.$$

Note that the fourth equality depends on Fubini's Theorem (Theorem 3.1), while the other equalities depend on the definitions given in this proof. So

$$T_\alpha(Uh_V) = T_\alpha \left(T_{\alpha v} \left(f \otimes T_{\alpha v} h_V \right) - h_V \right)$$
$$= T_\alpha \left(T_{\alpha v} \left(f \otimes T_{\alpha v} h_V \right) \right) - T_\alpha h_V = 0.$$

Hence $T_\alpha(Uh_V) = 0$ for arbitrary α, so $Uh_V = 0 \in L^2(\mathbb{R}^3)$, implying that the image of U is trivial. From the definition of U we can now see that for any $h \in V$ we have $h = f T_{\alpha v} h$. In other words, for any nonnegative real number r and any element x of the unit sphere S^2 we have

$$h(r, \theta, \phi) = f(r) \left(T_{\alpha v} h \right) (\theta, \phi).$$ □

With the first statement of Proposition 7.7, the $L^2(\mathbb{R}^3)$ model for the motion of the electron in the hydrogen atom implies a specific prediction. Since the nontrivial invariant irreducible subspaces correspond to the elementary states of the hydrogen atom (as we argued in Section 6.2), the proposition implies that every elementary state should have odd dimension.

Experimental evidence corroborates this prediction, up to a factor of two. The shells of the hydrogen atom have dimensions $2 = 2 \times 1$ for s-shells, $6 = 2 \times 3$ for p-shells, $10 = 2 \times 5$ for d-shells, and so on. The accepted physical model that correctly predicts the dimensions of the shells of the electron in the hydrogen atom attributes this factor of two to the *spin* of the electron. We discuss spin in more detail (and more precision) in Chapter 10.

The second statement of Proposition 7.7, corrected by a factor of two for spin, predicts that we should find elementary states of every dimension of the form $2(2\ell + 1)$ where ℓ is a nonnegative integer. This statement cannot be proved experimentally, as it involves an infinite number of states. Yet it is suggestive, especially in hindsight. It is a basic premise of the universally accepted current model of the hydrogen atom. In a similar vein, consider the following corollary of Proposition 7.5.

Proposition 7.8 *The subrepresentation*

$$\bigoplus_{\ell=0}^{\infty} \left(\mathcal{I} \otimes \mathcal{Y}^{\ell} \right)$$

of $L^2(\mathbb{R}^3)$ spans $L^2(\mathbb{R}^3)$. In other words,

$$\left(\bigoplus_{\ell=0}^{\infty} \left(\mathcal{I} \otimes \mathcal{Y}^{\ell} \right) \right)^{\perp} = 0.$$

We leave the proof to the reader in Exercise 7.8.

Proposition 7.8 is appealing when we recall that $\ell = 0$ corresponds to the s-shells, $\ell = 1$ corresponds to p-shells, etc., and the $\mathcal{I}$ consists of radial parts of wave functions. Thus the 0th summand $\mathcal{I} \otimes \mathcal{Y}^0$ corresponds to all combinations of s-shell states, the $\ell = 1$ summand corresponds to all combinations of p states, and so forth.

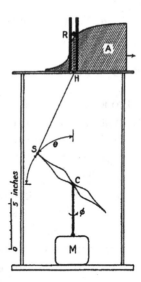

Figure 7.3. When set in motion and photographed, this machine could create images of the probability functions used to model electrons. In particular, it could create images of the spherical harmonics [Wh, Fig. 5].

Note that none of the results of this section mention energy, so that we cannot even predict a certain number of shells at or below a certain energy level. In contrast, the $so(4)$ symmetry of the hydrogen atom, presented in Section 8.6), make predictions about energy levels.

This mathematical model for the probability densities of various electron orbits allowed physicists to develop visualization tools. For example, in 1931, long before computer visualizations were possible, an article in *Physics Review* [Wh] featured a mechanical device (see Figure 7.3) designed to create images of the shapes of the electron orbitals (see Figure 7.4). There are many pictures of electron orbitals available on the internet. See for example [Co].

The results of this section, even with their limitations, are the punch line of our story, the "particularly beautiful goal" promised in the preface. Now is a perfect time for the reader to take a few moments to reflect on the journey. We have studied a significant amount of mathematics, including approximations in vector spaces of functions, representations, invariance, isomorphism, irreducibility and tensor products. We have used some big theorems, such as the Stone–Weierstrass Theorem, Fubini's Theorem and the Spectral Theorem. Was it worth it? And, putting aside any aesthetic pleasure the reader may have experienced, was it worth it from the experimental point of view? In other words, are the predictions of this section worth the effort of building the mathematical machinery?

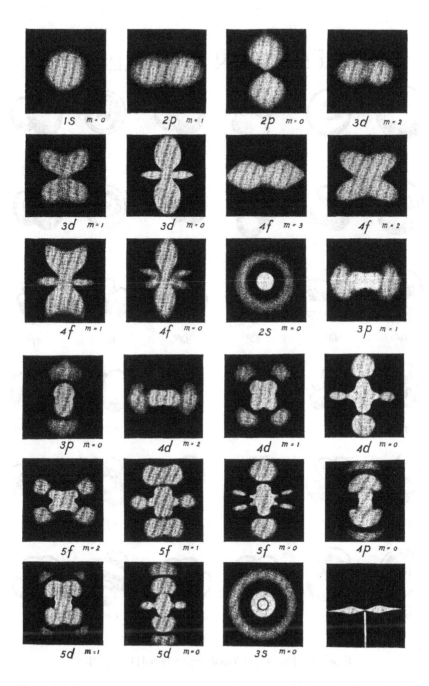

Figure 7.4. Some images created using the machine shown in Figure 7.3 [Wh, Fig. 6].

ELECTRON ORBITALS

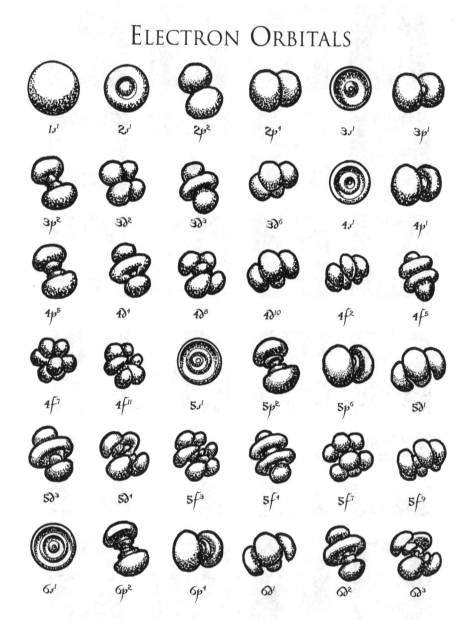

Figure 7.5. Electron orbitals drawn by Tweed [Tw, p. 58].

Before giving a final answer to these questions, the reader should appreciate that this story of the hydrogen atom is only one application of representation theory to quantum physics. The results of this section are not a quirky corner of accidental relevance. Whenever there are equivalent observers of a quantum system, there is room for representation theory. For example, the representation theory of finite groups makes predictions about the spectroscopy of molecules and lattices with symmetry. The representation theory of the Poincaré group predicts that elementary particles in spacetime should be distinguished by a continuous nonnegative parameter (mass) and a discrete nonnegative parameter (spin). The author hopes that our story of the hydrogen atom has given the reader a meaningful taste of one of the great ideas of 19th- and 20th-century mathematical physics.

7.4   Exercises

Exercise 7.1 *In Section 7.1 there is an informal proof that the Laplacian restricted to $\mathcal{P}_3^\ell$ is surjective onto $\mathcal{P}_3^{\ell-2}$ for any nonnegative integer ℓ. Turn this informal proof into a formal proof by induction.*

Exercise 7.2 *Calculate the rank of the restricted Laplacian by finding bases for $\mathcal{P}_3^\ell$ and $\mathcal{P}_3^{\ell-2}$ in which the matrix of the restricted Laplacian is upper triangular. (A matrix M is* upper triangular *if $M_{ij} = 0$ whenever $i > j$.)*

Exercise 7.3 *Write the polynomial $x^2 + y^2$ in $L^2(S^2)$ as a sum of harmonic polynomials.*

Exercise 7.4 *Illustrate Proposition 7.3 by finding a basis of $\mathcal{P}_3^2$ consisting of five harmonic polynomials and one polynomial with a factor of r^2. Find a basis of $\mathcal{P}_3^3$ consisting of seven harmonic polynomials and three polynomials with a factor of r^2.*

Exercise 7.5 (For students of the Fourier transform) *Suppose $f \in L^2(\mathbb{R}^3)$ is twice differentiable. Let $\hat{f}$ denote the Fourier transform of f. Consider the function g defined by*

$$g(x, y, z) := (x^2 + y^2 + z^2)\hat{f}(x, y, z).$$

Show that $g \in L^2(\mathbb{R}^3)$ and that $g = \widehat{(\nabla^2 f)}$.

Exercise 7.6 *The following statements are generalizations of statements proved in Proposition 7.7.*

Suppose (G, V_1, ρ_1) and (G, V_2, ρ_2) are two representations of the same group G.

1. *If both V_1 and V_2 are invariant, show that the representation $V_1 \otimes V_2$ is invariant.*

2. *If both V_1 and V_2 are irreducible, show that the representation $V_1 \otimes V_2$ is irreducible.*

Exercise 7.7 (For students of measure theory) *Prove rigorously that all the claims of the last paragraph of the proof of Proposition 7.5 are true. For example, show that if $q \in L^2(\mathbb{R}^3)$, then q_R is a well-defined element of $L^2(B_R)$.*

Exercise 7.8 (Proposition 7.8) *Prove Proposition 7.8. (Hint: use Definition 2.6 and Proposition 7.5.)*

8

The Algebra *so*(4) Symmetry
of the Hydrogen Atom

We began studying physics together, and Sandro was surprised when I tried to
explain to him some of the ideas that at the time I was confusedly cultivating.
That the nobility of Man, acquired in a hundred centuries of trial and error, lay
in making himself the conqueror of matter, and that I had enrolled in chemistry
because I wanted to remain faithful to this nobility. That conquering matter is
to understand it, and understanding matter is necessary to understanding the
universe and ourselves: and that therefore Mendeleev's Periodic Table, which
just during those weeks we were laboriously learning to unravel, was poetry,
loftier and more solemn than all the poetry we had swallowed down in liceo;
and come to think of it, it even rhymed!
— Primo Levi, *The Periodic Table* [Le, p. 41]

In Chapter 7 we predicted the dimensions of the various shells of the hydrogen atom — the *s*-shells, *p*-shells, and so forth. Except for a missing factor
of two, our predictions match experimental data. However, so far, we made
no predictions about dimensions of *energy levels* for the hydrogen atom. Because the energy is invariant under rotations, our model predicts that every
shell, whether *s*, *p*, *d* or *f*, should lie within one energy level. In fact, something much more remarkable is true of the energy levels of the hydrogen atom.
To a fair amount of precision, most energy levels contain more than one shell.
Here is the pattern:

Energy level	shells	dimension
Lowest	s	2
2nd and 3rd lowest	s, p	8
4th and 5th lowest	s, p, d	18
6th and 7th lowest	s, p, d, f	32

Connoisseurs of the periodic table will notice that the dimensions are equal to the lengths of the rows of the periodic table. Number pattern hounds will notice that the dimensions are twice squares: $2 = 2 \times 1^2, 8 = 2 \times 2^2, 18 = 2 \times 3^2$ and $32 = 2 \times 4^2$. Unlike the dimensions of the shells, which apply to any spherically symmetric quantum system, these dimensions are a consequence of the particular functional form of the Schrödinger operator. The formula $(x^2 + y^2 + z^2)^{-1}$ in the Coulomb potential allows us to find a *hidden symmetry*, that is, an extra symmetry that does not correspond to the spatial symmetry. We will find that every energy eigenspace of the hydrogen atom Hamiltonian must be a representation of the *Lie algebra so*(4). What's more, we will obtain a list of allowable energy levels and, for each allowable energy level, one and only one corresponding irreducible representation of *so*(4). The dimensions of these representations (multiplied by 2 to account for spin) will give us the numbers of elements in each row of the periodic table.

8.1   Lie Algebras

In this section we introduce Lie algebras, Lie algebra homomorphisms and Lie algebra Cartesian sums. In the examples we introduce all the Lie algebras we will need in our study of the hydrogen atom.

Definition 8.1 *A real Lie algebra is a real vector space* $\mathfrak{g}$ *with a bracket operation* $[\cdot, \cdot] \colon \mathfrak{g} \times \mathfrak{g} \to \mathfrak{g}$ *satisfying (for all* $A, B, C \in \mathfrak{g}$ *and* $r, s \in \mathbb{R}$*):*

1. *Asymmetry:* $[A, B] = -[B, A];$

2. *Linearity in both slots:* $[rA + sB, C] = r[A, C] + s[B, C]$ *and* $[A, rB + sC] = r[A, B] + s[A, C];$

3. *The Jacobi identity:* $[A, [B, C]] + [B, [C, A]] + [C, [A, B]] = 0.$

The bracket $[\cdot, \cdot]$ *is called the* Lie bracket.

The Lie bracket is sometimes called the *commutator*. If $[A, B] = 0$, then we say that A *commutes with* B or A and B *commute*.

For example, recall the algebra **Q** of quaternions introduced in Section 1.5. Consider the (real!) three-dimensional subspace $\mathfrak{g}_\mathbf{Q}$ spanned by $\mathbf{i}, \mathbf{j}, \mathbf{k}$. Any element of $\mathfrak{g}_\mathbf{Q}$ can be written $x\mathbf{i} + y\mathbf{j} + z\mathbf{k}$. The usual multiplication of quaternions followed by projection Π onto the subspace $\mathfrak{g}_\mathbf{Q}$ (along the *real line*, i.e., the set of real scalar multiples of 1), is a Lie bracket on $\mathfrak{g}_\mathbf{Q}$. First we show asymmetry:

$$
\begin{aligned}
[x\mathbf{i} + y\mathbf{j} + z\mathbf{k}, \tilde{x}\mathbf{i} + \tilde{y}\mathbf{j} + \tilde{z}\mathbf{k}] &:= \Pi\left((x\mathbf{i} + y\mathbf{j} + z\mathbf{k})(\tilde{x}\mathbf{i} + \tilde{y}\mathbf{j} + \tilde{z}\mathbf{k})\right) \\
&= (y\tilde{z} - z\tilde{y})\mathbf{i} + (z\tilde{x} - x\tilde{z})\mathbf{j} + (x\tilde{y} - y\tilde{x})\mathbf{k} \\
&= -\left((\tilde{y}z - \tilde{z}y)\mathbf{i} + (\tilde{z}x - \tilde{x}z)\mathbf{j} + (\tilde{x}y - \tilde{y}x)\mathbf{k}\right) \\
&= -[\tilde{x}\mathbf{i} + \tilde{y}\mathbf{j} + \tilde{z}\mathbf{k}, x\mathbf{i} + y\mathbf{j} + z\mathbf{k}].
\end{aligned}
$$

Linearity follows from the distributive law and the linearity of the projection. To prove the Jacobi identity, set

$$
\begin{aligned}
A &:= x\mathbf{i} + y\mathbf{j} + z\mathbf{k} \\
\tilde{A} &:= \tilde{x}\mathbf{i} + \tilde{y}\mathbf{j} + \tilde{z}\mathbf{k} \\
\hat{A} &:= \hat{x}\mathbf{i} + \hat{y}\mathbf{j} + \hat{z}\mathbf{k},
\end{aligned}
$$

where we temporarily free the symbol ˆ from its loyalty to the Fourier transform. Then

$$
\begin{aligned}
[\hat{A}, [A, \tilde{A}]] &= [\hat{x}\mathbf{i} + \hat{y}\mathbf{j} + \hat{z}\mathbf{k}, [x\mathbf{i} + y\mathbf{j} + z\mathbf{k}, \tilde{x}\mathbf{i} + \tilde{y}\mathbf{j} + \tilde{z}\mathbf{k}]] \\
&= [\hat{x}\mathbf{i} + \hat{y}\mathbf{j} + \hat{z}\mathbf{k}, (y\tilde{z} - z\tilde{y})\mathbf{i} + (z\tilde{x} - x\tilde{z})\mathbf{j} + (x\tilde{y} - y\tilde{x})\mathbf{k}] \\
&= (\hat{y}(x\tilde{y} - y\tilde{x}) - \hat{z}(z\tilde{x} - x\tilde{z}))\mathbf{i} + (\hat{z}(y\tilde{z} - z\tilde{y}) - \hat{x}(x\tilde{y} - y\tilde{x}))\mathbf{j} \\
&\quad + (\hat{x}(z\tilde{x} - x\tilde{z}) - \hat{y}(y\tilde{z} - z\tilde{y}))\mathbf{k}.
\end{aligned}
$$

Two more calculations of this genre tell us that the coefficient of $\mathbf{i}$ in the expression $[\hat{A}, [A, \tilde{A}]] + [A, [\tilde{A}, \hat{A}]] + [A, [\tilde{A}, \hat{A}]]$ is

$$
\begin{aligned}
(\hat{y}(x\tilde{y} - y\tilde{x}) - \hat{z}(z\tilde{x} - x\tilde{z})) &+ (y(\tilde{x}\hat{y} - \tilde{y}\hat{x}) - z(\tilde{z}\hat{x} - \tilde{x}\hat{z})) \\
&+ (\tilde{y}(\hat{x}y - \hat{y}x) - \tilde{z}(\hat{z}x - \hat{x}z)) = 0,
\end{aligned}
$$

by cancellation of like terms. Similar calculations show that the coefficients of $\mathbf{j}$ and $\mathbf{k}$ are also zero; as a result, we have $[\hat{A}, [A, \tilde{A}]] + [A, [\tilde{A}, \hat{A}]] + [A, [\tilde{A}, \hat{A}]] = 0$, the Jacobi identity.

The calculation of the other two coefficients (of $\mathbf{j}$ and $\mathbf{k}$) are more than just similar to the calculation of the coefficient of $\mathbf{i}$. Notice that our formulas for A, $\tilde{A}$ and $\hat{A}$ are *cyclic*: if we take a formula and replace each $\mathbf{i}$ by $\mathbf{j}$,

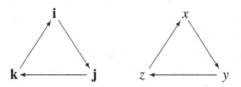

Figure 8.1. A mnemonic for cyclic calculations.

each **j** by **k** and each **k** by **i**, while replacing each x by y, each y by z and each z by x, we get an equivalent formula back. If we replace any statement proven from cyclic formulas by the analogous statement (obtained by making the replacements given above), we get another provable statement. Thus "the coefficient of **i** is 0" implies "the coefficient of **j** is 0", which in turn implies "the coefficient of **k** is 0." This kind of *cyclic calculation* will play a large role in this chapter. See Figure 8.1 for a mnemonic.

A host of examples of Lie algebras can be constructed by considering vector subspaces of a Lie algebra. Not every vector subspace is a Lie algebra, but any subspace *closed under the bracket* is a Lie algebra. That is, if a subspace V satisfies $[A, B] \in V$ for every $A, B \in V$, then V is a Lie algebra. In such a case we can say that V *inherits the Lie algebra structure* from the larger Lie algebra. We call V a *Lie subalgebra* of the larger Lie algebra. A nice big Lie algebra with many interesting subalgebras is the set of $n \times n$ matrices with complex entries. This set has a Lie bracket defined by

$$[A, B] := AB - BA. \tag{8.1}$$

This Lie algebra is usually denoted $g\ell(n, \mathbb{C})$ and is sometimes called the *general linear (Lie) algebra over the complex numbers*. Although this algebra is naturally a complex vector space, for our purposes we will think of it as a real Lie algebra, so that we can take real subspaces.[1] We encourage the reader to check the three criteria for a Lie bracket (especially the Jacobi identity) by direct calculation.

One Lie subalgebra of $gl(2, \mathbb{C})$ is the *special unitary algebra*,

$$su(2) := \left\{ A \in gl(2, \mathbb{C}) : A + A^* = 0, \operatorname{Tr} A = 0 \right\}.$$

---

[1]Because the bracket operation is complex linear in each slot, it is also a *complex Lie algebra*.

It is indeed a real subspace, since both the conditions on A are linear. Also, it is closed under the Lie bracket: if $A, B \in su(2)$ then

$$
\begin{aligned}
[A, B] + [A, B]^* &= AB - BA + (AB)^* - (BA)^* \\
&= A(B + B^*) - (A + A^*)B^* \\
&\quad - (B + B^*)A + B^*(A + A^*) \\
&= 0.
\end{aligned}
$$

Finally, we have

$$
\mathrm{Tr}[A, B] = \mathrm{Tr}(AB) - \mathrm{Tr}(BA) = 0,
$$

where the last equality follows from a general property of the trace (given in Proposition 2.8). So the real vector space $su(2)$ is closed under the Lie bracket, and hence it is a Lie algebra. It is not hard to see that

$$
su(2) = \left\{ \begin{pmatrix} iX & Y + iZ \\ -Y + iZ & -iX \end{pmatrix} : X, Y, Z \in \mathbb{R} \right\}.
$$

Any linear transformation A satisfying the condition $A + A^* = 0$ can be called *anti-Hermitian* or *skew-Hermitian*.

The name $su(2)$ suggests that this algebra might be related to the group $SU(2)$, and indeed it is. We can think of the Lie algebra $su(2)$ as the vector space of possible velocities (at the identity element I) of particles moving inside the Lie group $SU(2)$. Physicists sometimes call these velocities *infinitesimal elements*. In other (more mathematical) words, we can think of the Lie algebra $su(2)$ as the set[2] of derivatives at I of differentiable curves in the Lie group $SU(2)$. Consider a trajectory (a.k.a. a moving particle or a curve) inside the group $SU(2)$. That is, consider a function

$$
g(t) = \begin{pmatrix} u(t) + ix(t) & -y(t) + iz(t) \\ y(t) + iz(t) & u(t) - ix(t) \end{pmatrix}
$$

of time t taking values inside the group $SU(2)$. The functions u, x, y, z are real-valued and satisfy $u(t)^2 + x(t)^2 + y(t)^2 + z(t)^2 = 1$ for every t. Suppose that $u(0) = 1$ and $x(0) = y(0) = z(0) = 0$; then $g(0) = I \in SU(2)$. The constraint on $u(t), x(t), y(t), z(t)$ can be differentiated at $t = 0$ to yield $2u'(0) = 0$. So every derivative $g'(0)$ is of the form

$$
\begin{pmatrix} iX & -Y + iZ \\ Y + iZ & -iX \end{pmatrix},
$$

---

[2]Students of differential geometry may recognize this set as the tangent space to the manifold $SU(2)$ at the point I.

where $X, Y, Z \in \mathbb{R}$.

To prove the converse that every matrix of this form arises as a velocity in $SU(2)$, it is useful to prove a Spectral Theorem for $su(2)$:

Proposition 8.1 (Spectral Theorem for *su*(2)) *Consider an element A of $su(2)$. Then there is a real nonnegative number λ and a matrix $M \in SU(2)$ such that*

$$M^*AM = \begin{pmatrix} i\lambda & 0 \\ 0 & -i\lambda \end{pmatrix}, \tag{8.2}$$

where M^* denotes the conjugate transpose of M.

Note that because M is unitary we can rewrite the conclusion as

$$A = M \begin{pmatrix} i\lambda & 0 \\ 0 & -i\lambda \end{pmatrix} M^{-1}. \tag{8.3}$$

The reader may wish to compare this Spectral Theorem to Proposition 4.4.

Proof. To find the eigenvalues of A, we consider its characteristic polynomial. Then we use eigenvectors to construct the matrix M.

The characteristic polynomial of the matrix

$$A = \begin{pmatrix} iX & -Y+iZ \\ Y+iZ & -iX \end{pmatrix}$$

is $\xi^2 + (X^2 + Y^2 + Z^2)$. Note that either $X^2 + Y^2 + Z^2 = 0$, in which case there is a double root at $\xi = 0$, or $X^2 + Y^2 + Z^2 > 0$, in which case the roots are $\xi = \pm i\sqrt{X^2 + Y^2 + Z^2}$. In the first case, we have $X = Y = Z = 0$, and any $M \in SU(2)$ satisfies the requirements of the theorem (with $\lambda := 0$).

So suppose that $X^2 + Y^2 + Z^2 > 0$. Set $\lambda := \sqrt{X^2 + Y^2 + Z^2}$. Then $i\lambda \neq -i\lambda$. We will build the matrix M from eigenvectors of A. By the definition of eigenvalues, there are nonzero vectors $v, w \in \mathbb{C}^2$ such that $Av = i\lambda v$ and $Aw = -i\lambda w$. Without loss of generality we may assume that $\|v\| = \|w\| = 1$. Since λ is real, it follows from

$$\langle w, v \rangle = \langle Aw, Av \rangle = \langle -i\lambda w, i\lambda v \rangle = -\lambda^2 \langle w, v \rangle$$

that $\langle w, v \rangle = 0$. Define a two-by-two matrix whose columns are v and w:

$$\tilde{M} := \begin{pmatrix} v & w \end{pmatrix}.$$

The matrix $\tilde{M}$ is almost, but not quite, the matrix we need. We have

$$\tilde{M}^* A \tilde{M} = \begin{pmatrix} v^* \\ w^* \end{pmatrix} A \begin{pmatrix} v & w \end{pmatrix} = \begin{pmatrix} v^* \\ w^* \end{pmatrix} \begin{pmatrix} i\lambda v & -i\lambda w \end{pmatrix}$$

$$= \begin{pmatrix} i\lambda & 0 \\ 0 & -i\lambda \end{pmatrix},$$

as desired, but it is possible that $\tilde{M}$ is not in $SU(2)$. We do have

$$\tilde{M}^* \tilde{M} = \begin{pmatrix} \langle v, v \rangle & \langle v, w \rangle \\ \langle w, v \rangle & \langle w, w \rangle \end{pmatrix} = I,$$

but there is no guarantee that $\det \tilde{M} = 1$. A slight modification yields a matrix in $SU(2)$. The calculation above shows that $\tilde{M}$ is invertible and that $\left| \det \tilde{M} \right|^2 = \det \tilde{M}^* \det \tilde{M} = 1$. Hence there must be a complex number γ such that $|\gamma| = 1$ and $\gamma^2 \det \tilde{M} = 1$. Set $M := \gamma \tilde{M}$. Then M satisfies all our conditions:

$$M^* A M = \gamma^* \tilde{M}^* A \tilde{M} \gamma = \tilde{M}^* A \tilde{M} = \begin{pmatrix} i\lambda & 0 \\ 0 & -i\lambda \end{pmatrix},$$

$M^* M = \tilde{M}^* \gamma^* \gamma \tilde{M} = \tilde{M}^* \tilde{M} = I$ and $\det M = \gamma^2 \det \tilde{M} = 1$. So M is an element of $SU(2)$ and satisfies the requirements of the theorem.   □

As a corollary, we can show that every matrix in the algebra $su(2)$ is a velocity at the identity in the group $SU(2)$.

Proposition 8.2 *Suppose $A \in su(2)$. Then for any $t \in \mathbb{R}$ we have $\exp(tA) \in SU(2)$. Furthermore, the derivative of the function $\exp(tA)$ with respect to t at $t = 0$ is A.*

Proof. First consider the special case

$$A = \begin{pmatrix} iX & 0 \\ 0 & -iX \end{pmatrix},$$

where $X \in \mathbb{R}$. Then

$$\exp(tA) = \begin{pmatrix} e^{itX} & 0 \\ 0 & e^{-itX} \end{pmatrix} \in SU(2),$$

since $e^{-iX} = (e^{iX})^*$ and $\left| e^{iX} \right|^2 + 0^2 = 1$. We can calculate the derivative entry by entry:

$$\frac{d}{dt} \begin{pmatrix} e^{itX} & 0 \\ 0 & e^{-itX} \end{pmatrix} = \begin{pmatrix} iX e^{itX} & 0 \\ 0 & -iX e^{-itX} \end{pmatrix}.$$

Evaluating at $t = 0$ we obtain A.

Now consider the general case. Suppose A is an arbitrary element of $su(2)$. Then by the Spectral Theorem (Proposition 8.1) there is a nonnegative real number λ and a matrix $M \in SU(2)$ such that

$$A = M \begin{pmatrix} i\lambda & 0 \\ 0 & -i\lambda \end{pmatrix} M^{-1}. \tag{8.4}$$

It follows that for any real number t we have

$$tA = M \begin{pmatrix} it\lambda & 0 \\ 0 & -it\lambda \end{pmatrix} M^{-1}. \tag{8.5}$$

Hence, by part 3 of Proposition 1.4 we have

$$\exp tA = M \left(\exp t \begin{pmatrix} i\lambda & 0 \\ 0 & -i\lambda \end{pmatrix} \right) M^{-1}.$$

Since each of the three matrices on the right-hand side is in $SU(2)$, so is $\exp tA$.

It remains to differentiate $\exp tA$ with respect to t. Because M and M^{-1} are constant with respect to t, we can apply the calculation above to find that the derivative of $\exp tA$ with respect to t, evaluated at $t = 0$, is

$$M \begin{pmatrix} i\lambda & 0 \\ 0 & -i\lambda \end{pmatrix} M^{-1} = A. \tag{8.6}$$

$\square$

So the Lie algebra $su(2)$ is indeed the set of derivatives at I of differentiable curves in the Lie group $SU(2)$. This situation is common enough to merit its own terminology: we say that $su(2)$ is *the Lie algebra associated to $SU(2)$*, or, more succinctly, that $su(2)$ is *the Lie algebra of $SU(2)$*. To avoid confusion in oral discussions, one can refer to "the algebra $su(2)$" or "little $su(2)$" and "the group $SU(2)$" or "big $SU(2)$." Readers interested in the general theory of Lie groups and Lie algebras should note that there is a unique Lie algebra associated to any given Lie group [Wa, Proposition 3.7]; however, a Lie algebra does not determine a Lie group uniquely, as the reader may show in Exercise 8.5.

The Lie algebras $so(3)$ and $so(4)$ will be useful to us. For any natural number n one can define

$$so(n) := \left\{ A \in gl(n, \mathbb{C}) : A + A^T = 0 \text{ and all entries of } A \text{ are real} \right\}.$$

It is easy to see that $so(n)$ is a real vector subspace of $gl(n, \mathbb{C})$. Also, if $A, B \in so(n)$, then all entries of the matrix $[A, B]$ are real and

$$
\begin{aligned}
[A, B] + [A, B]^T &= AB - BA + (AB)^T - (BA)^T \\
&= AB - BA + B^T A^T - A^T B^T \\
&= A(B + B^T) - (A + A^T)B^T \\
&\quad - (B + B^T)A + B^T(A + A^T) \\
&= 0.
\end{aligned}
$$

So for any n, the real vector space $so(n)$ is a Lie algebra. For example,

$$
so(3) = \left\{ \begin{pmatrix} 0 & -Z & Y \\ Z & 0 & -X \\ Y & -X & 0 \end{pmatrix} : X, Y, Z \in \mathbb{R} \right\}.
$$

a vector space of real dimension 3.

The notions of *Lie algebra homomorphism* and *Lie algebra isomorphism* will be important to us.

Definition 8.2 *Suppose $\mathfrak{g}_1$ and $\mathfrak{g}_2$ are Lie algebras with bracket operations $[\cdot, \cdot]_1$ and $[\cdot, \cdot]_2$, respectively. Suppose T is a linear transformation from $\mathfrak{g}_1$ to $\mathfrak{g}_2$. Then T is a* Lie algebra homomorphism *if it respects the Lie bracket, i.e., if*

$$
[TA, TB]_2 = T([A, B]_1)
$$

for every $A, B \in \mathfrak{g}_1$. If T is injective and surjective, then T is a Lie algebra isomorphism; *in this case we write $\mathfrak{g}_1 \cong \mathfrak{g}_2$ and we say that $\mathfrak{g}_1$ is isomorphic to $\mathfrak{g}_2$.*

To define a Lie algebra homomorphism, it suffices to define it on basis elements of $\mathfrak{g}_1$ and check that the commutation relations are satisfied. Because the homomorphism is linear, it is defined uniquely by its value at basis elements. Because the bracket is linear, if the brackets of basis elements satisfy the equality in Definition 8.7, then any linear combination of basis elements will satisfy equality in Definition 8.7.

For example, the three-dimensional subspace $\mathfrak{g}_Q$ of the quaternions introduced above is isomorphic to $su(2)$, and both are isomorphic to $so(3)$. Define

$T_1 \colon \mathfrak{g}_Q \to su(2)$ by

$$T_1(\mathbf{i}) = \frac{1}{2}\begin{pmatrix} i & 0 \\ 0 & -i \end{pmatrix}$$

$$T_1(\mathbf{j}) = \frac{1}{2}\begin{pmatrix} 0 & 1 \\ -1 & 0 \end{pmatrix}$$

$$T_1(\mathbf{k}) = \frac{1}{2}\begin{pmatrix} 0 & i \\ i & 0 \end{pmatrix}.$$

The reader should check that Definition 8.7 is satisfied and notice that those factors of $1/2$ are necessary. Thus T_1 is a Lie algebra homomorphism. To see that it is an isomorphism, note that the matrices on the three right-hand sides of the defining equations for T_1 form a basis of $su(2)$. Similarly, defining $T_2 \colon \mathfrak{g}_Q \to so(3)$ by

$$T_2(\mathbf{i}) = \begin{pmatrix} 0 & 0 & 0 \\ 0 & 0 & -1 \\ 0 & 1 & 0 \end{pmatrix}$$

$$T_2(\mathbf{j}) = \begin{pmatrix} 0 & 0 & 1 \\ 0 & 0 & 0 \\ -1 & 0 & 0 \end{pmatrix}$$

$$T_2(\mathbf{k}) = \begin{pmatrix} 0 & -1 & 0 \\ 1 & 0 & 0 \\ 0 & 0 & 0 \end{pmatrix}$$

yields an isomorphism of Lie algebras. Readers should check that T_2 is injective, surjective, and respects the brackets. Finally, we can conclude that $T_2 \circ T_1^{-1}$ is injective, surjective, and respects the brackets, so $su(2)$ is isomorphic to $so(3)$.

Not all three-dimensional Lie algebras are isomorphic. For example, consider $\mathbb{R}^3$ with the *trivial Lie bracket*: define $[v, w] := 0$ for all $v, w \in \mathbb{R}^3$. Then for any Lie algebra homomorphism $T \colon su(2) \to \mathbb{R}^3$ we have

$$T(\mathbf{k}) = [T(\mathbf{i}), T(\mathbf{j})] = 0,$$

so T is not injective, and hence is not an isomorphism of Lie algebras. In fact, by a cyclic argument, $T(\mathbf{i}) = T(\mathbf{j}) = 0$ as well, so the only Lie algebra homomorphism from $su(2)$ to the commutative algebra $\mathbb{R}^3$ is the trivial homomorphism.

Another example of a three-dimensional Lie algebra is the *Heisenberg algebra*. This algebra consists of the set

$$\mathcal{H} := \left\{ \begin{pmatrix} 0 & p & r \\ 0 & 0 & q \\ 0 & 0 & 0 \end{pmatrix} : p, q, r \in \mathbb{R} \right\},$$

with the usual matrix Lie bracket.

When two or more Lie algebras are isomorphic, it is common practice to call them "equal" or "the same." For example, we will refer to $\mathfrak{g}_Q$, $so(3)$ and $su(2)$ as the "same" algebra and use the shorthand **i** for $T_2(\mathbf{i})$ or $T_1(\mathbf{i})$, etc., when the context precludes confusion. This Lie algebra shows up in yet another guise in many physics texts, where one encounters triples of operators, say, $\hat{\mathbf{J}}_x$, $\hat{\mathbf{J}}_y$, $\hat{\mathbf{J}}_z$, satisfying commutation relations

$$\left[\hat{\mathbf{J}}_x, \hat{\mathbf{J}}_y\right] = i\hbar\hat{\mathbf{J}}_z, \quad \left[\hat{\mathbf{J}}_y, \hat{\mathbf{J}}_z\right] = i\hbar\hat{\mathbf{J}}_x, \quad \left[\hat{\mathbf{J}}_z, \hat{\mathbf{J}}_x\right] = i\hbar\hat{\mathbf{J}}_y.$$

In physics applications these operators are usually *Hermitian*, i.e., they satisfy $\langle Hv, w \rangle = \langle v, Hw \rangle$ for all vectors v and w. We can define an isomorphism of Lie algebras by

$$T_3(\mathbf{i}) := \frac{1}{i\hbar}\hat{\mathbf{J}}_x, \quad T_3(\mathbf{j}) := \frac{1}{i\hbar}\hat{\mathbf{J}}_y, \quad T_3(\mathbf{k}) := \frac{1}{i\hbar}\hat{\mathbf{J}}_z.$$

Note that this definition yields the correct bracket relations. For example,

$$[T_3(\mathbf{i}), T_3(\mathbf{j})] = \frac{1}{(i\hbar)^2}\left[\hat{\mathbf{J}}_x, \hat{\mathbf{J}}_y\right] = \frac{1}{i\hbar}\hat{\mathbf{J}}_z = T_3(\mathbf{k}).$$

Such triples of operators are often called "angular momentum operators" or "generators of angular momentum." Sometimes they are indeed related to actual mechanical angular momentum; more often, the label "angular momentum" is the physicists' way of saying that the operators satisfy the commutation relations given above.

In our analysis of the Lie algebra $so(4)$ we will use the Cartesian sum of Lie algebras.

Definition 8.3 *Suppose $\mathfrak{g}_1$ and $\mathfrak{g}_2$ are Lie algebras, with brackets $[\cdot, \cdot]_1$ and $[\cdot, \cdot]_2$, respectively. Then the Cartesian sum $\mathfrak{g}_1 \oplus \mathfrak{g}_2$ of vector spaces is a Lie algebra with bracket operation defined by*

$$[(A_1, A_2), (B_1, B_2)] := ([A_1, B_1]_1, [A_2, B_2]_2).$$

We will find it useful to know that $so(4)$ is isomorphic to $su(2) \oplus su(2)$.

Proposition 8.3 *There is a Lie algebra isomorphism from $su(2) \oplus su(2)$ to $so(4)$.*

Proof. First we define an isomorphism $S: \mathfrak{g}_Q \oplus Q \to so(4)$ of Lie algebras by

$$
S(\mathbf{i}, 0) := \frac{1}{2} \begin{pmatrix} 0 & 1 & 0 & 0 \\ -1 & 0 & 0 & 0 \\ 0 & 0 & 0 & -1 \\ 0 & 0 & 1 & 0 \end{pmatrix} \qquad S(0, \mathbf{i}) := \frac{1}{2} \begin{pmatrix} 0 & -1 & 0 & 0 \\ 1 & 0 & 0 & 0 \\ 0 & 0 & 0 & -1 \\ 0 & 0 & 1 & 0 \end{pmatrix}
$$

$$
S(\mathbf{j}, 0) := \frac{1}{2} \begin{pmatrix} 0 & 0 & 1 & 0 \\ 0 & 0 & 0 & 1 \\ -1 & 0 & 0 & 0 \\ 0 & -1 & 0 & 0 \end{pmatrix} \qquad S(0, \mathbf{j}) := \frac{1}{2} \begin{pmatrix} 0 & 0 & -1 & 0 \\ 0 & 0 & 0 & 1 \\ 1 & 0 & 0 & 0 \\ 0 & -1 & 0 & 0 \end{pmatrix}
$$

$$
S(\mathbf{k}, 0) := \frac{1}{2} \begin{pmatrix} 0 & 0 & 0 & 1 \\ 0 & 0 & -1 & 0 \\ 0 & 1 & 0 & 0 \\ -1 & 0 & 0 & 0 \end{pmatrix} \qquad S(0, \mathbf{k}) := \frac{1}{2} \begin{pmatrix} 0 & 0 & 0 & -1 \\ 0 & 0 & -1 & 0 \\ 0 & 1 & 0 & 0 \\ 1 & 0 & 0 & 0 \end{pmatrix}.
$$

To confirm that this is an isomorphism of Lie algebras, note first that S is a well-defined linear transformation (by Proposition 2.3). Then check that it is a homomorphism of Lie algebras by checking all bracket relations between the matrices above. We leave this verification mostly to the reader, giving just one example:

$$
[S(\mathbf{j}, 0), S(\mathbf{k}, 0)] = \left[\frac{1}{2} \begin{pmatrix} 0 & 0 & 1 & 0 \\ 0 & 0 & 0 & 1 \\ -1 & 0 & 0 & 0 \\ 0 & -1 & 0 & 0 \end{pmatrix}, \frac{1}{2} \begin{pmatrix} 0 & 0 & 0 & 1 \\ 0 & 0 & -1 & 0 \\ 0 & 1 & 0 & 0 \\ -1 & 0 & 0 & 0 \end{pmatrix} \right]
$$

$$
= \frac{1}{2} \begin{pmatrix} 0 & 1 & 0 & 0 \\ -1 & 0 & 0 & 0 \\ 0 & 0 & 0 & -1 \\ 0 & 0 & 1 & 0 \end{pmatrix} = S(\mathbf{i}, 0) = S([\mathbf{j}, \mathbf{k}]).
$$

Because the six matrices above form a vector space basis of the Lie algebra $so(4)$, the Lie algebra homomorphism S is injective and surjective, i.e., it is an isomorphism.

Because S and T_1 are both isomorphisms of Lie algebras, so is

$$S \circ (T_1 \oplus T_1)^{-1} : su(2) \oplus su(2) \to so(4)$$
$$(x, y) \mapsto S(T^{-1}x, T^{-1}y).$$

Thus the Lie algebra $su(2) \oplus su(2)$ is isomorphic to the Lie algebra $so(4)$. □

Lie algebras are "infinitesimal" versions of Lie groups. Because symmetries of physical systems give rise to Lie groups, we can think of Lie algebras as *infinitesimal symmetries*. For a very physical presentation of the Lie algebra $so(3)$, see the section entitled "Infinitesimal Rotation" in Goldstein's mechanics textbook [Go]. While Lie groups can have rich nonlinear global structure, Lie algebras are linear spaces and are therefore often easier to work with. Yet the representation theory of Lie algebras is almost as powerful as the representation theory of Lie groups, as we will see below: finding a representation of the Lie algebra $so(4)$ on a space of physically interesting states of the Schrödinger operator will yield a very strong prediction about the dimensions of the shells of the hydrogen atom.[3]

8.2   Representations of Lie Algebras

Like Lie groups, Lie algebras have representations. In this section we define and discuss these representations. In the examples we develop facility calculating with partial differential operators. Finally, we prove Schur's Lemma along with two propositions used to construct subrepresentations.

Suppose V is a complex vector space. Let $g\ell(V)$ denote the vector space[4] of all complex linear transformations from V to V. Then we can define a Lie bracket on $g\ell(V)$ by $[A, B] := AB - BA$.

Definition 8.4 *A Lie algebra homomorphism ρ from a Lie algebra $\mathfrak{g}$ to $g\ell(V)$ is called a* representation *of $\mathfrak{g}$ on V.*

By analogy with our notation for group representations, we denote a representation by a triple $(\mathfrak{g}, V, \rho)$ or, when the rest is clear from context, simply by V or ρ. As for groups, we define homomorphisms and isomorphisms of representations.

---

[3]Fock's analysis, using Lie groups instead of algebras, is stronger, as it implies Proposition 8.14 rather than relying on it. See Chapter 9.

[4]Not to be confused with the group $\mathcal{GL}(V)$ of all invertible linear transformations from V to V, which is not a vector space.

Definition 8.5 *Suppose* $(\mathfrak{g}, V, \rho)$ *and* $(\mathfrak{g}, \tilde{V}, \tilde{\rho})$ *are two representations of one Lie algebra* $\mathfrak{g}$*. Suppose* $T: V \rightarrow \tilde{V}$ *is a linear transformation such that for any* $v \in V$ *and any* $A \in \mathfrak{g}$ *we have*

$$T \circ \rho(A) = \tilde{\rho}(A) \circ T.$$

Then we say that T *is a* homomorphism of (Lie algebra) representations. *If in addition* T *is injective and surjective then we say that* T *is an* isomorphism of (Lie algebra) representations *and that* ρ *is* isomorphic to $\tilde{\rho}$*.*

Partial differential operators will play a large role in the examples of Lie algebra representations that concern us. Hence it behooves us to consider partial derivative calculations carefully. Consider a simple example:

$$\partial_x(x\partial_y) = \partial_y + x\partial_x\partial_y \neq x\partial_x\partial_y = (x\partial_y)\partial_x. \tag{8.7}$$

To understand the first equality, we apply each term to a function $f(x, y, z)$. Correct application of the product rule for derivatives yields, for example,

$$\begin{aligned}
\partial_x(x\partial_y)f(x, y, z) &= \partial_x\left(x\frac{\partial f}{\partial y}(x, y, z)\right) \\
&= (\partial_x(x))\left(\frac{\partial f}{\partial y}(x, y, z)\right) + x\partial_x\left(\frac{\partial f}{\partial y}(x, y, z)\right) \\
&= \left(\frac{\partial f}{\partial y}(x, y, z)\right) + x\left(\frac{\partial^2 f}{\partial_x\partial_y}(x, y, z)\right).
\end{aligned}$$

As a general rule, when a calculation with differential operators proves mysterious, it is often helpful to apply the operators in question to an arbitrary function. This example shows that composition of partial differential operators is not commutative. The point is that when one variable is used both for differentiation and in a coefficient, the product rule for multiplication yields an extra term.

Beware of confusing the composition of a differential operator and a *multiplication operator*, such as the operator x that takes an arbitrary function $f(x, y, z)$ to the function $xf(x, y, z)$, with the application of the differential operator to a function. For example, if $g(x, y, z)$ is a function to which we wish to apply the differential operator ∂_y, we might write

$$\partial_y(g(x, y, z)) = \frac{\partial g}{\partial y}(x, y, z),$$

where each side of the equation is to be understood as a function. On the other hand, if we wish to compose the operator ∂_y with the multiplication operator taking any $f(x, y, z)$ to $g(x, y, z)f(x, y, z)$, we might write

$$\partial_y(g(x, y, z)) = \frac{\partial g}{\partial y}(x, y, z) + g(x, y, z)\partial_y,$$

where each side of the equation is to be understood as an operator on functions. We will adopt the standard practice of putting the burden on the reader to choose the correct calculation. See Exercise 8.13.

There is a natural representation of the Lie algebra $so(3)$ using partial differential operators on $L^2(\mathbb{R}^3)$. We can define the three basic *angular momentum operators* as linear transformations on $L^2(\mathbb{R}^3)$ as follows:

$$\mathbf{L_i} := z\partial_y - y\partial_z,$$
$$\mathbf{L_j} := x\partial_z - z\partial_x,$$
$$\mathbf{L_k} := y\partial_x - x\partial_y.$$

Some readers may rightly object that these partial differential operators are undefined on many elements of $L^2(\mathbb{R}^3)$, namely, functions that are not sufficiently differentiable. To define these operators precisely, we let $W^\infty(\mathbb{R}^3)$ denote the (dense) subspace of infinitely differentiable functions[5] in $L^2(\mathbb{R}^3)$ all of whose derivatives are also in $L^2(\mathbb{R}^3)$; we define a function $\mathbf{L}: su(2) \to g\ell\left(W^\infty(\mathbb{R}^3)\right)$ by

$$\mathbf{L}(c_i\mathbf{i} + c_j\mathbf{j} + c_k\mathbf{k}) := c_i\mathbf{L_i} + c_j\mathbf{L_j} + c_k\mathbf{L_k}.$$

Physicists call $\mathbf{L}$ the *total angular momentum*. To check that it is a Lie algebra homomorphism, we must check that the Lie brackets behave properly. They

---

[5]Some readers may wonder why we make this restriction, especially if they have experience applying angular momentum operators to discontinuous physical quantities. It is possible, with some effort, to make mathematical sense of the angular momentum of a discontinuous quantity but, as the purposes of the text do not require the result, we choose not to make the effort. Compare spherical harmonics, which are effective because physicists know how to extrapolate from spherical harmonics to many cases of interest by taking linear combinations; likewise, dense subspaces are useful because mathematicians know how to extrapolate from dense subspaces to the desired spaces.

do: for any $f \in W^\infty(\mathbb{R}^3)$ we have

$$
\begin{aligned}
[\mathbf{L_i}, \mathbf{L_j}]f(x, y, z) &= \mathbf{L_i L_j} f(x, y, z) - \mathbf{L_j L_i} f(x, y, z) \\
&= (z\partial_y - y\partial_z)(x\partial_z - z\partial_x)f(x, y, z) \\
&\quad - (z\partial_x - x\partial_z)(y\partial_z - z\partial_y)f(x, y, z) \\
&= (y\partial_x - x\partial_y)f(x, y, z) \\
&= \mathbf{L_k} f(x, y, z),
\end{aligned}
$$

where the second-to-last equality follows from a careful calculation obeying the rules of differentiation. Hence $[\mathbf{L_i}, \mathbf{L_j}] = \mathbf{L_k}$. Because the set of defining formulas for the angular momentum operators is a cyclic formula we also have, by cyclic reasoning,[6] that $[\mathbf{L_j}, \mathbf{L_k}] = \mathbf{L_i}$ and $[\mathbf{L_k}, \mathbf{L_i}] = \mathbf{L_j}$.

We can define invariant subspaces, subrepresentations and irreducible representations exactly as we did for groups.

Definition 8.6 *Suppose* $\mathfrak{g}$ *is an arbitrary Lie algebra and* $(\mathfrak{g}, V, \rho)$ *is a Lie algebra representation. A subspace W of V is an* invariant subspace *for* ρ *if* $\rho(A)w \in W$ *for every* $A \in \mathfrak{g}$ *and every* $w \in W$. *If W is an invariant subspace for* ρ, *then the representation* $\rho_W : \mathfrak{g} \to g\ell(V)$ *defined by*

$$
\rho_W(A) := \rho(A)\Big|_W
$$

is called a subrepresentation *of* ρ. *If V and* {0} *are the only invariant subspaces of V, then we say that* $(\mathfrak{g}, V, \rho)$ *is an* irreducible representation.

All of the results of Section 6.1 apply, *mutatis mutandis*, to irreducible Lie algebra representations. For example, if T is a homomorphism of Lie algebra representations, then the kernel of T and the image of T are both invariant subspaces. This leads to Schur's Lemma for Lie algebra representations.

Proposition 8.4 (Schur's Lemma) *Suppose* $(\mathfrak{g}, V_1, \rho_1)$ *and* $(\mathfrak{g}, V_2, \rho_2)$ *are irreducible representations of the Lie algebra* $\mathfrak{g}$. *Suppose that* $T : V_1 \to V_2$ *is a homomorphism of representations. Then there are only two possible cases:*

- *The function T is the zero function.*

- *The representations* $(\mathfrak{g}, V_1, \rho_1)$ *and* $(\mathfrak{g}, V_2, \rho_2)$ *are isomorphic (and T is an isomorphism).*

---

[6]Not to be confused with circular reasoning!

The proof is identical to the proof of Proposition 6.2.

We end this section with two useful tools for finding subrepresentations. The first is the Lie algebra analog to Proposition 5.2.

Proposition 8.5 *Suppose* $\mathfrak{g}$ *is a Lie algebra and* $(\mathfrak{g}, V, \rho)$ *is a Lie algebra representation. Suppose* $T : V \to V$ *commutes with* ρ. *Then each eigenspace of* T *is an invariant space of the representation* ρ.

We will use this proposition in Propositions 8.11 and 8.13.

Proof. Suppose v is an eigenvector of T with eigenvalue λ. Then for every $A \in \mathfrak{g}$ we have

$$T\rho(A)v = \rho(A)Tv = \lambda\rho(A)v,$$

so $\rho(A)v$ is in the eigenspace of T with eigenvalue λ. Hence the eigenspace of T with eigenvalue λ is an invariant space of the representation ρ. Since λ was arbitrary, this concludes the proof.   □

The image of a homomorphism of representations is always a subrepresentation.

Proposition 8.6 *Suppose* $(\mathfrak{g}, V_1, \rho_1)$ *and* $(\mathfrak{g}, V_2, \rho_2)$ *are two Lie algebra representations. Suppose* $T : V_1 \to V_2$ *is a homomorphism of representations. Then the image of* V_1 *under* T *is a subrepresentation of* V_2.

We will use this proposition in Proposition 8.13.

Proof. It suffices to show that Image(T) is an invariant space for ρ_2. Suppose v_2 lies in the image of T. Then there exists an element v_1 of V_1 such that $v_2 = Tv_1$. It follows that for any $A \in \mathfrak{g}$ we have

$$\rho_2(A)v_2 = \rho_2(A)Tv_1 = T\rho_1(A)v_1 \in \text{Image}(T).$$

We conclude that Image(T) is an invariant space for ρ_2.   □

The results of this section indicate strong similarities between Lie group representations and Lie algebra representations; however, there are important differences. While every Lie group representation corresponds to a Lie algebra representation, the converse is not true. For instance, while there are only odd-dimensional irreducible representations of the Lie group $SO(3)$ (Proposition 6.16), we will see in the next section that there are representations of the Lie algebra $so(3)$ of both even and odd order. This is one indication of a very fundamental asymmetry between groups and algebras. The fact that there are no infinite-dimensional irreducible representations of the Lie group $SO(3)$ on complex scalar product spaces (see discussion at the end of Section 6.6) while there are infinite-dimensional irreducible representations of the Lie algebra $so(3)$ on complex scalar product spaces (see Exercise 8.10) is yet another manifestation of the asymmetry.

In a sense that can be made quite precise. Lie groups are global objects and Lie algebras are local objects. To put it another way, Lie algebras are infinitesimal versions of Lie groups. In our main examples, the representation of the Lie group $SO(3)$ on $L^2(\mathbb{R}^3)$ operates by rotations of functions, while the representation of the Lie algebra $so(3)$ operates by differential operators on functions, sometimes called "infinitesimal generators of rotations." A differential operator A is *local* in the sense that one can calculate $(Af)(x_0, y_0, z_0)$ from the values of f near the point (x_0, y_0, z_0). By contrast, if B is a nontrivial rotation operator, then the calculation of $(Bf)(x_0, y_0, z_0)$ requires information about the values of f at some fixed, nonzero distance from (x_0, y_0, z_0). While global objects can be localized (by zooming in on one feature), local objects cannot always be extended into global ones. The interplay between local and global concepts is important in many fields of mathematics.

8.3   Raising Operators, Lowering Operators and Irreducible Representations of $su(2)$

The goal of this section is to classify finite-dimensional irreducible representations of the Lie algebra $su(2)$. As in the classification of irreducible representations of the Lie group $SU(2)$, we will first introduce a family of irreducible representations arising from an action of the Lie algebra on polynomials of two variables and then show that these are the only finite-dimensional irreducible representations of $su(2)$ (up to isomorphism). The main technical tools are the "raising" and "lowering" operators, as well as the eigenvectors and eigenvalues of $\rho(\mathbf{i})$ for arbitrary representations ρ.

We will construct a family of irreducible representations of the Lie algebra $su(2)$ as subrepresentations of a single representation on $\mathcal{P}$, the vector space of complex-coefficient polynomials in two variables. Recall from Section 8.1 that we can think of the algebra $su(2)$ as the real vector space spanned by $\{\mathbf{i}, \mathbf{j}, \mathbf{k}\}$, with bracket defined by $[\mathbf{i}, \mathbf{j}] = \mathbf{k}$, $[\mathbf{j}, \mathbf{k}] = \mathbf{i}$ and $[\mathbf{k}, \mathbf{i}] = \mathbf{j}$. For any $c_\mathbf{i}, c_\mathbf{j}, c_\mathbf{k} \in \mathbb{R}$, define the function $\mathbf{U} \colon su(2) \to g\ell(\mathcal{P})$ by

$$\mathbf{U}(c_\mathbf{i}\mathbf{i} + c_\mathbf{j}\mathbf{j} + c_\mathbf{k}\mathbf{k}) := c_\mathbf{i}\mathbf{U}_\mathbf{i} + c_\mathbf{j}\mathbf{U}_\mathbf{j} + c_\mathbf{k}\mathbf{U}_\mathbf{k}, \qquad (8.8)$$

where

$$\mathbf{U_i} := i \left(x \partial_x - y \partial_y\right)/2,$$
$$\mathbf{U_j} := \left(x \partial_y - y \partial_x\right)/2,$$
$$\mathbf{U_k} := i \left(x \partial_y + y \partial_x\right)/2.$$

The reader should check the bracket relations.

Each operator preserves the degree of homogeneous polynomials. For example, applying $\mathbf{U_i}$ to a degree-n monomial yields a degree-n monomial: for any $k = 0, 1, \ldots, n$, we have

$$\mathbf{U_i}(x^k y^{n-k}) = \frac{i}{2}\left(xkx^{k-1}y^{n-k} - y(n-k)x^k y^{n-k-1}\right) = \frac{i(2k-n)}{2}x^k y^{n-k};$$

similarly we find

$$\mathbf{U_j}(x^k y^{n-k}) = \frac{1}{2}\left((n-k)x^{k+1}y^{n-k-1} - kx^{k-1}y^{n-k+1}\right),$$
$$\mathbf{U_k}(x^k y^{n-k}) = \frac{i}{2}\left((n-k)x^{k+1}y^{n-k-1} + kx^{k-1}y^{n-k+1}\right).$$

Hence $\mathbf{U_i}$, $\mathbf{U_j}$ and $\mathbf{U_k}$ preserve the degree of any monomial. Hence $\mathbf{U}$ preserved the degree of any polynomial and takes any homogeneous polynomial to another homogeneous polynomial. In other words, each space $\mathcal{P}^n$ of homogeneous polynomials of a particular degree n is a subrepresentation of $(sl(2), \mathcal{P}, \mathbf{U})$.

In fact, for any particular nonnegative integer n, the operators $\mathbf{U_i}$, $\mathbf{U_j}$ and $\mathbf{U_k}$ form an irreducible representation of $su(2)$ on $\mathcal{P}^n$. The proof of this fact uses the eigenvectors of

$$\mathbf{U_i} = \frac{i}{2}\left(x \partial_x - y \partial_y\right).$$

The first of the three calculations above shows that each monomial is an eigenvector for the operator $\mathbf{U_i}$. The eigenvalues of $\mathbf{U_i}$ on $\mathcal{P}^n$ are

$$-\frac{in}{2}, i - \frac{in}{2}, \ldots, \frac{in}{2} - i, \frac{in}{2},$$

as pictured in Figure 8.2. We will also use the *raising operator*[7] *for the representation* $\mathbf{U}$

$$\mathbf{X} := \mathbf{U_j} - i\mathbf{U_k} = x \partial_y$$

---

[7] The raising and lowering operators were introduced by Dirac in his book, *The Principles of Quantum Mechanics* [Di, Section 39].

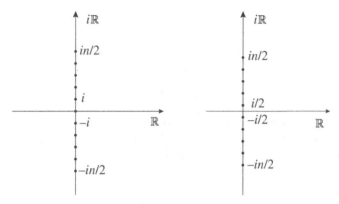

Figure 8.2. The eigenvalues of $\mathbf{U_i}$ on $\mathcal{P}^n$, namely, $-in/2, i - in/2, \ldots, in/2 - i, in/2$. The picture on the left is for even n; the one on the right is for odd n.

and the *lowering operator for the representation* $\mathbf{U}$

$$\mathbf{Y} := \mathbf{U_j} + i\mathbf{U_k} = -y\partial_x.$$

Note that

$$\mathbf{X}x^k y^{n-k} = (n-k)x^{k+1}y^{n-k-1}$$

for each integer $k = 0, \ldots, n$, so $\mathbf{X}$ "raises" the exponent of x in each term and "raises" the $\mathbf{U_i}$-eigenvalue from $i(2k - n)/2$ to $i(2k - n)/2 + i$ in the complex plane. Similarly we have $\mathbf{Y}x^k y^{n-k} = -kx^{k-1}y^{n-k+1}$, "lowering" the exponent of x and the $\mathbf{U_i}$-eigenvalue in each term. Because $\mathbf{X}$ and $\mathbf{Y}$ are complex linear combinations of the operators in the representation, $\mathbf{X}$ and $\mathbf{Y}$ preserve invariant subspaces. Suppose V is an invariant subspace of $\mathcal{P}^n$ and $v := \sum_{k=0}^{n} c_k x^k y^{n-k}$ is a nonzero vector in V. Let k_0 denote the smallest integer such that $c_{k_0} \neq 0$. Then

$$\mathbf{X}^{n-k_0} v = c_{k_0}(n - k_0)!x^n \neq 0.$$

So $x^n \in V$. But now we can apply the lowering operator to x^n and conclude that $x^{n-1}y \in V$. Repeating n times, we find that for any integer $m = 0, \ldots, n$ we have $x^{n-m}y^m \in V$. So $V = \mathcal{P}^n$. Hence $\mathcal{P}^n$ is an irreducible subrepresentation for the representation $\mathbf{U}$ of $su(2)$.

Up to isomorphism, these are the only finite-dimensional irreducible representations of the Lie algebra $su(2)$, as we shall show in Proposition 8.9 below. The proof of this proposition requires a more detailed understanding of the structure of the Lie algebra $su(2)$. In particular, we must construct raising and lowering operators for arbitrary representations of $su(2)$.

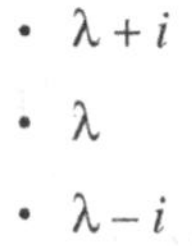

- $\lambda + i$
- λ
- $\lambda - i$

Figure 8.3. Raised and lowered eigenvalues.

Definition 8.7 *Suppose $(su(2), V, \rho)$ is a Lie algebra representation. Define the* raising operator *for ρ by*

$$\mathbf{X}_\rho := \rho(\mathbf{j}) - i\rho(\mathbf{k})$$

and the lowering operator *for ρ by*

$$\mathbf{Y}_\rho := \rho(\mathbf{j}) + i\rho(\mathbf{k}).$$

The next proposition details the relationship between $\mathbf{X}_\rho$, $\mathbf{Y}_\rho$ and $\rho(\mathbf{i})$. The eigenvalues of $\rho(\mathbf{i})$ play an important role.

Proposition 8.7 *Suppose $(su(2), V, \rho)$ is a Lie algebra representation. Then $[\mathbf{X}_\rho, \mathbf{Y}_\rho] = 2i\rho(\mathbf{i})$. Furthermore, if $v \in V$ is an eigenvector for $\rho(\mathbf{i})$ with eigenvalue λ, then*

$$\rho(\mathbf{i})(\mathbf{X}_\rho v) = (\lambda + i)\mathbf{X}_\rho v$$
$$\rho(\mathbf{i})(\mathbf{Y}_\rho v) = (\lambda - i)\mathbf{Y}_\rho v.$$

Note that $\lambda + i$ might not be an eigenvalue of $\rho(\mathbf{i})$, since $\mathbf{X}_\rho v$ might be 0. Similarly, $\lambda - i$ might not be an eigenvalue of $\rho(\mathbf{i})$. Still, Proposition 8.7 is often useful in constructing eigenvectors. For example, in Proposition 8.9 we will define an isomorphism of representations by mapping eigenvectors of $\rho(\mathbf{i})$ for some representation ρ to eigenvectors of $\mathbf{U_i}$. Note that the raising operator raises the eigenvalue (moving it upwards in the complex plane) while the lowering operator lowers the eigenvalue. See Figure 8.3.

Proof. The first statement follows from a calculation:

$$\begin{aligned}[\mathbf{X}_\rho, \mathbf{Y}_\rho] &= [\rho(\mathbf{j}) - i\rho(\mathbf{k}), \rho(\mathbf{j}) + i\rho(\mathbf{k})] \\ &= -i[\rho(\mathbf{k}), \rho(\mathbf{j})] + i[\rho(\mathbf{j}), \rho(\mathbf{k})] \\ &= 2i[\rho(\mathbf{j}), \rho(\mathbf{k})] = 2i\rho(\mathbf{i}).\end{aligned}$$

Now let $\lambda \in \mathbb{C}$ denote any eigenvalue of $\rho(\mathbf{i})$. Let v denote an eigenvector of $\rho(\mathbf{i})$ with eigenvalue λ. Then "raising" the vector v via $\mathbf{X}_\rho$ we find that

$$
\begin{aligned}
\rho(\mathbf{i})\left(\mathbf{X}_\rho v\right) &= \rho(\mathbf{i})\left(\left(\rho(\mathbf{j}) - i\rho(\mathbf{k})\right)v\right)\\
&= \left([\rho(\mathbf{i}), \rho(\mathbf{j})] - i[\rho(\mathbf{i}), \rho(\mathbf{k})] + \rho(\mathbf{j})\rho(\mathbf{i}) - i\rho(\mathbf{k})\rho(\mathbf{i})\right)v\\
&= \left(\rho(\mathbf{k}) + i\rho(\mathbf{j}) + \rho(\mathbf{j})\rho(\mathbf{i}) - i\rho(\mathbf{k})\rho(\mathbf{i})\right)v\\
&= (i + \lambda)\left(\rho(\mathbf{j}) - i\rho(\mathbf{k})\right)v\\
&= (i + \lambda)\left(\mathbf{X}_\rho v\right).
\end{aligned}
$$

Hence either $\mathbf{X}_\rho v = 0$ or $\mathbf{X}_\rho v$ is an eigenvector of $\rho(\mathbf{i})$ with eigenvalue $\lambda + i$. Replacing i by $-i$ and $\mathbf{X}_\rho$ by $\mathbf{Y}_\rho$ we find also that either $\mathbf{Y}_\rho v = 0$ or $\mathbf{Y}_\rho v$ is an eigenvector of $\rho(\mathbf{i})$ with eigenvalue $\lambda - i$. $\qquad\square$

Highest weight vectors are useful in the construction of irreducible subrepresentations, a key element of many proofs.

Definition 8.8 *Suppose* $(su(2), V, \rho)$ *is a finite-dimensional Lie algebra representation. Suppose* v_0 *is an eigenvector of* $\rho(\mathbf{i})$ *with the property that* $\mathbf{X}_\rho v_0 = 0$*. Then* v_0 *is a* highest weight vector *for the representation* ρ.

For example, in $\mathcal{P}^n$ the polynomial x^n is a highest weight vector. The next proposition shows how highest weight vectors generate irreducible subrepresentations.

Proposition 8.8 *Suppose* $(su(2), V, \rho)$ *is a finite-dimensional Lie algebra representation. Then there exists at least one highest weight vector for* ρ *in* V*. Suppose* v_0 *is a highest weight vector for* ρ*. Then there is a unique non-negative integer n such that* $\mathbf{Y}_\rho^n v_0 \neq 0$ *and* $\mathbf{Y}_\rho^{n+1} = 0$*. For any* $k = 0, \ldots, n$ *we have*

$$
\rho(\mathbf{i})\mathbf{Y}_\rho^k v_0 = \frac{i}{2}(n - 2k)\,\mathbf{Y}_\rho^k v_0. \tag{8.9}
$$

Furthermore,

$$
\left\{\mathbf{Y}_\rho^k v_0 : k = 0, \ldots, n\right\}
$$

is a basis for an irreducible subrepresentation W of V.

Note that we verified this proposition for the case $V = \mathcal{P}^n$ earlier in this section.

Proof. First we must show that V has at least one highest weight vector. Let v denote any eigenvector for $\rho(\mathbf{i})$ in V. Let λ denote the eigenvalue associated to v. Then for each $k \in \mathbb{N}$, by Proposition 8.7, either $\mathbf{X}_\rho^k v = 0$ or $\mathbf{X}_\rho^k v$ is an eigenvector for $\rho(\mathbf{i})$ with eigenvalue $\lambda + ik$. Since V is finite dimensional, the linear operator $\rho(\mathbf{i})$ can have only a finite number of distinct eigenvalues. Hence there must be a k such that $\mathbf{X}_\rho^k v = 0$ but $\mathbf{X}_\rho^{k-1} v \neq 0$. Because the vector $\mathbf{X}_\rho^{k-1} v$ is an eigenvector of $\rho(\mathbf{i})$ and lies in the kernel of $\mathbf{X}_\rho$, the vector $\mathbf{X}_\rho^{k-1} v$ is a highest weight vector for ρ.

Now we let v_0 denote a highest weight vector for ρ and construct the non-negative integer n. By Proposition 8.7 we know that if $\mathbf{Y}_\rho^k v_0$ is not an eigen-vector for $\rho(\mathbf{i})$, then $\mathbf{Y}_\rho^k v_0 = 0$. Since $\rho(\mathbf{i})$ has only a finite number of eigen-values, it follows that there must be a smallest nonnegative integer n such that $\mathbf{Y}_\rho^n v_0 \neq 0$ and $\mathbf{Y}_\rho^{n+1} v_0 = 0$. Proposition 8.7 also ensures that the eigenvalues are distinct; we conclude that the set

$$S := \left\{ \mathbf{Y}_\rho^k v_0 : k = 0, \dots, n \right\}$$

is linearly independent. Note that for any $m \geq 0$ we have

$$\mathbf{Y}_\rho^{n+1+m} v_0 = \mathbf{Y}_\rho^m \mathbf{Y}_\rho^{n+1} v_0 = 0,$$

so n is the unique nonnegative integer satisfying the conditions of the propo-sition.

Let W denote the subspace of V spanned by the set S. To show that W is a subrepresentation it suffices to show that W is invariant under $\rho(\mathbf{i})$, $\mathbf{X}_\rho$ and $\mathbf{Y}_\rho$, since $\rho(\mathbf{j}) = (\mathbf{Y}_\rho + \mathbf{X}_\rho)/2$ and $\rho(\mathbf{k}) = (\mathbf{Y}_\rho - \mathbf{X}_\rho)/2i$. Because each vector in S is an eigenvector for $\rho(\mathbf{i})$, the vector space W is invariant under $\rho(\mathbf{i})$. To see that W is invariant under $\mathbf{Y}_\rho$, note first that for any $k = 0, \dots, n-1$, we have

$$\mathbf{Y}_\rho \mathbf{Y}_\rho^k v_0 = \mathbf{Y}^{k+1} v_0 \in W.$$

For the case $k = n$ we have

$$\mathbf{Y}_\rho \mathbf{Y}_\rho^n v_0 = \mathbf{Y}_\rho^{n+1} v_0 = 0 \in W.$$

In either case we find that $\mathbf{Y}_\rho \mathbf{Y}_\rho^k v_0 \in W$. To see that W is also invariant under $\mathbf{X}_\rho$, we argue by induction on k that $\mathbf{X}_\rho(\mathbf{Y}_\rho^k v_0) \in W$. For the base case $(k = 0)$ we know from the definition of a highest weight vector that $\mathbf{X}_\rho v_0 = 0 \in W$. The inductive step is

$$\mathbf{X}_\rho \mathbf{Y}_\rho^k v_0 = \left(\mathbf{Y}_\rho \mathbf{X}_\rho + [\mathbf{X}_\rho, \mathbf{Y}_\rho] \right) \mathbf{Y}_\rho^{k-1} v_0$$
$$= \mathbf{Y}_\rho \left(\mathbf{X}_\rho \mathbf{Y}_\rho^{k-1} v_0 \right) + 2i\rho(\mathbf{i}) \mathbf{Y}_\rho^{k-1} v_0,$$

where we have used the first statement of Proposition 8.7. The first term lies in W by the inductive hypothesis and the fact that W is invariant under $\mathbf{Y}_\rho$;

the second term lies in W because $\mathbf{Y}_\rho^{k-1} v_0$ is an eigenvector for $\rho(\mathbf{i})$. Hence W is invariant under $\mathbf{X}_\rho$. So W is a nonempty invariant subspace for the representation ρ. Since S is linearly independent and spans W it is a basis for W.

Next we check the eigenvector condition, Equation 8.9. By the definition of a highest weight, v_0 is an eigenvector for $\rho(\mathbf{i})$. Let λ_0 denote the eigenvalue of $\rho(\mathbf{i})$ for the eigenvector v_0. Then (by an easy induction) it follows from Proposition 8.7 that the eigenvalue associated to $\mathbf{Y}_\rho^k v_0$ is $\lambda_0 - ik$. On the other hand, note that the trace of $\rho(\mathbf{i})$ on any finite-dimensional space is

$$\text{Tr}(\rho(\mathbf{i})) = \frac{1}{2i} \text{Tr}([X, Y]) = \frac{1}{2i} (\text{Tr}(XY) - \text{Tr}(YX)) = 0.$$

On W, we can express the trace explicitly in terms of the eigenvalues:

$$0 = \text{Tr}(\rho(\mathbf{i})) = \sum_{k=0}^{n} (\lambda_0 - ik) = (n+1)\lambda_0 - i\frac{n(n+1)}{2}.$$

It follows that $\lambda_0 = in/2$. Hence the eigenvalue corresponding to $\mathbf{Y}_\rho^k v_0$ is

$$i\left(\frac{n}{2} - k\right).$$

Finally we must show that W is irreducible. Suppose U is a nontrivial subrepresentation of W. We show that $\mathbf{Y}_\rho^n v_0 \in U$. Let u denote a nonzero vector in U. Expand u in the eigenbasis S:

$$u = \sum_{k=0}^{n} c_k \mathbf{Y}_\rho^k v_0 = c_0 v_0 + c_1 \mathbf{Y}_\rho v_0 + c_2 \mathbf{Y}_\rho^2 v_0 + \cdots + c_n \mathbf{Y}_\rho^n v_0.$$

Let k_0 denote the smallest k for which the coefficient $c_k \neq 0$. Then

$$\mathbf{Y}_\rho^n v_0 = \mathbf{Y}_\rho^{n-k_0} \mathbf{Y}_\rho^{k_0} v_0 = \frac{1}{c_{k_0}} \mathbf{Y}_\rho^{n-k_0} u \in U.$$

By an argument similar to the one used to construct W from v_0 we can find a subrepresentation $\tilde{U}$ of U containing $\mathbf{Y}_\rho^n v_0$. Without loss of generality we may assume $U = \tilde{U}$. We know that $\rho(\mathbf{i})$ has the eigenvalue $-in/2$ on U. Since U is a representation, we can consider the linear transformation $\rho(\mathbf{i}): U \to U$, and by an argument similar to one above we find that the eigenvalues of $\rho(\mathbf{i})$ on U must be

$$i\frac{n}{2}, i\left(\frac{n}{2} - 1\right), \ldots, -i\frac{n}{2}.$$

Hence $\dim U \geq \dim W$. But $U \subset W$, so $U = W$. Because U was an arbitrary nontrivial subrepresentation, it follows that W is irreducible. $\qquad\square$

Now we are ready to classify the finite-dimensional, irreducible Lie algebra representations of $su(2)$.

Proposition 8.9 *Suppose $(su(2), V, \rho)$ is a finite-dimensional irreducible Lie algebra representation. Set*

$$n := \dim V - 1.$$

Then $(su(2), V, \rho)$ is isomorphic to the representation $(su(2), \mathcal{P}^n, \mathbf{U})$.

In other words, the representations $\mathbf{U}$ of $su(2)$ as differential operators on homogeneous polynomials in two variables are essentially the only finite-dimensional irreducible representations, and they are classified by their dimensions. Unlike the Lie group $SO(3)$, the Lie algebra $su(2)$ has infinite-dimensional irreducible representations on complex scalar product spaces. See Exercise 8.10.

Proof. The main idea is to define an isomorphism by mapping the eigen-vectors of $\rho(\mathbf{i})$ to the monomials in $\mathcal{P}^n$. Choose any highest weight vector v_0 for the representation ρ. Because V is in irreducible representation of $su(2)$, the nontrivial subrepresentation W constructed in Proposition 8.8 must be all of V. Hence we have an explicit basis for V, namely,

$$S := \left\{ \mathbf{Y}_\rho^k v_0 : k = 0, \dots, n \right\}.$$

We can use the basis S to define a linear transformation $T \colon V \to \mathcal{P}^n$. This transformation will turn out to be an isomorphism of representations. Define $T \colon V \to \mathcal{P}^n$ by

$$T(\mathbf{Y}_\rho^k v_0) = \mathbf{Y}^k(y^n),$$

for each integer $k = 0, \dots, n$. Because T takes a basis to a basis, it follows from Exercise 2.17 that T is an isomorphism of vector spaces.

It remains to check that T is an isomorphism of representations. For the remaining condition of Definition 8.5, it suffices to check that for any $k = 0, \dots, n$ we have

$$T(\rho(\mathbf{q})\mathbf{Y}_\rho^k v_0) = \mathbf{U_q}(T(\mathbf{Y}_\rho^k v_0)) \tag{8.10}$$

$$T\left(\mathbf{Y}_\rho \left(\mathbf{Y}_\rho^k v_0\right)\right) = \mathbf{Y}\left(T\left(\mathbf{Y}_\rho^k v_0\right)\right) \tag{8.11}$$

$$T\left(\mathbf{X}_\rho \left(\mathbf{Y}_\rho^k v_0\right)\right) = \mathbf{X}\left(T\left(\mathbf{Y}_\rho^k v_0\right)\right). \tag{8.12}$$

For Equation 8.10 we have

$$T\left(\rho(\mathbf{i})\left(\mathbf{Y}_\rho^k v_0\right)\right) = (\lambda_0 + ik)\, T\left(\mathbf{Y}_\rho^k v_0\right) = \mathbf{U_i}\, T\left(\mathbf{Y}_\rho^k v_0\right).$$

Equation 8.11 follows easily from the definition of T:

$$T\left(\mathbf{Y}_\rho\left(\mathbf{Y}_\rho^k v_0\right)\right) = T\left(\mathbf{Y}_\rho^{k+1} v_0\right) = \mathbf{Y}^{k+1}(y^n) = \mathbf{Y}\left(\mathbf{Y}^k(y^n)\right)$$
$$= \mathbf{Y}\left(T(\mathbf{Y}_\rho^k v_0)\right),$$

where the second equals sign holds true even if $k = n$ because in that case both sides are 0. We can prove Equation 8.12 by induction on k. For the base case we find that

$$T\left(\mathbf{X}_\rho v_0\right) = T(0) = 0 = \mathbf{X}(y^n) = \mathbf{X}\, T(v_0).$$

The inductive step is:

$$T\left(\mathbf{X}_\rho(\mathbf{Y}_\rho^k v_0)\right) = T\left(\mathbf{X}_\rho \mathbf{Y}_\rho\left(\mathbf{Y}_\rho^{k-1} v_0\right)\right)$$
$$= T\left((\mathbf{Y}_\rho \mathbf{X}_\rho + [\mathbf{X}_\rho, \mathbf{Y}_\rho])\left(\mathbf{Y}_\rho^{k-1} v_0\right)\right)$$
$$= T\left(\mathbf{Y}_\rho \mathbf{X}_\rho\left(\mathbf{Y}_\rho^{k-1} v_0\right)\right) + T\left(2i\rho(\mathbf{i})\left(\mathbf{Y}_\rho^{k-1} v_0\right)\right)$$
$$= \mathbf{Y}\, T\left(\mathbf{X}_\rho\left(\mathbf{Y}_\rho^{k-1} v_0\right)\right) + 2i\mathbf{U_i}\, T\left(\left(\mathbf{Y}_\rho^{k-1} v_0\right)\right)$$
$$= \mathbf{YX}\, T\left(\mathbf{Y}_\rho^{k-1} v_0\right) + [\mathbf{X}, \mathbf{Y}]\, T\left(\mathbf{Y}_\rho^{k-1} v_0\right)$$
$$= \mathbf{XY}\, T\left(\mathbf{Y}_\rho^{k-1} v_0\right)$$
$$= \mathbf{X}\, T\left(\mathbf{Y}_\rho^k v_0\right).$$

The fourth equality follows from Equations 8.11 and 8.10, while the fifth uses the inductive hypothesis.

Hence T is an isomorphism of representations. So ρ is isomorphic to the restriction of $\mathbf{U}$ to $\mathcal{P}^n$.   □

Note that because the $\mathcal{P}^n$'s all have different dimensions, none is isomorphic to any other. Hence our list of finite-dimensional irreducible representations of $su(2)$ is complete and without repeats.

We encourage the reader to ponder the role of the raising operators ($\mathbf{X}$ and $\mathbf{X}_\rho$) and the lowering operators ($\mathbf{Y}$ and $\mathbf{Y}_\rho$) in the proofs in this section. Note that these operators do not live in the Lie algebra $su(2)$ itself; if we blindly apply the defining recipe to matrices in $su(2)$ we get, for example,

$$(\mathbf{j} - i\mathbf{k}) = \frac{1}{2}\begin{pmatrix} 0 & 1 \\ -1 & 0 \end{pmatrix} - \frac{i}{2}\begin{pmatrix} 0 & i \\ i & 0 \end{pmatrix} = \begin{pmatrix} 0 & 1 \\ 0 & 0 \end{pmatrix},$$

which is not an anti-Hermitian matrix, and hence is not an element of the Lie algebra $su(2)$. However, whenever we have a representation $(su(2), V, \rho)$ on a *complex* vector space V, we can define these operators on V. Defining raising and lowering operators ("the neatest trick in all of physics," according to at least one physicist [Roe]) is possible only on complex vector spaces, not on real vector spaces. This is but one example of a common pattern: study of the complex numbers $\mathbb{C}$ often sheds light on purely real phenomena.

The results of the current section, both the lowering operators and the classification, will come in handy in Section 8.4, where we classify the irreducible representations of $so(4)$. One can apply the classification of the irreducible representations of the Lie algebra $su(2)$ to the study of intrinsic spin, as an alternative to our analysis of spin in Section 10.4. More generally, raising and lowering operators are widely useful in the study of Lie algebra representations.

8.4 The Casimir Operator and Irreducible Representations of $so(4)$

The *Casimir operator* is a useful tool for identifying a representation of the Lie algebra $su(2)$. In this section we investigate Casimir operators and apply them to the classification of the finite-dimensional irreducible representations of the Lie algebra $so(4)$.

Definition 8.9 *Suppose* $(su(2), V, \rho)$ *is a Lie algebra representation. The* Casimir operator *for* ρ *is the linear transformation* $\mathbf{C} \colon V \to V$ *defined by*

$$\mathbf{C} := \rho(\mathbf{i})^2 + \rho(\mathbf{j})^2 + \rho(\mathbf{k})^2.$$

Like the raising and lowering operators, the Casimir operator does not correspond to any particular element of the Lie algebra $su(2)$. However, for any vector space V, both squaring and addition are well defined in the algebra $g\ell(V)$ of linear transformations. Given a representation, we can define the Casimir element of that representation.[8]

The main feature of the Casimir operator is that it commutes with every operator in the image of the representation.

---

[8] Generalizing this technique of applying algebraic operations legitimate in any $g\ell(V)$ but not necessarily in $\mathfrak{g}$, one can define "universal enveloping algebras." See Humphreys [Hu, Section 17.2] or Fulton and Harris [FH, Appendix C].

Proposition 8.10 *Suppose* $(su(2), V, \rho)$ *is a representation and* $\mathbf{C}$ *is its Casimir operator. Then* $\mathbf{C}$ *commutes with* ρ.

Proof. First note that $\mathbf{C}$ commutes with $\rho(\mathbf{i})$:

$$
\begin{aligned}
[\mathbf{C}, \rho(\mathbf{i})] &= [\rho(\mathbf{i})^2 + \rho(\mathbf{j})^2 + \rho(\mathbf{k})^2, \rho(\mathbf{i})] \\
&= \rho(\mathbf{j})^2\rho(\mathbf{i}) - \rho(\mathbf{i})\rho(\mathbf{j})^2 + \rho(\mathbf{k})^2\rho(\mathbf{i}) - \rho(\mathbf{i})\rho(\mathbf{k})^2 \\
&= \rho(\mathbf{j})[\rho(\mathbf{j}), \rho(\mathbf{i})] - [\rho(\mathbf{i}), \rho(\mathbf{j})]\rho(\mathbf{j}) \\
&\quad + \rho(\mathbf{k})[\rho(\mathbf{k}), \rho(\mathbf{i})] - [\rho(\mathbf{i}), \rho(\mathbf{k})]\rho(\mathbf{k}) \\
&= -\rho(\mathbf{j})\rho(\mathbf{k}) - \rho(\mathbf{k})\rho(\mathbf{j}) + \rho(\mathbf{k})\rho(\mathbf{j}) + \rho(\mathbf{j})\rho(\mathbf{k}) \\
&= 0.
\end{aligned}
$$

Cyclic reasoning implies that also $[\mathbf{C}, \rho(\mathbf{j})] = [\mathbf{C}, \rho(\mathbf{k})] = 0$. Because $\{\mathbf{i}, \mathbf{j}, \mathbf{k}\}$ is a basis for $su(2)$, it follows that $[\mathbf{C}, \rho(\mathbf{q})] = 0$ for any element $\mathbf{q} \in su(2)$. □

For example, consider the representation of $su(2)$ on polynomials in two variables defined by Equation 8.8. The Casimir operator for this representation is

$$
\begin{aligned}
\mathbf{C} &= \mathbf{U}_\mathbf{i}^2 + \mathbf{U}_\mathbf{j}^2 + \mathbf{U}_\mathbf{k}^2 \\
&= -\frac{1}{4}\left(x\partial_x - y\partial_y\right)^2 + \frac{1}{4}\left(x\partial_y - y\partial_x\right)^2 - \frac{1}{4}\left(x\partial_y + y\partial_x\right)^2 \\
&= -\frac{1}{4}\left(x^2\partial_x^2 + y^2\partial_y^2 + 3x\partial_x + 3y\partial_y + 2xy\partial_x\partial_y\right).
\end{aligned}
$$

This Casimir operator is constant on various interesting vector spaces of polynomials in two variables. Note that the restriction of the Casimir operator $\mathbf{C}$ to the space $\mathcal{P}^0$ of constant polynomials is the zero operator. The restriction to the space $\mathcal{P}^1$ of purely linear polynomials is multiplication by $-\frac{3}{4}$. Continuing on this theme, we find that

$$
\mathbf{C}x^2 = -\frac{1}{4}(2+6)x^2 = -2x^2
$$

$$
\mathbf{C}y^2 = -\frac{1}{4}(2+6)y^2 = -2y^2
$$

$$
\mathbf{C}xy = -\frac{1}{4}(3+3+2)xy = -2xy,
$$

so $\mathbf{C}$ is constant on $\mathcal{P}^2$ with value -2. These are three examples of a general phenomenon.

Proposition 8.11 *Suppose $(su(2), V, \rho)$ is a finite-dimensional irreducible Lie algebra representation. Then the Casimir operator is a scalar multiple of the identity on V.*

We encourage the reader to compare this proposition and its proof to the corresponding proposition for group representations (Proposition 6.3). The converse of Proposition 8.11 is false, as the reader is asked to show in Exercise 8.11.

Proof. Since V is a finite-dimensional complex vector space, $\mathbf{C}$ must have at least one eigenvalue λ. Define

$$W := \{v \in V : \mathbf{C}v = \lambda v\};$$

i.e, W is the eigenspace corresponding to λ. By Proposition 8.5, this subspace is invariant under ρ because $\mathbf{C}$ commutes with the representation by Proposition 8.10. But because λ is an eigenvalue for $\mathbf{C}$, the subspace W is not equal to $\{0\}$. Hence, since ρ is irreducible, we conclude by Schur's Lemma (Proposition 8.4) that $W = V$. So $\mathbf{C}v = \lambda I v$ for every $v \in V$. In other words, $\mathbf{C}$ is a scalar multiple of the identity. $\qquad\square$

Let us evaluate the Casimir operator $\mathbf{C}$ restricted to $\mathcal{P}^n$ for arbitrary n. By Proposition 8.11 it suffices to evaluate $\mathbf{C}$ on any one element of $\mathcal{P}^n$, say, x^n. We find that

$$\mathbf{C}x^n = -\frac{1}{4}(n^2 + 2n).$$

Hence on $\mathcal{P}^n$ we have

$$\mathbf{C} = -\frac{1}{4}(n^2 + 2n)I. \tag{8.13}$$

Since, by Proposition 8.9, each finite-dimensional irreducible representation of $su(2)$ is isomorphic to $\mathcal{P}^n$ for some n, it follows that the only possible value of the Casimir operator on a finite-dimensional representation is $-\frac{1}{4}(n^2 + 2n)$ for some n. In other words, the possible values are

$$0, -\frac{3}{4}, -2, -\frac{15}{4}, -6, \ldots.$$

Physicists may be more familiar with another description of this sequence of numbers: $-\ell(\ell + 1)$, where $\ell := \frac{n}{2}$ is the quantum number taking half-integer values $0, \frac{1}{2}, 1, \frac{3}{2}, \ldots$. The number ℓ is emphasized because it shows up in the eigenvalues of $\rho(\mathbf{i})$.

Proposition 8.12 *Suppose* $(su(2), V, \rho)$ *is a finite-dimensional Lie algebra representation. Suppose* $\mathbf{C} = -\ell(\ell + 1)I$ *on V for some nonnegative half-integer* ℓ. *Then the eigenvalues of* $\rho(\mathbf{i}) \colon V \to V$ *are*

$$\{-i\ell, -i\ell + i, \ldots, i\ell - i, i\ell\}.$$

Proof. Choose any eigenvector u of $\rho(\mathbf{i})$. Let λ denote the corresponding eigenvalue. For every $k \in \mathbb{N}$ the vector $\mathbf{X}_\rho^k u$ is either trivial or an eigenvector for $\rho(\mathbf{i})$ with eigenvalue $\lambda + ik$. Let k_0 denote the natural number such that

$$u_0 := \mathbf{X}_\rho^{k_0 - 1} u$$

is nontrivial and $\mathbf{X}_\rho u_0 = \mathbf{X}_\rho^{k_0} u = 0$. (Because V is finite dimensional, there must be such a k_0.) Then u_0 is a highest weight vector. Let W denote the irreducible subrepresentation spanned by

$$\left\{ \mathbf{Y}_\rho^k u_0 \colon k = 0, \ldots, n \right\},$$

whose existence is guaranteed by Proposition 8.8. By Proposition 8.9, the representation W is isomorphic to the representation $\mathcal{P}^n$. By Equation 8.13 we know that

$$-\ell(\ell + 1)I = \mathbf{C} = -\frac{n^2 + 2n}{4} I.$$

There are two solutions for n in terms of ℓ, but only one is a nonnegative integer. We conclude that $n = 2\ell$. So W is isomorphic to $\mathcal{P}^{2\ell}$. Hence the eigenvalues of $\rho(\mathbf{i})$ on W must be the same as the eigenvalues of $\mathbf{U_i}$ on $\mathcal{P}^{2\ell}$. These eigenvalues are shown in Figure 8.2. In terms of ℓ, the eigenvalue corresponding to u_0 is $i\ell$. Hence we find that the eigenvalue associated to our arbitrary eigenvector u is

$$\lambda = i\ell - ik_0 + i.$$

Note that because the dimension of V is $n + 1$, the definition of k_0 ensures that $1 \le k_0 \le n + 1 = 2\ell + 1$. Hence

$$\lambda \in \{i\ell, i\ell - i, \ldots, -i\ell + i, -i\ell\},$$

which proves the proposition. $\square$

Next we use Casimir operators to classify the irreducible representations of $su(2) \oplus su(2)$. This classification will use the natural representation on a tensor product.

Definition 8.10 *Suppose* $(\mathfrak{g}_1, V_1, \rho_1)$ *and* $(\mathfrak{g}_2, V_2, \rho_2)$ *are two representations of a Lie algebra* $\mathfrak{g}$. *Then the* tensor product *of the two representations is*

$$(\mathfrak{g}_1 \oplus \mathfrak{g}_2, V_1 \otimes V_2, \rho_1 \otimes I + I \otimes \rho_2),$$

where

$$(\rho_1 \otimes I + I \otimes \rho_2)(A, B) := \rho_1(A) \otimes I + I \otimes \rho_2(B)$$

for any $A \in \mathfrak{g}_1$ *and* $B \in \mathfrak{g}_2$.

If we think of a Lie algebra as the space of derivatives of a Lie group at the identity, then the expression $\rho_1(\mathbf{q}) \otimes I + I \otimes \rho_2(\mathbf{p})$ looks like the product rule for derivatives. We leave it to the reader (in Exercise 8.12) to show that $\rho_1 \otimes I + I \otimes \rho_2$ satisfies the definition of a Lie algebra representation.

The next proposition classifies finite-dimensional irreducible representations of $so(4)$. Recall from Proposition 8.3 that $so(4) \cong su(2) \oplus su(2)$, so the representations of the two Lie algebras must be identical. Hence it suffices to classify the finite-dimensional irreducible representations of $su(2) \oplus su(2)$.

Proposition 8.13 *Suppose* $(su(2) \oplus su(2), V, \rho)$ *is a finite-dimensional irreducible representation. Then there are irreducible representations*

$$\left(su(2), W_1, \rho_1\right) \text{ and } \left(su(2), W_2, \rho_2\right)$$

such that the representation $(su(2) \oplus su(2), V, \rho)$ *is isomorphic to the Lie algebra representation*

$$\left(su(2) \oplus su(2), W_1 \otimes W_2, \rho_1 \otimes I + I \otimes \rho_2\right).$$

Like the proof of Proposition 8.9, the proof of this proposition uses the technology of raising operators, lowering operators and weights.

Proof. First we introduce some notation. We will write arbitrary elements of $su(2) \oplus su(2)$ as $(\mathbf{q}, \mathbf{p})$, where $\mathbf{q}, \mathbf{p} \in su(2)$. Note that by the definition of the Cartesian sum of Lie algebras we have $[(\mathbf{q}, 0), (0, \mathbf{p})] = 0$ for all $\mathbf{q}, \mathbf{p} \in su(2)$.

Next we use Casimirs to find a vector w that is a highest-weight vector for both

$$\rho_1 := \rho\Big|_{su(2) \oplus 0} \qquad \text{and} \qquad \rho_2 := \rho\Big|_{0 \oplus su(2)}.$$

Set

$$\mathbf{C}_1 := \rho(\mathbf{i}, 0)^2 + \rho(\mathbf{j}, 0)^2 + \rho(\mathbf{j}, 0)^2$$
$$\mathbf{C}_2 := \rho(0, \mathbf{i})^2 + \rho(0, \mathbf{j})^2 + \rho(0, \mathbf{j})^2.$$

In other words, the operator $\mathbf{C}_1$ is the Casimir operator for the representation ρ_1 and $\mathbf{C}_2$ is the Casimir operator for ρ_2. We want to show that the Casimir operators $\mathbf{C}_1$ and $\mathbf{C}_2$ are scalar multiples of the identity on V. To this end, note that for any $\mathbf{q} \in su(2)$ we have $[\mathbf{C}_1, \rho(\mathbf{q}, 0)] = 0$ by Proposition 8.10, while for any $\mathbf{p} \in su(2)$ we have

$$[\mathbf{C}_1, \rho(0, \mathbf{p})] = [\rho(\mathbf{i}, 0)^2 + \rho(\mathbf{j}, 0)^2 + \rho(\mathbf{j}, 0)^2, \rho(0, \mathbf{p})] = 0,$$

since $[(\mathbf{q}, 0), (0, \mathbf{p})] = 0$ for any $\mathbf{q} \in su(2)$. Hence for any element $(\mathbf{q}, \mathbf{p})$ of $su(2) \oplus su(2)$ we have

$$[\mathbf{C}_1, \rho(\mathbf{q}, \mathbf{p})] = [\mathbf{C}_1, \rho(\mathbf{q}, 0)] + [\mathbf{C}_1, \rho(0, \mathbf{p})] = 0.$$

So $\mathbf{C}_1$ commutes with ρ. It follows from Proposition 8.5 that each eigenspace of $\mathbf{C}_1$ is an invariant space for the representation ρ. Because ρ is irreducible, we conclude that $\mathbf{C}_1$ has only one eigenspace, namely, all of V. Hence $\mathbf{C}_1$ must be a scalar multiple of the identity on V. Similarly, $\mathbf{C}_2$ must be a scalar multiple of the identity on V. By Proposition 8.9 and Equation 8.13, we know that the Casimir operators can take on only certain values on finite-dimensional representations, so we can choose nonnegative half-integers ℓ_1 and ℓ_2 such that $\mathbf{C}_1 = -\ell_1(\ell_1 + 1)$ and $\mathbf{C}_2 = -\ell_2(\ell_2 + 1)$.

Set

$$U := \{u \in V : \rho(\mathbf{i}, 0)u = i\ell_1 u, \rho(0, \mathbf{i})u = i\ell_2 u\}.$$

Since $\mathbf{C}_1 = -\ell_1(\ell_1 + 1)$ on V, Proposition 8.12 implies that the eigenspace of $\rho(\mathbf{i}, 0): V \to V$ for the eigenvalue $i\ell_1$ is not empty. Furthermore, since $\rho(\mathbf{i}, 0)$ commutes with every operator of the form $\rho(0, \mathbf{q})$, the $i\ell_1$-eigenspace of $\rho(\mathbf{i}, 0)$ is invariant under the restriction of ρ to $0 \oplus su(2)$ and hence (again by Proposition 8.12) the $i\ell_2$-eigenspace of $\rho(0, \mathbf{i})$ restricted to the eigenspace of $\rho(\mathbf{i}, 0)$ is not empty. Hence U is not empty. Let w denote any nonzero element of U.

Next we define irreducible representations

$$(su(2), W_1, \rho_1) \quad \text{and} \quad (su(2), W_2, \rho_2)$$

such that $\rho(\mathbf{q}, \mathbf{p}) = \rho_1(\mathbf{q}) \otimes I + I \otimes \rho_2(\mathbf{p})$ for any $\mathbf{q}, \mathbf{p} \in Q$. Let $\mathbf{Y}_1$ and $\mathbf{Y}_2$ denote the lowering operators for the representations ρ_1 and ρ_2, respectively. In other words, define

$$\mathbf{Y}_1 := \rho(\mathbf{j}, 0) + i\rho(\mathbf{k}, 0), \qquad \mathbf{Y}_2 := \rho(0, \mathbf{j}) + i\rho(0, \mathbf{k}).$$

Let W_1 denote the span of the set

$$S_1 := \left\{\mathbf{Y}_1^k w : k = 0, \ldots, 2\ell_1\right\},$$

and let W_2 denote the span of the set

$$S_2 := \left\{ \mathbf{Y}_2^k w : k = 0, \ldots, 2\ell_2 \right\},$$

By Proposition 8.8, both W_1 and W_2 are irreducible.

Next we define a linear transformation $T : W_1 \otimes W_2 \to V$ by

$$T\left(\left(\mathbf{Y}_1^{k_1} w \right) \otimes \left(\mathbf{Y}_2^{k_2} w \right) \right) := \mathbf{Y}_1^{k_1} \mathbf{Y}_2^{k_2} w, \tag{8.14}$$

for any $k_1 = 0, \ldots, 2\ell_1$ and any $k_2 = 0, \ldots, 2\ell_2$. We will show that T is the desired isomorphism of representations.

First we prove that T is a homomorphism of representations. It suffices to check the basis vectors of $W_1 \otimes W_2$. For $k_1 = 0, \ldots, 2\ell_1$ and $k_2 = 0, \ldots, 2\ell_2$ and arbitrary $(\mathbf{q}, \mathbf{p}) \in su(2) \oplus su(2)$ we have

$$\rho(\mathbf{q}, \mathbf{p}) T\left(\mathbf{Y}_1^{k_1} w \otimes \mathbf{Y}_2^{k_2} w \right) = \rho(\mathbf{q}, \mathbf{p}) \mathbf{Y}_1^{k_1} \mathbf{Y}_2^{k_2} w$$

$$= \rho(\mathbf{q}, 0) \mathbf{Y}_1^{k_1} \mathbf{Y}_2^{k_2} w + \mathbf{Y}_1^{k_1} \rho(0, \mathbf{p}) \mathbf{Y}_2^{k_2} w$$

$$= T\left(\rho_1(\mathbf{q}) \mathbf{Y}_1^{k_1} w \otimes \mathbf{Y}_2^{k_2} w + \mathbf{Y}_1^{k_1} w \otimes \rho_2(\mathbf{p}) \mathbf{Y}_2^{k_2} w \right)$$

$$= T\left((\rho_1(\mathbf{q}) \otimes I + I \otimes \rho_2(\mathbf{p})) \mathbf{Y}_1^{k_1} w \otimes \mathbf{Y}_2^{k_2} w \right).$$

Hence T is a homomorphism of representations.

To show that T is injective, it suffices to show that the image of the basis

$$\left\{ (\mathbf{Y}_1^{k_1} w) \otimes (\mathbf{Y}_2^{k_2} w) : k_1 = 0, \ldots, 2\ell_1; k_2 = 0, \ldots, 2\ell_2 \right\}$$

is linearly independent in V. By the definition of the linear transformation T, this image is

$$S := \left\{ \mathbf{Y}_1^{k_1} \mathbf{Y}_2^{k_2} w : k_1 = 0, \ldots, 2\ell_1; k_2 = 0, \ldots, 2\ell_2 \right\}.$$

Suppose the scalars $\left\{ c_{k_1 k_2} \in \mathbb{C} : k_1 = 0, \ldots, 2\ell_1; k_2 = 0, \ldots, 2\ell_2 \right\}$ satisfy

$$0 = \sum_{k_1=0}^{2\ell_1} \sum_{k_2=0}^{2\ell_2} c_{k_1 k_2} \mathbf{Y}_1^{k_1} \mathbf{Y}_2^{k_2} w.$$

Note that for each fixed k_1, the vector $\sum_{k_2=0}^{2\ell_2} c_{k_1 k_2} \mathbf{Y}_1^{k_1} \mathbf{Y}_2^{k_2} w$ is an eigenvector for $\rho(\mathbf{i}, 0)$ with eigenvector $i(\ell_1 - k_1)$. Because these eigenvalues are distinct, the equation above implies that for each k_1 we have

$$0 = \sum_{k_2=0}^{2\ell_2} c_{k_1 k_2} \mathbf{Y}_1^{k_1} \mathbf{Y}_2^{k_2} w.$$

But for each k_2 we know that the vector $\mathbf{Y}_2^{k_2} w$ is an eigenvector for $\rho(0, \mathbf{i})$ with eigenvalue $i(\ell_2 - k_2)$. Because these eigenvalues are distinct, it follows that $c_{k_1 k_2} = 0$ for each k_1, k_2. Hence the linear transformation T is injective.

It remains to show that T is surjective. We apply Proposition 8.6 to see that Image(T) is a subrepresentation of $(su(2) \oplus su(2), V, \rho)$. Since Image($T$) is not trivial and V is irreducible, it follows that $V = $ Image(T), i.e., that T is surjective onto V. This completes the proof that $T: W_1 \otimes W_2 \rightarrow V$ is an isomorphism of representations.   □

In this section we have used the Casimir operator of the Lie algebra $su(2)$ to help us classify irreducible representations of $so(4)$. This is one glimpse of the power of the Casimir operator, whose most important feature is that it commutes with the image under the representation of the Lie algebra. Casimir operators play an important role in the representation theory of many different Lie algebras. As we will see in Section 8.6, the Schrödinger Hamiltonian operator for the hydrogen atom has $so(4)$ symmetry. We will use both the Casimir operator and our classification of the irreducible representations of $so(4)$ to make predictions about the hydrogen atom.

8.5   Bound States of the Hydrogen Atom

In this section we discuss the *bound states* of the hydrogen atom. These are states where the electron stays with the nucleus. In contrast, an electron with lots of energy could simply speed past the nucleus without getting trapped. Such an *unbound* electron does not stop long enough form a coherent atom; hence in our study of the atom, it makes sense to study only the bound states.

At long last, it is time to appeal to the Schrödinger operator

$$\mathbf{H} := -\frac{\hbar^2}{2\mathbf{m}} \left(\partial_x^2 + \partial_y^2 + \partial_z^2 \right) - \frac{\mathbf{e}^2}{\sqrt{x^2 + y^2 + z^2}},$$

where $\mathbf{e}$ is the charge of the electron. The function $e^2/\sqrt{x^2 + y^2 + z^2}$ is called the *Coulomb potential*. Note that the Schrödinger operator is a cyclic formula, as is the Coulomb potential. Experiments show that the Schrödinger operator can be used to completely determine the spatial behavior of the electron in a (nonrelativistic) hydrogen atom in many situations. Although the model is not perfect (for example, it does not correctly predict relativistic effects or the microfine splitting of the spectral lines of hydrogen), it yields useful, correct predictions for many experiments.

The Schrödinger operator can be used to make predictions about measurements of the energy of the electron in a hydrogen atom. For example, suppose $\phi \in L^2(\mathbb{R}^3)$ satisfies the *Schrödinger eigenvalue equation*

$$\mathbf{H}\phi = E\phi \tag{8.15}$$

for some real number E. We will find it convenient to recall the Laplacian operator $\nabla^2 := \partial_x^2 + \partial_y^2 + \partial_z^2$ and write the Schrödinger eigenvalue equation explicitly as

$$-\frac{\hbar^2}{2\mathbf{m}}\left(\nabla^2\phi\right)(x, y, z) - \frac{e^2}{\sqrt{x^2 + y^2 + z^2}}\phi(x, y, z) = E\phi(x, y, z). \tag{8.16}$$

Consider an electron in the state corresponding to ϕ. If we measure the energy of such an electron, we are sure to obtain the energy value E. For this reason the eigenvalues of the Schrödinger operator are known as *energy levels* or *energy eigenvalues*, and the corresponding eigenfunctions are called *energy eigenstates*. More generally, consider an electron in a state that is a superposition of energy eigenstates:

$$\phi = \sum_E c_E \phi_E,$$

where $\sum_E |c_E|^2 = 1$ and for each E the function ϕ_E is an eigenfunction corresponding to the eigenvalue E and c_E is a complex number. Consider measuring the energy of such an electron. The probability that the measured energy will be E is the number $|c_E|^2 \in [0, 1]$. We are particularly interested in the vector space spanned by eigenstates corresponding to negative energy values.[9] These are known as *bound states* because they are "bound" to the nucleus of the hydrogen atom — they do not have enough energy to escape. The bound states form a vector subspace of $L^2(\mathbb{R}^3)$.

---

[9]We are sweeping an issue under the rug here. What we really want to study is the vector space of states whose energy is sure to be negative when measured. In fact, in the case of this particular operator (the Schrödinger operator with the Coulomb potential), the vector space of states sure to have negative energy is precisely equal to the span of the negative eigenstates. Proving this equality requires subtle techniques of functional analysis. To get a glimpse of the issue, see Exercise 8.16. In the language of physics, the problem is that there may be plane-wave eigenfunctions; in the language of mathematics, the problem is that there may be continuous spectrum. Again, this issue is moot in the case of negative energy for the Schrödinger operator, where the only solutions whose energy is sure to be measured negative are (finite or countably infinite) linear combinations of *bona fide* eigenfunctions in $L^2(\mathbb{R}^3)$.

Why is zero is the cutoff between the bound and unbound states? Energy, after all, can be measured only relatively. One can measure energy differences physically, but adding an overall constant to an energy function never changes the physical predictions. For example, in order to define potential energy in the study of classical mechanical motion under the influence of gravity, one must pick an arbitrary reference height. For our Schrödinger operator, the fact that the Coulomb potential increases toward zero as $x^2 + y^2 + z^2$ gets large fixes zero as the sensible cutoff. Physically, a particle with energy greater than zero has enough energy to escape the Coulomb potential well, and so is unbound. On the other hand, a particle with energy less than zero is not likely to climb out of the potential well; in other words, such a particle is bound to the nucleus.

Proposition 8.14 *Each negative eigenvalue E of the Schrödinger operator has a finite number of linearly independent eigenfunctions.*

The proof depends on Proposition A.3 of Appendix A, which ensures that all $L^2(\mathbb{R}^3)$ solutions of the Schrödinger equation can be approximated by linear combinations of solutions where the radial and angular variables have been separated.

Proof. First we will show that only a finite number of solutions are of the form $\alpha \otimes Y_{\ell,m}$, for $\alpha \in \mathcal{I}$ and $Y_{\ell,m}$ a spherical harmonic function. Then we will apply Proposition A.3 to conclude that these solutions span the space of all square-integrable solutions.

Fix an eigenvalue E. Suppose we have a solution to the eigenvalue equation for the Schrödinger operator in the given form. I.e, suppose we have a function $\alpha \in \mathcal{I}$ and a spherical harmonic function $Y_{\ell,m}$ such that

$$\left(-\frac{\hbar^2}{2\mathbf{m}} \left(\nabla_r^2 + \nabla_{\theta,\phi}^2 \right) - \frac{e^2}{r} - E \right) \alpha(r) Y_{\ell,m}(\theta, \phi) = 0,$$

where

$$\nabla_r^2 := \partial_r^2 + \frac{2}{r}\partial_r,$$

$$\nabla_{\theta,\phi}^2 := \frac{1}{r^2}\partial_\theta^2 + \frac{\cos\theta}{r^2 \sin\theta}\partial_\theta + \frac{1}{r^2 \sin^2\theta}\partial_\phi^2.$$

After dividing by $\alpha(r)Y_{\ell,m}(\theta, \phi)$, rearranging and applying Equation 1.13, we obtain

$$\frac{\hbar^2}{2\mathbf{m}}\alpha''(r) + \frac{\hbar^2}{r\mathbf{m}}\alpha'(r) + \left(\frac{e^2}{r} + E - \frac{\ell(\ell+1)\hbar^2}{2\mathbf{m}r^2} \right) \alpha(r) = 0. \qquad (8.17)$$

Not every solution α to this equation will correspond to an $L^2(\mathbb{R}^3)$ eigenfunction of the Schrödinger operator. In order for $\alpha(r)Y_{\ell,m}(\theta, \phi)$ to be square integrable, the integral

$$\int_0^\infty |\alpha(r)|^2 r^2 dr \tag{8.18}$$

must converge. It turns out that if $\ell \geq 1$ and $\ell > \sqrt{\mathbf{m}e^2}/\hbar\sqrt{-2E}$, then there is no solution α that makes the integral converge. Note that Equation 8.17 is a second-order linear ordinary differential equation with a regular singular point at $r = 0$. It is well known (see, e.g., Simmons [Sim, Section 30]), that every solution of such an equation can be written in the form

$$\alpha(r) = r^K \sum_{j=0}^\infty c_j r^j$$

on some neighborhood of $r = 0$, for some $K \in \mathbb{R}$ and some $c_0 \neq 0$. Without loss of generality, we can take $c_0 = 1$: because the equation is linear, dividing by c_0 yields another solution. Because power series converge uniformly on any closed subset of their domains of convergence, we can switch[10] the order of differentiation and summation after plugging the series expression for α into Equation 8.17 to obtain

$$\frac{\hbar^2}{2\mathbf{m}} (K(K-1) + 2K - \ell(\ell+1)) r^{K-2} + \text{higher order terms } = 0.$$

Hence we have

$$(K - \ell)(K + \ell + 1) = K(K-1) + 2K - \ell(\ell+1) = 0,$$

which holds only if $K = \ell$ or if $K = -\ell - 1$.

Neither of these solutions give convergence of Integral 8.18. Consider first the solution with $K = -\ell - 1$. Near $r = 0$ we have

$$|\alpha(r)|^2 r^2 \sim r^{2K-2},$$

so Integral 8.18 will converge at the lower limit only if $2K + 2 > -1$, i.e., only if $K > -3/2$. But we have assumed that $\ell \geq 1$, so $K = -\ell - 1 \leq -2 < -3/2$. So the solution with $K = -\ell - 1$ does not correspond to a square integrable eigenfunction of the Schrödinger operator. On the other

---

[10]The point is that it is not always possible to switch infinite summations and differentiations. In a rigorous mathematical proof, such manipulations must be carefully justified.

hand, if we take $K = \ell$, we have problems with convergence as r goes to ∞. A straightforward maximization calculation (see Exercise 8.17) shows that if $\ell > \sqrt{m}\mathbf{e}^2/\hbar\sqrt{-2E}$, then for every $r > 0$ we have

$$\frac{\mathbf{e}^2}{r} + E - \frac{\ell(\ell + 1)\hbar^2}{2\mathbf{m}r^2} < 0.$$

It follows that at any critical point r_0, i.e., any point such that $\alpha'(r_0) = 0$, the real numbers $\alpha(r_0)$ and $\alpha''(r_0)$ must have the same sign. Hence there are no local maxima of α at points where the value of α is positive. Near the origin ($r = 0$) we have $\alpha(r) \sim r^\ell$, so there must be a point r_1 such that $\alpha'(r_1) > 0$ and $\alpha(r_1) > 0$. Hence for all $r > r_1$, we have $\alpha(r) > \alpha(r_1)$; otherwise there would have to be a local maximum between r_1 and r, in a region where α is positive. So Integral 8.18 cannot converge at the upper limit. In other words, α does not yield an $L^2(\mathbb{R}^3)$-eigenfunction of the Schrödinger operator either.

We have shown that if $\ell \geq 1$ and $\ell > \sqrt{m}\mathbf{e}^2/\hbar\sqrt{-2E}$, then there is no eigenfunction in $L^2(\mathbb{R}^3)$ of the Schrödinger operator with eigenvalue E. Since ℓ must be a nonnegative integer, it follows that for any fixed $E < 0$ there are only a finite number of corresponding eigenfunctions.   □

Because of the spherical symmetry of physical space, any realistic physical operator (such as the Schrödinger operator) must commute with the angular momentum operators. In other words, for any $g \in SO(3)$ and any f in the domain of the Schrödinger operator $\mathbf{H}$ we must have $\mathbf{H} \circ \rho(g) = \rho(g) \circ \mathbf{H}$, where ρ denotes the natural representation of $SO(3)$ on $L^2(\mathbb{R}^3)$. In Exercise 8.15 we invite the reader to check that $\mathbf{H}$ does indeed commute with rotation. The commutation of $\mathbf{H}$ and the angular momentum operators is the *infinitesimal* version of the commutation with rotation; i.e., we can obtain the former by differentiating the latter. More explicitly, we differentiate the equation

$$\mathbf{H}\left(f\left(\begin{pmatrix} 1 & 0 & 0 \\ 0 & \cos\theta & -\sin\theta \\ 0 & \sin\theta & \cos\theta \end{pmatrix} \begin{pmatrix} x \\ y \\ z \end{pmatrix} \right) \right)$$
$$= (\mathbf{H}f)\left(\begin{pmatrix} 1 & 0 & 0 \\ 0 & \cos\theta & -\sin\theta \\ 0 & \sin\theta & \cos\theta \end{pmatrix} \begin{pmatrix} x \\ y \\ z \end{pmatrix} \right)$$

with respect to the real variable θ and evaluate at $\theta = 0$ to deduce that $\mathbf{H}\mathbf{L}_i f = \mathbf{L}_i \mathbf{H} f$. So $[\mathbf{H}, \mathbf{L}_i] = 0$ and, by a cyclic argument, $[\mathbf{H}, \mathbf{L}_j] = [\mathbf{H}, \mathbf{L}_k] = 0$ as well. By the linearity of the Lie algebra homomorphism $\mathbf{L}$, it follows that for any $\mathbf{q} \in su(2)$, we have $[\mathbf{H}, \mathbf{L_q}] = 0$. By Proposition 8.5 we conclude that $\mathbf{L_q}$ preserves the eigenspaces of $\mathbf{H}$. We can summarize by saying that the angular momentum operators restricted to a single eigenspace of the Schrödinger operator $\mathbf{H}$ form a representation of $su(2)$.

We can put these representations (one for each eigenvalue of the Schrödinger operator) together to form a representation of $su(2)$ on the vector space of bound states of the hydrogen atom. We will see in Section 8.6 that there is a physically natural representation of the larger Lie algebra $so(4) \cong su(2) \oplus su(2)$ on the set of bound states of the hydrogen atom.

8.6   The Hydrogen Representations of $so(4)$

We can use the representation theory of the Lie algebra $so(4)$ along with the stunning fact that there is a representation of $so(4)$ on the space of bound states of the Schrödinger operator with the Coulomb potential to make a satisfying prediction about the dimensions of the shells of the hydrogen atom and the energy levels of these shells.

In Section 8.5 we saw that the angular momentum operators commute with the Schrödinger operator. There is another, more obscure, set of operators commuting with the Schrödinger operator — by analogy with the classical two-body problem, these may be called the *Runge–Lenz operators*.[11] The Runge–Lenz operators are defined only on the bound states of the Schrödinger operator, and their useful properties depend on the explicit functional form of the Coulomb potential. Amazingly enough, on the vector space of bound states we can combine the Runge–Lenz and angular momentum operators to form a representation of $so(4)$. The construction of the Runge–Lenz operators was first published by Pauli [P].

We consider one eigenspace of the Schrödinger operator at a time. Fix an eigenvalue $E < 0$ of the Schrödinger operator. Let V_E denote the eigenspace corresponding to E. From Proposition 8.5 we know that there is a representation of $su(2)$ on the eigenspace V_E. We will extend this to a representation of $su(2) \oplus su(2)$. To this end we introduce three more operators on V_E. Define

---

[11] For an introduction to the Runge–Lenz vectors in the classical context, see [Mi].

the Runge–Lenz operators:

$$\mathbf{R_i} := \frac{i\hbar}{\sqrt{-8mE}} \left(\mathbf{L_k}\partial_y + \partial_y\mathbf{L_k} - \mathbf{L_j}\partial_z - \partial_z\mathbf{L_j} + \frac{2me^2 x}{\hbar^2\sqrt{x^2 + y^2 + z^2}} \right)$$

$$\mathbf{R_j} := \frac{i\hbar}{\sqrt{-8mE}} \left(\mathbf{L_i}\partial_z + \partial_z\mathbf{L_i} - \mathbf{L_k}\partial_x - \partial_x\mathbf{L_k} + \frac{2me^2 y}{\hbar^2\sqrt{x^2 + y^2 + z^2}} \right)$$

$$\mathbf{R_k} := \frac{i\hbar}{\sqrt{-8mE}} \left(\mathbf{L_j}\partial_x + \partial_x\mathbf{L_j} - \mathbf{L_i}\partial_y - \partial_y\mathbf{L_i} + \frac{2me^2 z}{\hbar^2\sqrt{x^2 + y^2 + z^2}} \right).$$

Note that this collection of formulas is cyclic.

To see how these operators will play a role in building our representation of $su(2) \oplus su(2)$, we calculate their Lie brackets. In this section we give formulas without proof, leaving the details for Section 8.7. First we note that

$$\left[\mathbf{R_i}, \mathbf{R_j} \right] = \mathbf{L_k} \tag{8.19}$$

on the E-eigenspace of the Schrödinger operator. The calculation relies on the fact that the operator $\mathbf{H} - E$ is 0 on the eigenspace. By cyclic reasoning, we have also $\left[\mathbf{R_j}, \mathbf{R_k} \right] = \mathbf{L_i}$ and $[\mathbf{R_k}, \mathbf{R_i}] = \mathbf{L_j}$. Next we have, as a purely algebraic consequence of the definitions,

$$[\mathbf{L_i}, \mathbf{R_j}] = [\mathbf{R_i}, \mathbf{L_j}] = \mathbf{R_k}, \tag{8.20}$$

and hence, cyclically, we have $[\mathbf{L_j}, \mathbf{R_k}] = [\mathbf{L_k}, \mathbf{R_i}] = \mathbf{R_j}$ and $[\mathbf{R_j}, \mathbf{L_k}] = [\mathbf{R_k}, \mathbf{L_i}] = \mathbf{R_j}$. Another algebraic calculation yields

$$[\mathbf{L_i}, \mathbf{R_i}] = 0 \tag{8.21}$$

and hence $[\mathbf{L_j}, \mathbf{R_j}] = [\mathbf{L_k}, \mathbf{R_k}] = 0$.

We will find it helpful to know that

$$\mathbf{R} \cdot \mathbf{L} := \mathbf{R_i}\mathbf{L_i} + \mathbf{R_j}\mathbf{L_j} + \mathbf{R_k}\mathbf{L_k} = 0. \tag{8.22}$$

Similarly, we have

$$\mathbf{L} \cdot \mathbf{R} := \mathbf{L_i}\mathbf{R_i} + \mathbf{L_j}\mathbf{R_j} + \mathbf{L_k}\mathbf{R_k} = 0. \tag{8.23}$$

The proofs of these two equalities are algebraic computations and do not require the Schrödinger eigenvalue equation.

However, we will need the Schrödinger eigenvalue equation to calculate

$$\mathbf{L}^2 + \mathbf{R}^2 := \mathbf{L} \cdot \mathbf{L} + \mathbf{R} \cdot \mathbf{R} = 1 + \frac{me^4}{2E\hbar^2}. \qquad (8.24)$$

Now we are ready to introduce the representation of $so(4)$. Define

$$\mathbf{A_i} := (\mathbf{L_i} + \mathbf{R_i})/2 \qquad\qquad \mathbf{B_i} := (\mathbf{L_i} - \mathbf{R_i})/2$$
$$\mathbf{A_j} := (\mathbf{L_j} + \mathbf{R_j})/2 \qquad\qquad \mathbf{B_j} := (\mathbf{L_j} - \mathbf{R_j})/2$$
$$\mathbf{A_k} := (\mathbf{L_k} + \mathbf{R_k})/2 \qquad\qquad \mathbf{B_k} := (\mathbf{L_k} - \mathbf{R_k})/2.$$

It follows easily from Equations 8.19 and 8.20 and the fact that $\mathbf{L}$ is a representation that

$$[\mathbf{A_i}, \mathbf{A_j}] = \frac{1}{4}\left([\mathbf{L_i}, \mathbf{L_j}] + [\mathbf{R_i}, \mathbf{R_j}] + [\mathbf{L_i}, \mathbf{R_j}] + [\mathbf{R_i}, \mathbf{L_j}]\right)$$
$$= \frac{1}{4}(2\mathbf{L_k} + 2\mathbf{R_k}) = \mathbf{A_k}$$

and likewise $[\mathbf{A_j}, \mathbf{A_k}] = \mathbf{A_i}$ and $[\mathbf{A_k}, \mathbf{A_i}] = \mathbf{A_j}$. So the $\mathbf{A}$'s form a representation of $su(2)$. We will call this the *diagonal $su(2)$ representation*, referring to the *diagonal subgroup* $\{(\mathbf{q}, \mathbf{q}): \mathbf{q} \in su(2)\}$ inside $su(2) \oplus su(2)$. Similarly, we have

$$[\mathbf{B_i}, \mathbf{B_j}] = \frac{1}{4}\left([\mathbf{L_i}, \mathbf{L_j}] + [\mathbf{R_i}, \mathbf{R_j}] - [\mathbf{L_i}, \mathbf{R_j}] - [\mathbf{R_i}, \mathbf{L_j}]\right)$$
$$= \frac{1}{4}(2\mathbf{L_k} - 2\mathbf{R_k}) = \mathbf{B_k},$$

In addition, each $\mathbf{A}$ commutes with each $\mathbf{B}$. For example,

$$[\mathbf{A_i}, \mathbf{B_i}] = \frac{1}{4}\left([\mathbf{L_i}, \mathbf{L_i}] - [\mathbf{R_i}, \mathbf{R_i}] + [\mathbf{R_i}, \mathbf{L_i}] - [\mathbf{L_i}, \mathbf{R_i}]\right) = 0,$$

by Equations 8.21 and

$$[\mathbf{A_i}, \mathbf{B_j}] = [\mathbf{L_i}, \mathbf{L_j}] - [\mathbf{R_i}, \mathbf{R_j}] - [\mathbf{R_i}, \mathbf{L_j}] + [\mathbf{L_i}, \mathbf{R_j}]$$
$$= \mathbf{L_k} - \mathbf{L_k} - \mathbf{R_k} + \mathbf{R_k} = 0$$

by Equations 8.19 and 8.20. So we have a representation of $so(4)$.

Let us calculate the value of the Casimir operator for each of the representations of $su(2)$. Because $\mathbf{L} \cdot \mathbf{R} = \mathbf{R} \cdot \mathbf{L} = 0$, we have

$$\mathbf{A}^2 = \mathbf{B}^2 = \frac{1}{4}\left(\mathbf{L}^2 + \mathbf{R}^2\right) = \frac{1}{4}\left(1 + \frac{me^4}{2E\hbar^2}\right).$$

Recall that the value of the Casimir operator determines an irreducible representation of $su(2)$. From Section 8.4, we know that the value of the Casimir must be $-\frac{1}{4}(n^2 + 2n)$, where n is a nonnegative integer. So

$$-(n^2 + 2n) = 1 + \frac{me^4}{2E\hbar^2}$$

and hence each eigenvalue E of the Schrödinger operator must be of the form

$$E = \frac{-me^4}{2\hbar^2(n + 1)^2} \tag{8.25}$$

for some nonnegative integer n. We remind the reader that **m** denotes the mass of the electron and **e** is the charge on the electron. Note that among the consequences of Equation 8.25 is the fact that the only *bona fide* eigenspaces for the Schrödinger operator are those corresponding to negative energy levels.

Furthermore, if we fix a nonnegative integer n, the eigenspace corresponding to the eigenvalue $E = -(\mathbf{me}^4)/2(n + 1)^2$ must be made up only of irreducible representations (of $so(4)$) isomorphic to $\mathcal{P}^n \otimes \mathcal{P}^n$. In particular, because the dimension of the eigenspace is finite (by Proposition 8.14) the dimension of the eigenspace must be an integer multiple of the dimension $(n + 1)^2$ of $\mathcal{P}^n \otimes \mathcal{P}^n$. Thus the lowest possible eigenvalue is $-\mathbf{me}^4/2$ and the dimension of its eigenspace must be divisible by 1, while the second lowest possible eigenvalue is $-\mathbf{me}^4/8$, and the dimension of its eigenspace must be divisible by 4, and so on. The actual dimension of the eigenspace can be determined experimentally. We collect the results in a table (Figure 8.4). To

n	Energy level (eigenvalue)	Dimension of $\mathcal{P}^n$	Dimension of electronic shell (dimension of eigenspace)
0	$-\mathbf{me}^4/2$	1	2
1	$-\mathbf{me}^4/8$	4	8
2	$-\mathbf{me}^4/18$	9	18
3	$-\mathbf{me}^4/32$	16	32
etc.			

Figure 8.4. A comparison of theory with experiment.

put it another way, our representation-theoretic calculation has predicted that the dimensions of the irreducible representations of $so(4)$ should divide the multiplicities of the corresponding energy levels of the hydrogen atom. But this is true: the multiplicities are

$$2 = 2 \times \mathbf{1}, \ 8 = 2 \times \mathbf{4}, \ 18 = 2 \times \mathbf{9} \ \text{and} \ 32 = 2 \times \mathbf{16}.$$

It is not an accident that the numbers 2, 8, 18, 32 are the lengths of the rows of
the periodic table. See Sections 1.3 and 1.4. As in Section 7.3, the prediction
is off by a factor of two. The factor of two is due to the spin of the electron.
See Section 11.4.

In this section we have presented the celebrated $so(4)$ symmetry of the
hydrogen atom. Thus the hydrogen atom has more symmetry than is evident
from its spatial symmetry. Is this a happy accident or a sign of a deeper symmetry in our world? The author does not know. The fourth dimension here is
an abstract theoretical construct, not a physical reality. However — and this is
one of the main points of this text — the abstract symmetry has real physical
consequences.

8.7   The Heinous Details

In this section we collect the calculations omitted in the previous section.
They are straightforward but tedious. Some details have been left to the reader
as exercises.

All calculations in this section involve operators on functions, as explained
in Section 8.1. Thus, for example, $[\mathbf{L_i}, y] = z$, since for any differentiable
function f of three real variables we have

$$[\mathbf{L_i}, y]f = \mathbf{L_i}(yf(x, y, z)) - y\mathbf{L_i}f(x, y, z)$$
$$= (z\partial_y - y\partial_z)(yf(x, y, z)) - (yz\partial_y - y^2\partial_z)f(x, y, z)$$
$$= zf(x, y, z).$$

We will find the following shorthand for part of the Runge–Lenz operators
helpful. Set

$$\mathbf{M_i} := \left(\mathbf{L_k}\partial_y + \partial_y\mathbf{L_k} - \mathbf{L_j}\partial_z - \partial_z\mathbf{L_j}\right)$$
$$= 2\left(y\partial_x\partial_y - x\partial_y^2 + z\partial_x\partial_z - x\partial_z^2 + \partial_x\right).$$

Thus $\mathbf{R_i}$ is the sum of the differential operator $\mathbf{M_i}$ and a multiplication operator. The cyclic versions of this formula define $\mathbf{M_j}$ and $\mathbf{M_k}$.

First we prove Equation 8.19: $[\mathbf{R_i}, \mathbf{R_j}] = \mathbf{L_k}$. With the help of Exercises 8.18 and 8.20 we find

$$[\mathbf{R_i}, \mathbf{R_j}] = \frac{\hbar^2}{8mE}\left[\mathbf{M_i} + \frac{2me^2}{\hbar^2}\frac{x}{\sqrt{x^2+y^2+z^2}}, \mathbf{M_j} + \frac{2me^2}{\hbar^2}\frac{y}{\sqrt{x^2+y^2+z^2}}\right]$$

$$= \frac{\hbar^2}{8mE}\left([\mathbf{M_i}, \mathbf{M_j}] + \frac{2me^2}{\hbar^2}\left(\left[\mathbf{M_i}, \frac{y}{\sqrt{x^2+y^2+z^2}}\right] + \left[\frac{x}{\sqrt{x^2+y^2+z^2}}, \mathbf{M_j}\right]\right)\right)$$

$$= \frac{1}{E}\left(\frac{\hbar^2}{2m}\mathbf{L_k}\nabla^2 + e^2\mathbf{L_k}\frac{1}{\sqrt{x^2+y^2+z^2}}\right).$$

If we restrict our attention to one energy level of the Schrödinger operator, then the Schrödinger eigenvalue equation (Equation 8.16) holds so that

$$\frac{\hbar^2}{2m}\mathbf{L_k}\nabla^2 + e^2\mathbf{L_k}\frac{1}{\sqrt{x^2+y^2+z^2}} = E\mathbf{L_k},$$

and hence

$$[\mathbf{R_i}, \mathbf{R_j}] = \mathbf{L_k}.$$

Next we verify Equation 8.20. A relatively straightforward calculation yields $[\mathbf{L_i}, \mathbf{M_j}] = \mathbf{M_k}$. By Exercise 8.14 we know that the function

$$1/\sqrt{x^2+y^2+z^2}$$

commutes with the angular momentum operator $\mathbf{L_i}$. Hence by the product rule we have

$$\left[\mathbf{L_i}, \frac{y}{\sqrt{x^2+y^2+z^2}}\right] = \left[\mathbf{L_i}, \frac{1}{\sqrt{x^2+y^2+z^2}}\right]y + \frac{1}{\sqrt{x^2+y^2+z^2}}[\mathbf{L_i}, y]$$

$$= \frac{1}{\sqrt{x^2+y^2+z^2}}[\mathbf{L_i}, y]$$

$$= \frac{z}{\sqrt{x^2+y^2+z^2}}.$$

Putting these together we have

$$[\mathbf{L_i}, \mathbf{R_j}] = \frac{i}{2\sqrt{-E}}\left(\frac{\hbar}{\sqrt{2m}}[\mathbf{L_i}, \mathbf{M_j}] + \frac{\sqrt{2m}e^2}{\hbar}\left[\mathbf{L_i}, \frac{y}{\sqrt{x^2+y^2+z^2}}\right]\right)$$

$$= \frac{i}{2\sqrt{-E}}\left(\frac{\hbar}{\sqrt{2m}}\mathbf{M_k} + \frac{\sqrt{2m}e^2}{\hbar}\frac{z}{\sqrt{x^2+y^2+z^2}}\right) = \mathbf{R_k}.$$

A similar calculation (left to the reader as Exercise 8.21) shows that

$$[\mathbf{R_i}, \mathbf{L_j}] = \mathbf{R_k}.$$

To see that Equation 8.22 is true, note that

$$
\begin{aligned}
\mathbf{M_i L_i} &= \left(2x\partial_y^2 - 2y\partial_x\partial_y - 2z\partial_x\partial_z + 2x\partial_z^2 - 2\partial_x\right)\left(y\partial_z - z\partial_y\right) \\
&= 2(xy\partial_y^2\partial_z - yz\partial_x\partial_z^2) + 2(yz\partial_x\partial_y^2 - xz\partial_y\partial_z^2) \\
&\quad + 4(z\partial_x\partial_y - y\partial_x\partial_z) + 2(xy\partial_z^3 - xz\partial_y^3) \\
&\quad + (2z^2\partial_x\partial_y\partial_z - 2y^2\partial_x\partial_y\partial_z).
\end{aligned}
$$

If we add the three different cyclic versions of the first parenthesized pair of terms in this last expression we get

$$(xy\partial_y^2\partial_z - yz\partial_x\partial_z^2) + (yz\partial_z^2\partial_x - zx\partial_y\partial_x^2) + (zx\partial_x^2\partial_y - xy\partial_z\partial_y^2) = 0.$$

The sums of the cyclic versions of the other pairs of terms are also equal to zero. Hence

$$\mathbf{M_i L_i} + \mathbf{M_j L_j} + \mathbf{M_k L_k} = 0. \tag{8.26}$$

Also, we have

$$
\begin{aligned}
&\frac{x}{\sqrt{x^2 + y^2 + z^2}}\mathbf{L_i} + \frac{y}{\sqrt{x^2 + y^2 + z^2}}\mathbf{L_j} + \frac{z}{\sqrt{x^2 + y^2 + z^2}}\mathbf{L_k} \\
&= \frac{1}{\sqrt{x^2 + y^2 + z^2}}\left(xz\partial_y - xy\partial_z + yx\partial_z - yz\partial_x + zy\partial_x - zx\partial_y\right) = 0.
\end{aligned}
$$

Since $\mathbf{R_i}$ is a linear combination of $\mathbf{M_i}$ and $x/\sqrt{x^2 + y^2 + z^2}$, etc., we conclude that Equation 8.22 holds:

$$\mathbf{R_i L_i} + \mathbf{R_j L_j} + \mathbf{R_k L_k} = 0.$$

A similar argument shows that Equations 8.21 and 8.23 hold true. First we have

$$
\begin{aligned}
\mathbf{L_i M_i} &= \left(y\partial_z - z\partial_y\right)\left(2x\partial_y^2 - 2y\partial_x\partial_y - 2z\partial_x\partial_z + 2x\partial_z^2 - 2\partial_x\right) \\
&= 2(xy\partial_y^2\partial_z - yz\partial_x\partial_z^2) + 2(xy\partial_z^3 - xz\partial_y^3) \\
&\quad + 2(yz\partial_x\partial_y^2 - xz\partial_y\partial_z^2) + 2(z^2 - y^2)\partial_x\partial_y\partial_z \\
&\quad + 4(z^2\partial_x\partial_y - y^2\partial_x\partial_z) = \mathbf{M_i L_i},
\end{aligned}
$$

so by Equation 8.26 we know that $L_iM_i + L_jM_j + L_kM_k = 0$. Second, we have

$$L_i \frac{x}{\sqrt{x^2 + y^2 + z^2}} = \frac{1}{\sqrt{x^2 + y^2 + z^2}} L_i x$$

$$= \frac{1}{\sqrt{x^2 + y^2 + z^2}} (y\partial_z - z\partial_x)x$$

$$= \frac{1}{\sqrt{x^2 + y^2 + z^2}} (xy\partial_z - xz\partial_x)$$

$$= \frac{x}{\sqrt{x^2 + y^2 + z^2}} L_i$$

$$= \frac{1}{\sqrt{x^2 + y^2 + z^2}} (xz\partial_y - xy\partial_z),$$

whose cyclic version is equal to 0. We conclude that $[L_i, R_i] = 0$ and

$$L_iR_i + L_jR_j + L_kR_k = 0.$$

To verify Equation 8.24, we first calculate the operator

$$\mathbf{L}^2 := L_i^2 + L_j^2 + L_k^2$$
$$= (z\partial_y - y\partial_z)^2 + (x\partial_z - z\partial_x)^2 + (y\partial_x - x\partial_y)^2$$
$$= (y^2 + z^2)\partial_x^2 + (x^2 + y^2)\partial_z^2 + (x^2 + z^2)\partial_y^2$$
$$\quad - 2yz\partial_y\partial_z - 2xz\partial_x\partial_z - 2xy\partial_x\partial_y - 2x\partial_x - 2y\partial_y - 2z\partial_z$$
$$= (x^2 + y^2 + z^2)\nabla^2 - (x\partial_x + y\partial_y + z\partial_z)^2 - (x\partial_x + y\partial_y + z\partial_z).$$

To calculate $\mathbf{R}^2$, begin by noting that

$$4E R_i^2 = \left(\frac{\hbar}{\sqrt{2m}} M_i + \frac{\sqrt{2m}e^2 x}{\hbar\sqrt{x^2 + y^2 + z^2}} \right)^2.$$

For clarity, we will calculate the squared terms and the cross-terms separately. For the first squared term we have (up to a constant)

$$\frac{1}{4}M_i^2 = \left(x\partial_y^2 - y\partial_x\partial_y - z\partial_x\partial_z + x\partial_z^2 - \partial_x \right)^2$$
$$= x^2\partial_y^4 + y^2\partial_x^2\partial_y^2 + y\partial_x^2\partial_y + z^2\partial_x^2\partial_z^2 + z\partial_x^2\partial_z + x^2\partial_z^4 + \partial_x^2$$
$$\quad - 2xy\partial_x\partial_y^3 - 2x\partial_x\partial_y^2 - y\partial_y^3 - 2xz\partial_x\partial_y^2\partial_z - z\partial_y^2\partial_z + 2x^2\partial_y^2\partial_z^2$$
$$\quad - 2x\partial_x\partial_y^2 - \partial_y^2 + 2yz\partial_x^2\partial_y\partial_z - 2xy\partial_x\partial_y\partial_z^2 - y\partial_y\partial_z^2 + 2y\partial_x^2\partial_y$$
$$\quad - 2xz\partial_x\partial_z^3 - z\partial_z^3 - 2x\partial_x\partial_z^2 + 2z\partial_x^2\partial_z - 2x\partial_x\partial_z^2 - \partial_z^2.$$

Adding and subtracting the terms $x^2\partial_x^4 + y^2\partial_x^2\partial_y^2 + z^2\partial_x^2\partial_z^2$ and regrouping terms, we find that

$$\frac{1}{4}\mathbf{M}_i^2 = \left(\partial_x^2 - \partial_y^2 - \partial_z^2\right)$$

$$+ \left(x^2\partial_y^4 + x^2\partial_z^4 + y^2\partial_x^2\partial_y^2 + z^2\partial_x^2\partial_z^2 + 2x^2\partial_y^2\partial_z^2 x^2\partial_x^4 + y^2\partial_x^2\partial_y^2 + z^2\partial_x^2\partial_z^2\right)$$

$$- \left(x^2\partial_x^4 + y^2\partial_x^2\partial_y^2 + z^2\partial_x^2\partial_z^2 + 2xy\partial_x\partial_y^3 + 2xy\partial_x\partial_y\partial_z^2\right.$$

$$\left. + 2xz\partial_x\partial_z^3 + z\partial_z^3 + y\partial_y\partial_z^2 + x\partial_x\partial_z^2\right)$$

$$- \left(y\partial_y^3 + z\partial_y^2\partial_z + 2x\partial_x\partial_y^2\right)$$

$$+ \left(2yz\partial_x^2\partial_y\partial_z - 2xz\partial_x\partial_y^2\partial_z - 2x\partial_x\partial_y^2 + 2y\partial_x^2\partial_y\right.$$

$$\left. - 3x\partial_x\partial_z^2 + 3z\partial_x^2\partial_z + y\partial_x^2\partial_y\right).$$

Straightforward but tedious algebra and cyclic arguments then lead to

$$\frac{1}{4}(\mathbf{M}_i^2 + \mathbf{M}_j^2 + \mathbf{M}_k^2) = \left(1 - \mathbf{L}^2\right)\nabla^2.$$

Hence the cyclic version of the first squared term of $4E\mathbf{R}_i^2$ is

$$\frac{2\hbar^2}{m}(1 - \mathbf{L}^2)\nabla^2. \tag{8.27}$$

The second squared term is

$$\frac{2me^4x^2}{\hbar^2(x^2 + y^2 + z^2)}$$

and its cyclic version is

$$\frac{2me^4}{\hbar^2}. \tag{8.28}$$

To calculate the sum of the cross terms we consider

$$\left(\mathbf{M}_i\frac{x}{\sqrt{x^2 + y^2 + z^2}} + \frac{x}{\sqrt{x^2 + y^2 + z^2}}\mathbf{M}_i\right)$$

$$= \left[\mathbf{M}_i, \frac{x}{\sqrt{x^2 + y^2 + z^2}}\right] + 2\frac{x}{\sqrt{x^2 + y^2 + z^2}}\mathbf{M}_i$$

$$= \frac{1}{\sqrt{x^2 + y^2 + z^2}}\left[\mathbf{M}_i, x\right]$$

$$+ x\left[\mathbf{M}_i, \frac{1}{\sqrt{x^2 + y^2 + z^2}}\right] + 2\frac{x}{\sqrt{x^2 + y^2 + z^2}}\mathbf{M}_i.$$

Let us calculate the cyclic versions of these three terms. First, use Exercise 8.19 to see that

$$\frac{1}{\sqrt{x^2 + y^2 + z^2}} \left([\mathbf{M_i}, x] + [\mathbf{M_j}, y] + [\mathbf{M_k}, z] \right)$$

$$= \frac{2}{\sqrt{x^2 + y^2 + z^2}} \left(3 + 2x\partial_x + 2y\partial_y + 2z\partial_z \right).$$

Next, we have, also with the help of Exercise 8.19, that

$$\left[x\mathbf{M_i} + y\mathbf{M_j} + z\mathbf{M_k}, \frac{1}{\sqrt{x^2 + y^2 + z^2}} \right]$$

$$= \frac{2}{(\sqrt{x^2 + y^2 + z^2})^3} \left(x^2 + y^2 + z^2 + x^2 y\partial_y + y^2 z\partial_z + z^2 x\partial_x + x^2 z\partial_z \right.$$

$$\left. + y^2 x\partial_x + z^2 y\partial_y - x^2 z\partial_z - xy^2\partial_x - yz^2\partial_y - x^2 y\partial_y - y^2 z\partial_z - xz^2\partial_x \right)$$

$$= \frac{2}{\sqrt{x^2 + y^2 + z^2}}.$$

Finally, we have

$$\frac{2}{\sqrt{x^2 + y^2 + z^2}} \left(x\mathbf{M_i} + y\mathbf{M_j} + z\mathbf{M_k} \right)$$

$$= \frac{4}{\sqrt{x^2 + y^2 + z^2}} \left(2xy\partial_x\partial_y + 2yz\partial_y\partial_z + 2xz\partial_x\partial_z \right.$$

$$\left. - x^2\partial_y^2 - x^2\partial_z^2 - y^2\partial_x^2 - y^2\partial_z^2 - z^2\partial_x^2 - z^2\partial_y^2 + x\partial_x + y\partial_y + z\partial_z \right)$$

$$= \frac{4}{\sqrt{x^2 + y^2 + z^2}} \left((x\partial_x + y\partial_y + z\partial_z)^2 + x\partial_x \right).$$

Adding these three results we obtain

$$\frac{4}{\sqrt{x^2 + y^2 + z^2}} \left(2 + 2x\partial_x + 2y\partial_y + 2z\partial_z + 2xy\partial_x\partial_y + 2yz\partial_y\partial_z \right.$$

$$\left. + 2xz\partial_x\partial_z - x^2\partial_y^2 - x^2\partial_z^2 - y^2\partial_x^2 - y^2\partial_z^2 - z^2\partial_x^2 - z^2\partial_y^2 \right).$$

It follows that the cyclic version of the sum of the cross terms is

$$\frac{4e^2(\mathbf{L}^2 - 1)}{\sqrt{x^2 + y^2 + z^2}}. \tag{8.29}$$

Thus $\mathbf{R}^2$ is equal to $\frac{1}{4E}$ times the sum of Formulas 8.27, 8.28 and 8.29. Recalling Formula 8.27 for $\mathbf{L}^2$ and collecting terms, we find

$$\mathbf{L}^2 + \mathbf{R}^2$$

$$= \mathbf{L}^2 + \frac{1}{4E} \left(\frac{2\hbar^2}{\mathbf{m}}(\mathbf{L}^2 - 1)\nabla^2 + \frac{2\mathbf{m}e^4}{\hbar^2} + (\mathbf{L}^2 - 1)\frac{4e^2}{\sqrt{x^2 + y^2 + z^2}} \right)$$

$$= \mathbf{L}^2 \left(1 + \frac{\hbar^2}{2\mathbf{m}E}\nabla^2 + \frac{1}{E}\frac{e^2}{\sqrt{x^2 + y^2 + z^2}} \right)$$

$$- \frac{1}{4E} \left(\frac{2\hbar^2}{\mathbf{m}}\nabla^2 + \frac{4e^2}{\sqrt{x^2 + y^2 + z^2}} \right) + \frac{\mathbf{m}e^4}{2E\hbar^2}.$$

If we are on an eigenspace for the eigenvalue $E < 0$ for the Schrödinger operator

$$-\frac{\hbar^2}{2\mathbf{m}}\nabla^2 - \frac{e^2}{\sqrt{x^2 + y^2 + z^2}},$$

then this expression reduces to

$$\mathbf{L}^2 + \mathbf{R}^2 = 1 + \frac{\mathbf{m}e^4}{2E\hbar^2}.$$

Congratulations! By working through this section you have verified the celebrated $so(4)$ symmetry of the hydrogen atom. A pause for celebration would be quite appropriate.

8.8 Exercises

Exercise 8.1 *Check that the quantity*

$$\frac{\mathbf{m}e^4}{2\hbar^2}$$

has the units of energy.

Exercise 8.2 *Find all Lie subalgebras of $\mathfrak{g}_Q$.*

Exercise 8.3 *Show that the Heisenberg Lie algebra $\mathcal{H}$ is isomorphic to neither the Lie algebra $su(2)$ nor the trivial three-dimensional Lie algebra.*

Exercise 8.4 *Show that $so(3)$ is the Lie algebra associated to the Lie group $SO(3)$.*

Exercise 8.5 *Show that the Lie group $SO(3)$ is not isomorphic to the Lie group $SU(2)$. (Hint: consider the center of each group, i.e., the set of group elements that commute with every other element.) Note that we have shown in the text that the Lie algebra $su(2)$ is isomorphic to the Lie algebra $so(3)$. Conclude that Lie algebras do not uniquely determine Lie groups.*

Exercise 8.6 *For any natural number n the Lie algebra $so(n)$ is defined by*

$$so(n) := \left\{ A \in g\ell\left(\mathbb{R}^n\right): A + A^* = 0, \operatorname{Tr} A = 0 \right\}.$$

Show that the dimension of $so(n)$ as a real vector space is the triangle number $n(n-1)/2$.

Exercise 8.7 *Show that every group element $g \in SU(2)$ is of the form $\exp M$ for some algebra element $M \in su(2)$.*

Suppose $(su(2), V, \rho)$ is a finite-dimensional Lie algebra representation of $su(2)$. Define a function $\sigma : SU(2) \to \mathcal{GL}(V)$ by

$$\sigma(X) := \exp(\rho(M)),$$

where $\exp M = X$. Show that σ is well defined, that the image of σ indeed lies in $\mathcal{GL}(V)$ and that σ is a group representation. (Remark: The finite dimensionality of V is necessary to assure convergence of the exponential of $\rho(M)$.) Readers familiar with the definition of the exponential map on an arbitrary Lie algebra should prove the corresponding generalization.

Exercise 8.8 *Suppose V is a vector space. Is the vector space $g\ell(V)$ a group under composition of linear transformations?*

Exercise 8.9 *Construct a Lie algebra representation $(su(2), V, \rho)$ with two highest weight vectors v_0 and v_1 such that their corresponding eigenvalues λ_0 and λ_1 (respectively) are not equal.*

Exercise 8.10 *In this exercise we construct infinite-dimensional irreducible representations of the Lie algebra $su(2)$. Suppose λ is a complex number such that $\lambda \neq in$ for any nonnegative integer n. Consider a countable set $S := \{v_0, v_1, v_2, \dots\}$ and let V denote the complex vector space of finite linear combinations of elements of S. Show that V can be made into a complex* *scalar product space. Define three linear operators $A, X, Y : V \to V$ by setting, for any nonnegative integer k,*

$$A v_k := \left(\frac{\lambda}{2} - ik \right) v_k$$

$$X v_k := \begin{cases} (\lambda - ik + i) v_{k-1}, & k \neq 0 \\ 0, & k = 0 \end{cases}$$

$$Y v_k := (ik + i) v_{k+1}.$$

Next show that the linear transformation $\rho : su(2) \to g\ell(V)$ defined by

$$\rho(\mathbf{i}) := A,$$

$$\rho(\mathbf{j}) := \frac{1}{2} (X + Y),$$

$$\rho(\mathbf{k}) := \frac{i}{2} (X - Y)$$

is a representation of $su(2)$. Finally, show that V is infinite dimensional and irreducible. (Hint: For irreducibility, show that for every subrepresentation W must contain the vector v_0.)

Exercise 8.11 *Find an example of a reducible (i.e., not irreducible) representation $(su(2), V, \rho)$ such that the Casimir operator is a scalar multiple of the identity on V. (This implies that the converse of Proposition 8.11 is false.)*

Exercise 8.12 *Show that the function $\rho_1 \otimes I + I \otimes \rho_2$ from Definition 8.10 satisfies the definition of a Lie algebra representation.*

Exercise 8.13 *Show that $[\partial_y, g(x, y, z)] = \frac{\partial g}{\partial y}$. Is this an equation of functions or an equation of operators?*

Exercise 8.14 *Suppose $f \in \mathcal{I}$, i.e., suppose that $f \in L^2(\mathbb{R}^3)$ and f is invariant under rotations. Show that f is a cyclic formula. Show that f commutes with the angular momentum operators, i.e., show that $[\mathbf{L_i}, f] = [\mathbf{L_j}, f] = [\mathbf{L_k}] = 0$.*

Exercise 8.15 (Used in Section 8.5) *Show that the operator $\mathbf{H}$ commutes with the natural representation of $SO(3)$ on $L^2(\mathbb{R}^3)$.*

Exercise 8.16 *Consider a free quantum particle in one dimension, i.e., consider the system whose state space is $L^2(\mathbb{R})$ and whose energy operator is*

$-(\hbar^2/2\mathbf{m})\nabla^2 = -(\hbar^2/2\mathbf{m})\partial^2$. *Show that this operator has no eigenfunctions in $L^2(\mathbb{R})$. On the other hand, consider the function*

$$\psi(x) = i \int_1^2 e^{i\omega x} d\omega.$$

Show that $\psi \in L^2(\mathbb{R})$ and that any energy measurement of a particle in the state ψ will yield a positive result; in fact, it will yield a result in the interval $[\hbar^2/2\mathbf{m}, 2\hbar^2/\mathbf{m}]$.

Exercise 8.17 *Suppose $E < 0$ and $\mathbf{e} > 0$. Show that if*

$$\ell > \sqrt{\mathbf{me}^2}/\hbar\sqrt{-2E},$$

then for every $r > 0$ the quantity

$$\frac{\mathbf{e}^2}{r} + E - \frac{\ell(\ell+1)\hbar^2}{2\mathbf{m}r^2}$$

is negative.

Exercise 8.18 (Used in Section 8.7) *In this exercise, both equations are equations of operators. Show that for any natural number n,*

$$\left[\partial_x, \frac{1}{\left(\sqrt{x^2+y^2+z^2}\right)^n}\right] = \frac{-nx}{\left(\sqrt{x^2+y^2+z^2}\right)^{n+2}}.$$

Show also that

$$\left[\frac{x}{\sqrt{x^2+y^2+z^2}}, \frac{y}{\sqrt{x^2+y^2+z^2}}\right] = 0.$$

Exercise 8.19 (Used in Section 8.7) *In this exercise, all equations are equations of operators. Show that*

$$\left[\mathbf{M_i}, \frac{1}{\sqrt{x^2+y^2+z^2}}\right] = \frac{2}{(\sqrt{x^2+y^2+z^2})^3}\left(x - y\mathbf{L_k} + z\mathbf{L_j}\right),$$

that

$$[\mathbf{M_i}, x] = 2 + 2y\partial_y + 2z\partial_z,$$

and that

$$[\mathbf{M_i}, z] = 2z\partial_x - 4x\partial_z.$$

Exercise 8.20 (Used in Section 8.7) *Verify the following equations of operators:*

$$[\mathbf{M_i}, \mathbf{M_j}] = 4\mathbf{L_k}\nabla^2,$$

$$\left[\mathbf{M_i}, \frac{y}{\sqrt{x^2+y^2+z^2}}\right] + \left[\frac{x}{\sqrt{x^2+y^2+z^2}}, \mathbf{M_j}\right] = 4\mathbf{L_k}\frac{1}{\sqrt{x^2+y^2+z^2}}.$$

Exercise 8.21 (Used in Section 8.7) *Verify the following equations of operators:*

$$[\mathbf{R_i}, \mathbf{L_j}] = \mathbf{R_k}$$
$$[\mathbf{R_i}, \mathbf{L_i}] = 0.$$

Exercise 8.22 *Is there anything in group representations of $SU(2)$ or $SO(3)$ analogous to the Casimir operator for Lie algebra $su(2)$?*

9
The Group $SO(4)$ Symmetry of the Hydrogen Atom

But first I pray yow, of youre curteisye,
That ye n' arette it nat my vileynye,
Thogh that I pleynly speke in this mateere,
To telle yow hir wordes and hir cheere,
Ne thogh I speke hir wordes proprely.
For this ye knowen al so wel as I,
Whoso shal telle a tale after a man,
He moot reherce as ny as evere he kan
Everich a word, if it be in his charge,
Al speke he never so rudeliche and large,
Or ellis he moot telle his tale untrewe,
Or feyne thyng, or fynde wordes newe.
He may nat spare, althogh he were his brother;
He moot as wel seye o word as another.
— Geoffrey Chaucer, *The Canterbury Tales* [Ch, lines 725–38]

In this chapter we present Fock's construction of a representation of the group $SO(4)$ on $L^2(\mathbb{R}^3)$, the phase space of the hydrogen atom. This representation commutes with the Schrödinger operator (otherwise known as the energy operator), and hence it is a physical symmetry of the hydrogen atom.

In one sense the group $SO(4)$ symmetry is no better than the algebra $so(4)$ symmetry, as both lead to the same conclusion about the dimensions of elementary states of the hydrogen atom. However, the group symmetry is more powerful than the algebra symmetry. From a strictly logical point of view, one can deduce the algebra symmetry from the group symmetry, but not vice versa. More impressively, the group symmetry yields a different proof of the finite dimension of the energy levels (Proposition 8.14).

Since the group symmetry is more powerful, it is not surprising that it requires stronger analytical technology. Instead of developing this technology, we put the burden on the reader to find it elsewhere. This chapter begins with prerequisites for Fock's argument in Section 9.1. We omit many proofs, instead giving a sketch of some key ingredients and references for the necessary ideas and techniques. In Section 9.2 we translate the original article by Fock.

9.1   Preliminaries

Fock's argument rests on the theory of the Fourier transform. In particular, he uses the momentum-space version of the Schrödinger equation. We let $\hat{f}$ denote the Fourier transform of $f \in L^2(\mathbb{R}^3)$.

Proposition 9.1 *Suppose* $f \in L^2(\mathbb{R}^3)$ *satisfies the position-space Schrödinger equation (Equation 8.16). Then the Fourier transform $\hat{f}$ of f satisfies the* momentum-space Schrödinger equation

$$-\frac{\hbar^2}{2\mathbf{m}}|p|^2\hat{f}(p) - \mathbf{e}^2\sqrt{\frac{2}{\pi}}\int_{\mathbb{R}^3}\frac{\hat{f}(\tilde{p})d\tilde{p}}{|p - \tilde{p}|^2} = \hat{f}(p).$$

If $\hat{f}$ satisfies the momentum-space Schrödinger equation then f satisfies the position-space Schrödinger equation.

The proof is a straightforward application of the fundamental properties of the Fourier transform, namely, its linearity, and how it intertwines differentiation, multiplication and convolution. This material is available in any introduction to Fourier transforms; for example, see [DyM, Chapter 2]. The only tricky part is the calculation of the Fourier transform of the Coulomb potential. See Exercise 9.3.

Some of Fock's terminology may be mysterious to the modern reader. In particular, *degenerate* energy levels are energy eigenvalues whose eigenspaces are reducible (i.e., not irreducible) representations.

In four dimensions, as in three dimensions, the restrictions of homogeneous harmonic polynomials of degree n to the unit sphere are called *spherical harmonic functions of degree n*. The analysis in four dimensions proceeds much as it did in three dimensions, although the dimension counts change.

Definition 9.1 *Let $\mathbb{H}_4^n$ denote the complex vector space of homogeneous harmonic polynomials of degree n in four variables. Let $\mathcal{Y}_4^n$ denote the complex* *vector space spanned by the restrictions of elements of* $\mathbb{H}_4^n$ *to the unit sphere* S^3 *in* $\mathbb{R}^4$. *Finally, define*

$$\mathcal{Y}_4 := \bigcup_{n=0}^{\infty} \mathcal{Y}_4^n.$$

Anticipating Fock's notation, let us call our variables x_1, x_2, x_3 and x_4. An example of a homogeneous harmonic polynomial of degree four is

$$(x_1 + ix^2)^2(x_3 - ix^4)^2;$$

note that

$$\nabla^2(x_1 + ix^2)^2(x_3 - ix^4)^2$$
$$= \left(\partial_{x_1}^2 + \partial_{x_2}^2 + \partial_{x_3}^2 + \partial_{x_4}^2\right)(x_1 + ix^2)^2(x_3 - ix^4)^2 = 0.$$

The results of Chapter 7 can be modified to the four-dimensional case.

Proposition 9.2 *For any nonnegative integer n, the natural representation of* $SO(4)$ *on* $\mathbb{H}_4^n$ *is irreducible. The dimension of this representation is* $(n+1)^2$.

The key to the proof of Proposition 9.2 is a classification (analogous to Proposition 6.16) of the representations of $SO(4)$, along with a tool (analogous to Proposition 6.17) for identifying representations. Most of the work has been done in Proposition 8.13; to get information about group representations from the classification of Lie algebra representations requires the insight that any Lie group representation on a vector space V induces a Lie algebra representation on V, obtained by differentiating the group representation on paths through the origin of the group. In other words, any Lie group representation has *infinitesimal generators*; the infinitesimal generators of the group representation form the algebra representation. See, for example, Bröcker and tom Dieck [BtD, Section II.9].

Proposition 9.3 *For any nonnegative integer n, the natural map (restriction) from* $\mathbb{H}_4^n$ *to* $\mathcal{Y}_4^n$ *is an isomorphism of complex vector spaces. The set* $\mathcal{Y}_4$ *spans the complex scalar product space* $L^2(S^3)$.

The proofs in Chapter 7 apply to Proposition 9.3 as well, *mutatis mutandis*.

Fock uses a stereographic projection from the three-sphere S^3 to Euclidean space $\mathbb{R}^3$. To serve his purposes, this projection must depend on a parameter; he calls the parameter α.

Definition 9.2 *Suppose α is a strictly positive real number. Define $S_\alpha : S^3 \to \mathbb{R}^3$ by*

$$S_\alpha : \begin{pmatrix} x_1 \\ x_2 \\ x_3 \\ x_4 \end{pmatrix} \mapsto \frac{\alpha}{1 + x_4} \begin{pmatrix} x_1 \\ x_2 \\ x_3 \end{pmatrix} \in \mathbb{R}^3.$$

The reader can check in Exercise 9.1 that S_α is invertible.

9.2   Fock's Original Article

In this section we present an English translation of V. Fock's original article, "Zur Theorie des Wasserstoffatoms" [F].

On the theory of the hydrogen atom.[1]

From **V. Fock** in Leningrad.

(Received August 5, 1935.)

   The Schrödinger equation for the hydrogen atom in momentum space is shown to be identical to the integral equation for the spherical harmonics of four-dimensional potential theory. Hence the transformation group of the hydrogen atom is the *four-dimensional* rotation group; this explains the degeneracies of the energy levels of hydrogen for the azimuthal quantum number ℓ. The consequences of the potential theory interpretation of the Schrödinger equation (such as the addition theorem) permit many physical applications. The method allows one to evaluate, almost without calculation, infinite sums appearing in the theory of the Compton effect on bound electrons and related problems. On the foundation of a simplified model of the atom one can hope to build explicit expressions for the density matrix in momentum space, for atom form factors, for the shielding potentials, and so on.

   It has long been known that the the energy levels of the hydrogen atom are degenerate with respect to the azimuthal quantum number ℓ; one speaks occasionally of an "accidental" degeneracy. But any degeneracy of eigenvalues is linked to the transformation group of the relevant equation: e.g., the degeneracy with respect to the magnetic quantum number m is allied to the usual rotation group. However, until now, the group corresponding to the "accidental" degeneracy of the hydrogen levels was unknown.

---

[1]Lecture given on February 8, 1935, in the theory seminar at Leningrad University. Compare V. Fock, *Bull. de l'ac. des sciences de l'URSS*, 1935, no. 2, 169.

In this work we will show that this group is equivalent to the four-dimensional rotation group.

1. It is known that the Schrödinger equation in momentum space takes the form of an integral equation:

$$\frac{1}{2m}\mathfrak{p}^2\psi(\mathfrak{p}) - \frac{Ze^2}{2\pi^2h}\int\frac{\psi(\mathfrak{p}')(d\mathfrak{p}')}{|\mathfrak{p}-\mathfrak{p}'|^2} = E\psi(\mathfrak{p}), \tag{9.1}$$

where $(d\mathfrak{p}') = dp'_x dp'_y dp'_z$ denotes the volume element in momentum space. Next we look at the point spectrum and let p_0 denote the mean quadratic momentum

$$p_0 = \sqrt{-2mE}. \tag{9.2}$$

We want to divide the components of the momentum vector by p_0 and think of the result as coordinates on a hyperplane, which we project stereographically onto the unit sphere in four-dimensional Euclidean space. The Cartesian coordinates on the sphere are

$$\left.\begin{aligned}
\xi &= \frac{2p_0 p_x}{p_0^2 + p^2} = \sin\alpha\sin\theta\cos\phi, \\[2mm]
\eta &= \frac{2p_0 p_y}{p_0^2 + p^2} = \sin\alpha\sin\theta\sin\phi, \\[2mm]
\zeta &= \frac{2p_0 p_z}{p_0^2 + p^2} = \sin\alpha\cos\phi, \\[2mm]
\chi &= \frac{p_0^2 - p^2}{p_0^2 + p^2} = \cos\alpha.
\end{aligned}\right\} \tag{9.3}$$

The angles α, θ, ϕ are spherical coordinates on the sphere; clearly θ and ϕ are the usual spherical coordinates on momentum space. The surface element on the unit sphere

$$d\Omega = \sin^2\alpha\,d\alpha\sin\theta\,d\theta\,d\phi \tag{9.4}$$

is related to the volume element in momentum space via

$$(d\mathfrak{p}) = dp_x dp_y dp_z = p^2 dp\sin\theta\,d\theta\,d\phi = \frac{1}{8p_0^2}(p_0^2 + p^2)^3\,d\Omega. \tag{9.5}$$

Let us define the abbreviation

$$\lambda = \frac{Zme^2}{hp_0} = \frac{Zme^2}{h\sqrt{-2mE}} \tag{9.6}$$

and introduce a new function

$$\Psi(\alpha, \theta, \phi) = \frac{\pi}{\sqrt{8}} p_0^{-5/2} (p_0^2 + p^2)^2 \phi(\mathfrak{p}). \tag{9.7}$$

Then the Schrödinger equation (9.1) can be written

$$\Psi(\alpha, \theta, \phi) = \frac{\lambda}{2\pi^2} \int \frac{\Psi(\alpha', \theta', \phi') \, d\Omega'}{4 \sin^2 \frac{\omega}{2}}. \tag{9.8}$$

The denominator $4 \sin^2 \omega/2$ in the integrand is the square of the four-dimensional distance between the points α, θ, ϕ and α', θ', ϕ' on the sphere:

$$4 \sin^2 \frac{\omega}{2} = (\xi - \xi')^2 + (\eta - \eta')^2 + (\zeta - \zeta')^2 + (\chi - \chi')^2. \tag{9.9}$$

Thus the number ω is the arclength of the great circle arc connecting the two points. We have

$$\cos \omega = \cos \alpha \cos \alpha' + \sin \alpha \sin \alpha' \cos \gamma, \tag{9.10}$$

where $\cos \gamma$ has the usual meaning:

$$\cos \gamma = \cos \theta \cos \theta' + \sin \theta \sin \theta' \cos(\phi - \phi'). \tag{9.10*}$$

The constant factor in (9.7) is chosen so that the normalization condition for Ψ is satisfied:

$$\frac{1}{2\pi^2} \int |\Psi(\alpha, \theta, \phi)|^2 \, d\Omega = \int \frac{p_0^2 + p^2}{2p_0^2} |\psi(\mathfrak{p})|^2 \, (d\mathfrak{p}) = \int |\psi(\mathfrak{p})|^2 \, (d\mathfrak{p}) = 1. \tag{9.7*}$$

Since the surface of a four-dimensional sphere has the value $2\pi^2$, the function $\Psi = 1$ in particular satisfies this normalization condition.

2. We would now like to show that equation (9.8) is nothing but the integral equation for the four-dimensional spherical harmonic functions.

We set

$$x_1 = r\xi; \quad x_2 = r\eta; \quad x_3 = \zeta; \quad x_4 = r\chi \tag{9.11}$$

and consider the Laplace equation

$$\frac{\partial^2 u}{\partial x_1^2} + \frac{\partial^2 u}{\partial x_2^2} + \frac{\partial^2 u}{\partial x_3^2} + \frac{\partial^2 u}{\partial x_4^2} = 0. \tag{9.12}$$

The function

$$G = \frac{1}{2R^2} + \frac{1}{2R_1^2} \qquad (9.13)$$

with

$$R^2 = r^2 - 2rr' \cos\omega + r'^2; \quad R_1^2 = 1 - 2rr' \cos\omega + r^2 r'^2 \qquad (9.14)$$

can be seen as a "Green's Function of the Third Kind"; on the sphere this function satisfies the boundary condition

$$\frac{\partial G}{\partial r'} + G = 0 \quad \text{for} \quad r' = 1. \qquad (9.15)$$

A function $u(x_1, x_2, x_3, x_4)$ harmonic on the interior of the unit sphere can be expressed in terms of the boundary values of $\partial u/\partial r + u$ by Green's Theorem as follows:

$$u(x_1, x_2, x_3, x_4) = \frac{1}{2\pi^2} \int \left(\frac{\partial u}{\partial r'} + u \right)_{r'=1} G \, d\Omega'. \qquad (9.16)$$

For a harmonic polynomial of degree $n - 1$

$$u = r^{n-1} \Psi_n(\alpha, \theta, \phi) \quad (n = 1, 2, \ldots) \qquad (9.17)$$

one has

$$\left(\frac{\partial u}{\partial r} + u \right)_{r=1} = nu = n\Psi_n(\alpha, \theta, \phi). \qquad (9.18)$$

If one uses this expression in (9.16) and uses (9.13) and (9.14) for $r' = 1$, one finds that

$$r^{n-1} \Psi_n(\alpha, \theta, \phi) = \frac{n}{2\pi^2} \int \frac{\Psi_n(\alpha', \theta', \phi')}{1 - 2r \cos\omega + r^2} \, d\Omega'. \qquad (9.19)$$

This equation holds for $r = 1$ also, in which case it coincides with the Schrödinger equation (9.8) when the parameter λ is equal to the whole number n; it is

$$\lambda = \frac{Zme^2}{h\sqrt{-2mE}} = n, \qquad (9.20)$$

which clearly is the principal quantum number.

Thus we have shown that the Schrödinger equation (9.1) or (9.8) can be solved with four-dimensional spherical harmonic functions. At the same time the transformation group of the Schrödinger equation has been found: this group is obviously identical to the four-dimensional rotation group.

3. We choose the following representation for the four-dimensional spherical harmonics. We set

$$\Psi_{n\ell m}(\alpha, \theta, \phi) = \Pi_{\ell}(n, \alpha)Y_{\ell m}(\theta, \phi), \tag{9.21}$$

where ℓ and m have their usual meanings of the azimuthal and magnetic quantum numbers, respectively, and $Y_{\ell m}(\theta, \phi)$ denotes the usual spherical harmonic function normalized by

$$\frac{1}{4\pi} \int_0^\pi \int_0^{2\pi} |Y_{\ell m}(\theta, \phi)|^2 \sin\theta \, d\theta \, d\phi = 1. \tag{9.22}$$

For brevity, one sets

$$M_{\ell} = \sqrt{n^2(n^2 - 1) \cdots (n^2 - \ell^2)}; \tag{9.23}$$

then one can consider a function $\Pi_{\ell}(n, \alpha)$, normalized by the condition

$$\frac{2}{\pi} \int_0^\pi \Pi_{\ell}^2(n, \alpha) \sin^2 \alpha \, d\alpha = 1, \tag{9.24}$$

and defined by one of the two equations

$$\Pi_{\ell}(n, \alpha) = \frac{M_{\ell}}{\sin^{\ell+1} \alpha} \int_0^\alpha \cos n\beta \frac{(\cos\beta - \cos\alpha)^{\ell}}{\ell!} d\beta \tag{9.25}$$

or

$$\Pi_{\ell}(n, \alpha) = \frac{\sin^{\ell} \alpha}{M_{\ell}} \frac{d^{\ell+1}(\cos n\alpha)}{d(\cos\alpha)^{\ell+1}}. \tag{9.25*}$$

For $\ell = 0$ we have

$$\Pi_0(n, \alpha) = \frac{\sin n\alpha}{\sin\alpha}. \tag{9.26}$$

Note that the defining equations (9.25) and (9.25*) hold true also for complex values of n (the continuous spectrum). The function Π_{ℓ} satisfies the relations

$$-\frac{d\Pi_{\ell}}{d\alpha} + \ell \operatorname{ctg} \alpha \Pi_{\ell} = \sqrt{n^2 - (\ell + 1)^2} \Pi_{\ell+1} \tag{9.27}$$

$$\frac{d\Pi_{\ell}}{d\alpha} + (\ell + 1) \operatorname{ctg} \alpha \Pi_{\ell} = \sqrt{n^2 - \ell^2} \Pi_{\ell-1}, \tag{27*}$$

which lead to the differential equation[2]

$$\frac{d^2\Pi_\ell}{d\alpha^2} + 2\,\text{ctg}\,\alpha\frac{d\Pi_\ell}{d\alpha} - \frac{\ell(\ell+1)}{\sin^2\alpha}\Pi_\ell + (n^2-1)\Pi_\ell = 0. \tag{9.28}$$

4. We proceed to establish the addition theorem for four-dimensional spherical harmonics. Equation (9.19) is an identity with respect to r. Expanding the integrand in powers of r

$$\frac{1}{1-2r\cos\omega+r^2} = \sum_{k=1}^{\infty} r^{k-1}\frac{\sin k\omega}{\sin\omega} \tag{9.29}$$

and setting the coefficients equal, one finds that

$$\frac{n}{2\pi^2}\int \Psi_n(\alpha',\theta',\phi')\frac{\sin k\omega}{\sin\omega}\,d\Omega' = \delta_{kn}\Psi_n(\alpha,\theta,\phi). \tag{9.30}$$

Now $n\frac{\sin n\omega}{\sin\omega}$, considered as a function of α', θ' and ϕ', is a four-dimensional spherical harmonic, which can be expanded in the $\Psi_{n\ell m}(\alpha',\theta',\phi')$'s. The coefficients of this expansion can be calculated from (9.30) (with $k=n$). In this way one finds the addition theorem

$$n\cdot\frac{\sin n\omega}{\sin\omega} = \sum_{\ell=0}^{n-1}\sum_{m=-\ell}^{+\ell}\bar{\Psi}_{n\ell m}(\alpha,\theta,\phi)\Psi_{n\ell m}(\alpha',\theta',\phi'). \tag{9.31}$$

Making use of the usual addition theorem for three-dimensional spherical harmonics and using the expression (9.21) for $\Psi_{n\ell m}$, one can rewrite (9.31) as

$$n\frac{\sin n\omega}{\sin\omega} = \sum_{\ell=0}^{\infty}\Pi_\ell(n,\alpha)\Pi_\ell(n,\alpha')(2\ell+1)P_\ell(\cos\gamma), \tag{9.32}$$

where P_ℓ denotes the Legendre polynomial and $\cos\gamma$ is defined by (9.10*). Here we have written the upper limit of the summation as $\ell=\infty$; we wish to indicate thereby that formula (9.32) is valid in this form also for complex values of n and α. If n is a whole number, the sum obviously ends with the term $\ell=n-1$.

---

[2]In his work on the wave equation of the Kepler problem in momentum space (ZS. f. Phys. **74**, 216, 1932), E. Hellras has derived a differential equation [Equations (9g) and (10b) in his article] which — after a simple transformation — can be understood as the differential equation of the four-dimensional spherical harmonics in stereographic projection. [With the gracious approval of E. Helleras, we correct the following misprints in his article: the number E that appears in the last term of his equations (9f) and (9g) should be multiplied by 4.]

5. We have given the geometric meaning of the integral equation (9.1) in the case of the point spectrum. In the case of the continuous spectrum ($E >$ 0) one must study, instead of the hypersphere, a two-sheeted hyperboloid in pseudo-Euclidean space. The region $0 < p < \sqrt{2mE}$ corresponds to one sheet and the region $\sqrt{2mE} < p < \infty$ corresponds to the other. In this case one can write the Schrödinger equation (9.1) as a system of two integral equations coupling the values of the desired function on the two sheets of the hyperboloid.

One can describe the state of affairs without reference to the fourth dimension as follows. In the case of the point spectrum the geometry of Riemann (constant positive curvature) reigns in momentum space, while in the case of the continuous spectrum the geometry of Lobatschewski (constant negative curvature) applies.

The geometrical meaning of the Schrödinger equation (9.1) is not as concrete in the case of the continuous spectrum as it is in the case of the point spectrum. Therefore, in applications it is better to derive formulas first for the point spectrum and only at the end allow the principal quantum number n to take pure imaginary values. This procedure allows one to see that the $\Pi_\ell(n, \alpha)$'s are analytic functions of n and α that, for pure imaginary values of n and α, differ from the corresponding functions of the continuous spectrum by only a constant factor.[3]

6. Now we will briefly indicate the problems that can be usefully treated with the above "geometric" theory of the hydrogen atom.[4] In many applications, such as the theory of the Compton effect in a bound electron[5] and in the inelastic matter theory of atoms[6] it is a question of determining the norm of the projection of a given function ϕ on the subspace of Hilbert space determined by the principal quantum number n.[7] This norm is defined by the sum

$$N = \int |P_n\phi|^2 \, d\tau = \sum_{\ell m} \left| \int \bar{\psi}_{n\ell m}\phi d\tau \right|^2. \tag{9.33}$$

---

[3]Compare V. Fock, *Foundations of Quantum Mechanics*, Leningrad 1932 (Russian).

[4]A more detailed treatment of these problems is planned and will appear in the Phys. ZS. d. Sowjetunion.

[5]G. Wentzel, ZS f. Phys. **58**, 348, 1929; F. Bloch, Phys. Rev. **46**, 674, 1934.

[6]H. Bethe, Ann. d. Phys. **5**, 325, 1930.

[7]J. v. Neumann, Mathematische Grundlagen der Quantenmechanik. Berlin, J. Springer, 1932.

The summation over ℓ usually poses great difficulties, especially when there is an infinite summation (continuous spectrum). Although the introduction of parabolic quantum numbers allows one to evaluate the sum in some cases, the calculations are still very complicated.

In comparison, if one uses the transformation group of the Schrödinger equation as well as the addition theorem (9.31) for the eigenfunctions, the summation is easy to carry out; the whole summation (9.33) is easier to calculate than one single term.

Our theory brings analogous simplifications to the calculation of the norm of the projection of an operator L on the nth subspace, that is, to the evaluation of the double sum

$$N(L) = \sum_{\ell m} \sum_{\ell' m'} \left| \bar{\psi}_{n\ell m} L \psi_{n\ell' m'} d\tau \right|^2 . \tag{9.34}$$

Expressions of the form (9.34) enter, for example, in the calculation of atom form factors, where the operator L has the form

$$L = e^{-\mathfrak{k} \frac{\partial}{\partial \mathfrak{p}}}; \quad L\psi(\mathfrak{p}) = \psi(\mathfrak{p} - \mathfrak{k}) \tag{9.35}$$

in momentum space. To evaluate (9.33) and (9.34) one uses the fact that these expressions are independent of the choice of orthogonal system $\psi_{n\ell m}$ on the subspace. An orthogonal substitution of the variables ξ, η, ζ, χ (four-dimensional rotation) introduces only a new orthogonal system, and so does not change the values of the sums (9.33) and (9.34). This rotation can be chosen so that the integrals in (9.33) and (9.34) simplify substantially or even vanish.[8] Thus one can, for example, essentially decompose the operator L defined by (9.35), which shifts the coordinate origin in momentum space, into a product of four-dimensional rotations, a reflection and a change of scale $p \to \lambda p$. This last operation gives rise to a sum that is much easier to calculate, as $\psi(\lambda p)$ has the same dependence on the angles θ and ϕ (usual spherical harmonics) as $\psi(p)$.

7. The projection $P_n \phi$ appearing in (9.33) of the function ϕ onto the subspace n of the Hilbert space is equal to

$$P_n \phi = \sum_{\ell m} \int \bar{\psi}_{n\ell m} \phi d\tau . \tag{9.36}$$

---

[8]In the expression (9.34) the $\psi_{n\ell m}$'s and the $\psi_{n\ell' m'}$'s can be replaced with two *different* rotations.

In momentum space the kernel of projection operator P_n has the form

$$\rho_n(\mathfrak{p}', \mathfrak{p}) = \sum_{\ell m} \bar{\psi}_{n\ell m}(\mathfrak{p}') \psi_{n\ell m}(\mathfrak{p}). \tag{9.37}$$

Here we can express the $\psi_{n\ell m}$'s in terms of four-dimensional spherical harmonics via (9.7). Because of the dependence on the principal quantum number, we now denote the mean quadratic momentum p_0 by p_n. So we have, instead of (9.7),

$$\Psi_{n\ell m}(\alpha, \theta, \phi) = \frac{\pi}{\sqrt{8}} p_n^{-5/2}(p_n^2 + p^2)^2 \psi_{n\ell m}(\mathfrak{p}). \tag{9.38}$$

Plugging (9.38) into (9.37) and using the addition theorem (9.31) one obtains

$$\rho_n(\mathfrak{p}', \mathfrak{p}) + \frac{8 p_n^5}{\pi^2 (p_n^2 + p^2)^2 (p_n^2 + p'^2)^2} \cdot n \frac{\sin n\omega}{\sin \omega} \tag{9.39}$$

and in the special case $\mathfrak{p}' = \mathfrak{p}$

$$\rho_n(\mathfrak{p}, \mathfrak{p}) + \frac{8 p_n^5 n^2}{\pi^2 (p_n^2 + p^2)^4}. \tag{9.40}$$

Hence the integral

$$4\pi \int_0^\infty \rho(\mathfrak{p}, \mathfrak{p}) p^2 dp = n^2 \tag{9.41}$$

equals the dimension of the subspace.

8. The great success of Bohr's model of Mendeleev's periodic table of the elements and the applicability of the Ritz formula for the energy levels show that treating the electron in an atom as if it were in a Coulomb field is a reasonable approximation.

It is therefore reasonable to consider the following model of the atom. The electrons in the atom can be assigned to "large strata": all electrons with principal quantum number n belong to the nth large stratum. Now electrons in the nth large stratum can be described only with hydrogen-like wave functions with the nuclear charge Z_n. Instead of Z_n one can introduce the mean quadratic momentum p_n, related to Z_n by

$$Z_n = np_n \frac{a}{\hbar} \quad (a \text{ hydrogen radius}). \tag{9.42}$$

Under these assumptions one can calculate the energy of an atom as a function of the nuclear charge Z and the parameter p_n and determine the value of

p_n from the minimum condition. Thus one can notice that under the given assumptions, the wave functions of the electrons in a large stratum are indeed orthogonal to one another, but not to the functions of another large stratum [sic]. Therefore it is consistent to neglect the exchange energy between electrons belonging to different large strata and to consider only the exchange energy inside each large stratum.

This procedure yields very satisfying results when applied to atoms with two large strata. For Na^+ ($Z = 11$) one finds, e.g., (in atomic units):

$$p_1 = 10.63; \quad p_2 = 3.45 \quad (Z = 11) \tag{9.43}$$

and for Al^{+++} ($Z = 13$) one finds that

$$p_1 = 12.62; \quad p_2 = 4.45 \quad (Z = 13). \tag{9.43*}$$

By this method one obtains a simple analytic expression for the shielding potential. With the above values of p_1 and p_2 this expression is hardly different from Hartree's "self-consistent field" calculated via incomparably difficult numerical techniques, and is even perhaps a bit more exact, as it lies between the "self-consistent field" with and without exchange in the case of the sodium atom.[9]

An analogous calculation went through for atoms with three large strata, namely, for Cu^+ ($Z = 29$) and for Zn^{++} ($Z = 30$). It gave

$$p_1 = 28.59; \quad p_2 = 10.64; \quad p_3 = 5.47 \quad (Z = 29) \tag{9.44}$$
$$p_1 = 29.59; \quad p_2 = 11.09; \quad p_3 = 5.84 \quad (Z = 30). \tag{9.44*}$$

The discrepancy between the shielding potential and the one calculated by Hartree is a bit bigger for Cu^+ (three strata) than for Na^+ and Al^{+++} (two strata), but the discrepancy does not surpass 1% of the entire value.

The exactness of the model proposed here seems — for atoms that are not too heavy — to satisfy fairly high standards.

To the extent that our model holds true, one can use the sum of the expressions (9.39) in the case of the large strata of the atom on hand for the density matrix of the atom in momentum space. But the knowledge of the density matrix allows one — as Dirac[10] especially has pointed out — to answer all questions about the atom, in particular the calculation of the atom form factors.

---

[9]Compare V. Fock and Mary Petrashen, Phys. ZS. d, Sowjetunion **6**, 368, 1934.
[10]P.A.M. Dirac, Proc. Cambr. Phil. Soc. **28**, 240, 1931, Nr. II.

As an example we cite here the atom form factor F_n for the nth biggest large stratum. In atomic units we have

$$F_n = \int e^{i\mathfrak{k}\cdot\mathfrak{r}}\rho_n(\mathfrak{r},\mathfrak{r})d\tau = \int \rho_n(\mathfrak{p}, \mathfrak{p} - \mathfrak{k})(d\mathfrak{p}). \tag{9.45}$$

If one plugs in the expression from (9.39), the integral is expressible in closed form. Abbreviating

$$x = \frac{4p_n^2 - k^2}{4p_n^2 + k^2} \tag{9.46}$$

one obtains

$$F_n = F_n(x) = \frac{1}{4n^2}T'_n(x)(1 + x)^2\{P'_n(x) + P'_{n-1}(x)\}, \tag{9.47}$$

where $T'_n(x)$ denote the derivative of the Tschebyschef polynomial

$$T_n(x) = \cos(n \arccos x) \tag{9.48}$$

and $P'_n(x)$ denotes the derivative of the Legendre polynomial $P_n(x)$. For $k = 0$ we have $x = 1$ and $F_n(1) = n^2$.

The sum of the expressions (9.40) over the large strata in the atom at hand is proportional to the charge density in momentum space. One can compare these quantities with the charge densities calculable from the Fermi-statistic model of the atom from which one sees that the latter model is less exact. For the atoms Ne ($Z = 10$) and Na$^+$ ($Z = 11$) one finds a good agreement for large p, while for small p (about $p < 2$ atomic units) the Fermi model gives charge density values that are much too high.

In conclusion, remark that our method, which on application to atoms with filled large strata yields exceptional simplifications, can probably be used as a foundation for handling atoms with large strata that are not full.

9.3   Exercises

Exercise 9.1 *Show that S_α^{-1} is given by the formula*

$$(p_1, p_2, p_3) \mapsto \frac{1}{\alpha^2 + |p|^2}(2\alpha p_1, 2\alpha p_2, 2\alpha p_3, \alpha^2 - |p|^2),$$

where $|p|^2 := p_1^2 + p_2^2 + p_3^2$. Check in particular that the image of a point $p \in \mathbb{R}^3$ under this function has length one in $\mathbb{R}^4$. Is S_α a linear transformation? Is S_α^{-1} a linear transformation?

Exercise 9.2 *Show that*

$$|S_\alpha(x)|^2 = \frac{(1 - x_4)\alpha^2}{1 + x_4}$$

and

$$\alpha^2 + |S_\alpha(x)|^2 = \frac{2\alpha^2}{1 + x_4}.$$

Exercise 9.3 (For students of the Fourier transform) *Calculate the Fourier transform of the Yukawa potential, $e^{-k|x|}/|x|$, where $k > 0$. Take a limit to show that the Fourier transform of $1/|x|$ is $4\pi/|p|^2$.*

10

Projective Representations and Spin

> Somewhere in the east: early morning: set off at dawn, travel round in front of
> the sun, steal a day's march on him. Keep it up for ever never grow a day older
> technically.
>
> — James Joyce, *Ulysses* [Joy, p. 57]

It is a bit of a lie to say, as we did in previous chapters, that complex scalar
product spaces are state spaces for quantum mechanical systems. Certainly
every nonzero vector in a complex scalar product space determines a quantum
mechanical state; however, the converse is not true. If two vectors differ only
by a phase factor, or if two vectors normalize to the same vector, then they will
determine the same physical state. This is one of the fundamental assumptions
of quantum mechanics. The quantum model we used in Chapters 2 through 9
ignored this subtlety. However, to understand spin we must face this issue.

10.1 Complex Projective Space

Mathematically, we collect the ambiguity of the phase factor into an equivalence relation (see Definition 1.3 of Section 1.7). In the current section we
introduce the necessary equivalence relation and use it to define *complex projective spaces*. We acquaint ourselves in some detail with the complex projective space $\mathbb{P}(\mathbb{C}^2)$. Finally, we show that linear transformations survive the
equivalence.

Suppose V is a complex scalar product space used in the study of a particular quantum mechanical system. (For example, consider $V = L^2(\mathbb{R}^3)$, the space used in the study of a mobile particle in $\mathbb{R}^3$.) If v and w are nonzero vectors in V, and if there is a nonzero complex number λ such that $v = \lambda w$, then v and w correspond to the same state of the quantum system: since $v = \lambda w$, we have

$$\frac{v}{\|v\|} = \frac{\lambda}{|\lambda|} \frac{w}{\|w\|},$$

so the normalized vectors $\frac{v}{\|v\|}$ and $\frac{w}{\|w\|}$ differ by a phase factor, namely the complex scalar $\frac{\lambda}{|\lambda|}$ of modulus one. So we define a physically natural equivalence relation on $V \setminus \{0\}$: we say that $v \sim w$ if and only if there is a (nonzero) complex scalar λ such that $v = \lambda w$. The proof that $\sim$ is an equivalence relation is identical to the argument necessary to resolve Exercise 1.23. One can think of the modulus $|\lambda|$ as the normalization factor and the directional part $\lambda/|\lambda|$ as the phase factor. Note that because $\lambda/|\lambda|$ lies on the unit circle, there is an $\alpha \in \mathbb{R}$ such that

$$\frac{\lambda}{|\lambda|} = e^{i\alpha}.$$

If we want to have a mathematical space in which each point corresponds to exactly one state of the quantum mechanical system, we must construct a space of equivalence classes.

Definition 10.1 *Suppose V is a complex vector space. We define the* projectivization *of V by*

$$\mathbb{P}(V) := V/\sim,$$

where $\sim$ is the equivalence relation defined above.

The set $\mathbb{P}(V)$ is sometimes called the *projective space over V, complex projective space* or, simply, *projective space*.

The simplest interesting complex projective space is $\mathbb{P}(\mathbb{C}^2)$. Let us write (c_0, c_1) for an element of $\mathbb{C}^2$. It is customary to denote the equivalence class of (c_0, c_1) by $[c_0 : c_1]$. For example,

$$[i : 1] = [1 : -i] = [2 + i : 1 - 2i] = \{(i\lambda, \lambda) \in \mathbb{C}^2 : 0 \neq \lambda \in \mathbb{C}\}.$$

The colon in the middle might remind you of the old-fashioned division sign, or ratio sign. The point is that whenever the ratios c_1/c_0 and b_1/b_0 are equal we have

$$\frac{b_0}{c_0}(c_0, c_1) = (b_0, b_1)$$

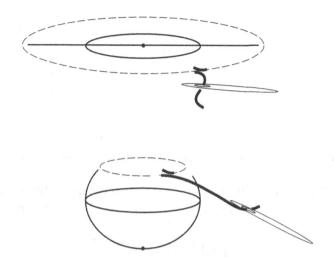

Figure 10.1. The drawstring approach to $\mathbb{P}(\mathbb{C}^2)$.

and hence $[c_0{:}c_1] = [b_0{:}b_1]$. In other words, the corresponding points in projective space are equal. Here b_0, b_1, c_0 and c_1 are all complex numbers. One might be tempted to conclude that the projective space $\mathbb{P}(\mathbb{C}^2)$ is the set of all possible ratios; but for a small technicality, one would be right. The technicality is that although it is common practice to say that "$1/0 = \infty$," division by zero is strictly illegal. In the rigorous mathematical treatment of projective space, we call the point $[0{:}1]$ the *point at infinity*. We accept the intuition of thinking of $[0{:}1]$ as infinite in some sense, but we also avoid the ambiguities that an undefined "∞" can create.

It turns out that $\mathbb{P}(\mathbb{C}^2)$ looks like the two-sphere S^2. To see this, think of $\mathbb{P}(\mathbb{C}^2)$ as $\mathbb{C} \cup \{[0{:}1]\}$, i.e., as a set of ratios including the infinite ratio $1/0$. Loosely speaking, one can imagine the complex numbers $\mathbb{C}$ as an infinite plane. Imagine sewing a drawstring into an infinitely large circle on the plane and then tightening it to form a sphere that is missing one point, the point at infinity. Put the point at infinity in and, voilà, it's a sphere. See Figure 10.1.

More precisely, we can use *stereographic projection* to find an injective, surjective function from the projective space $\mathbb{P}(\mathbb{C}^2)$ to the sphere S^2, via the plane of ratios. Stereographic projection is a function F from the xy-plane in $\mathbb{R}^3$ into the unit sphere in $\mathbb{R}^3$. We define

$$F(x, y) := \left(\frac{2x}{x^2 + y^2 + 1}, \frac{2y}{x^2 + y^2 + 1}, \frac{x^2 + y^2 - 1}{x^2 + y^2 + 1} \right). \qquad (10.1)$$

See Figure 10.2. Some properties stereographic projection are given in Exercise 10.5. In particular, the *north pole* $(0, 0, 1)$ is the only point omitted from the image; i.e., it is the only point on the sphere that does not corre-

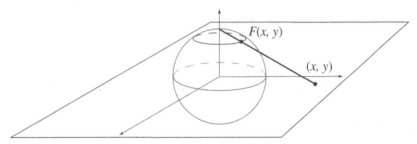

Figure 10.2. Stereographic projection. The formula for $F(x, y)$ is given in Equation 10.1.

spond to a point on the plane. We saw above that except for $[0{:}1]$, each point $[c_0{:}c_1] \in \mathbb{P}(\mathbb{C}^2)$ corresponds to the point $c_1/c_0 \in \mathbb{C}$, which corresponds to the point

$$\left(\Re\left(\frac{c_1}{c_0}\right), \Im\left(\frac{c_1}{c_0}\right) \right)$$

on the xy-plane. So the one-to-one correspondence between $\mathbb{P}(\mathbb{C}^2)$ and S^2 is given by

$$[c_0{:}c_1] \mapsto F\left(\Re\left(\frac{c_1}{c_0}\right), \Im\left(\frac{c_1}{c_0}\right) \right)$$

$$[0{:}1] \mapsto (0, 0, 1),$$

where F denotes stereographic projection. Note that $[1{:}0]$ is the south pole of the sphere, while $[0{:}1]$ is the north pole. For a more explicit formula, see Exercise 10.6.

The projective space $\mathbb{P}(\mathbb{C}^2)$ has many names. In mathematical texts it is often called *one-dimensional complex projective space*, denoted $\mathbb{CP}^1$. (Students of complex differential geometry may recognize that the space $\mathbb{P}(\mathbb{C}^2)$ is one-dimensional as a complex manifold: loosely speaking, this means that around any point of $\mathbb{P}(\mathbb{C}^2)$ there is a neighborhood that looks like an open subset of $\mathbb{C}$, and these neighborhoods overlap in a reasonable way.) In physics the space appears as the state space of a spin-1/2 particle. In computer science, it is known as a *qubit* (pronounced "cue-bit"), for reasons we will explain in Section 10.2. In this text we will use the name "qubit" because "$\mathbb{CP}^1$" has mathematical connotations we wish to avoid.[1]

---

[1]The most important of these connotations comes from complex geometry, where complex conjugation is not a natural function on $\mathbb{CP}^1$. In quantum mechanics, however, complex conjugation is a natural function. See Section 10.5.

For each natural number n, there is a projective space $\mathbb{P}(\mathbb{C}^{n+1})$, also known as $\mathbb{C}P^n$. Each element of $\mathbb{P}(\mathbb{C}^{n+1})$ is an equivalence class

$$[c_0{:}c_1 : \cdots : c_n] := \{\lambda(c_0, c_1, \ldots, c_n) : 0 \neq \lambda \in \mathbb{C}\},$$

where $c_0, c_1, \ldots, c_n$ are complex numbers. We can think of a large portion of these elements as a copy of $\mathbb{C}^n$: if $c_0 \neq 0$, then we have

$$[c_0{:}c_1 : \cdots : c_n] = [1{:}\frac{c_1}{c_0} : \cdots : \frac{c_n}{c_0}].$$

In other words, just as most of $\mathbb{P}(\mathbb{C}^2)$ corresponded to the complex plane because we could think of each equivalence class (except for $[0{:}1]$) as a bona fide ratio, each equivalence class in $\mathbb{P}(\mathbb{C}^{n+1})$ with $c_0 \neq 0$ corresponds to an n-tuple of ratios.

Can we extend our drawstring picture (Fig. 10.1) to an arbitrary $\mathbb{P}(\mathbb{C}^{n+1})$? We visualized $\mathbb{P}(\mathbb{C}^2)$ as a sphere, constructed by taking a plane and adding a point at infinity ($[0{:}1]$). For an arbitrary $\mathbb{P}(\mathbb{C}^{n+1})$, there is more than just one point with $c_0 = 0$. In fact, there is a whole $\mathbb{P}(\mathbb{C}^n)$ worth of them, as the reader may show in Exercise 10.4. In Section 10.4 we will see that $\mathbb{P}(\mathbb{C}^{n+1})$ is the state space for a particle of spin $(n + 1)/2$.

Whenever we consider a set of equivalence classes, it behooves us to ask what survives the equivalence. Note what does not survive: if $\dim V \geq 2$, the set $\mathbb{P}(V)$ is not a complex vector space: addition does not descend. For any element $v \in V \setminus \{0\}$, there must be a $w \in V \setminus \{0\}$ such that the set $\{v, w\}$ is linearly independent, by the assumption on dimension. By the definition of linear independence, it follows that for every $c \in \mathbb{C}$ we have

$$v + w \neq c(v + 2w).$$

In other words, while $v \sim v$ and $w \sim 2w$, it is not true that $v + w \sim v + 2w$. Hence the sum "$[v] + [w]$" is not well defined. Hence expressions such as

$$\frac{1}{\sqrt{2}}\big(|\phi\rangle + i\,|\psi\rangle\big),$$

which appear frequently in physics books, do not correspond to vector addition. In Section 10.3 we give a rigorous mathematical interpretation of such expressions.

However, the notion of a linear subspace descends to projective space.

Definition 10.2 *If W is a linear subspace of V, we define*

$$[W] := \{[w] : w \in W, w \neq 0\}.$$

Such a subset of $\mathbb{P}(V)$ is called a linear subspace *of $\mathbb{P}(V)$.*

Note that the empty set $\emptyset$ is a linear subspace of $\mathbb{P}(V)$, since $\emptyset = [\{0\}]$.

Any invertible linear transformation of V descends to a function from $\mathbb{P}(V)$ to itself that preserves subspaces. If the operator T were not invertible, there would be a nonzero v such that $Tv = 0$, in which case $[Tv] = \emptyset$ is not an element of $\mathbb{P}(V)$.

Proposition 10.1 *Suppose $T \colon V \to V$ is an invertible linear operator. Then a function $[T] \colon \mathbb{P}(V) \to \mathbb{P}(V)$ can be uniquely defined by requiring*

$$[T][v] := [Tv]$$

for all $v \in V$ such that $v \neq 0$. The function $[T]$ is called the projectivization *of T. This function preserves linear subspaces, i.e., if $[W]$ is an arbitrary linear subspace of $\mathbb{P}(V)$, then the image of $[W]$ under $[T]$ is also a linear subspace. Finally, a linear subspace W of V is an eigenspace of T for some nonzero eigenvalue λ if and only if $[W]$ consists entirely of fixed points of $[T]$.*

Proof. To show that $[T]$ is well defined, we must show that if $[w] = [v]$, then $[Tw] = [Tv]$. But $[w] = [v]$ if and only if there is a nonzero complex scalar c such that $w = cv$, in which case $Tw = cTv$ and hence $[Tw] = [Tv]$.

Now let $[W]$ denote an arbitrary linear subspace of $\mathbb{P}(V)$. Then the image of $[W]$ under $[T]$ is the set

$$\{[Tw] \colon w \in W\} = \mathbb{P}(\{Tw \colon w \in W\}).$$

Since T is a linear transformation, the image of W under T is a linear subspace of V, and hence the image of $[W]$ under $[T]$ is a linear subspaces of $\mathbb{P}(V)$.

Finally, if there is a complex number $\lambda \neq 0$ such that $Tw = \lambda_w w$ for each $w \in W$, then

$$[T][w] = [Tw] = [\lambda w] = [w]$$

so $[W]$ consists entirely of fixed points for $[T]$. On the other hand, suppose $[T][w] = [w]$ for any vector $w \in W$. Then for each $w \in W$ there is a complex number λ_w such that $Tw = \lambda w$. For any two linearly independent vector $w_1, w_2 \in W$ we have

$$\lambda_{w_1+w_2}(w_1 + w_2) = T(w_1 + w_2) = \lambda_{w_1} w_1 + \lambda_2 w_2,$$

and hence $\lambda_{w_1} = \lambda_{w_1+w_2} = \lambda_{w_2}$. It follows that every element of W is an eigenvector for T with eigenvalue λ_{w_1}. $\qquad\square$

For example, consider the linear transformation $T : \mathbb{C}^2 \rightarrow \mathbb{C}^2$ given by the matrix

$$\begin{pmatrix} 1 & 0 \\ 0 & e^{i\alpha} \end{pmatrix},$$

where α is real number. The projectivization $[T]$ satisfies, for any $[z_0{:}z_1] \in \mathbb{P}(\mathbb{C}^2)$,

$$[T]([z_0{:}z_1]) = [z_0{:}e^{i\alpha}z_1].$$

Notice that both the north pole ($[0{:}1]$) and the south pole ($[1{:}0]$) are fixed by $[T]$. This projectivization corresponds to the rotation of the two-sphere around the vertical axis through an angle of α.

Not every linear-subspace-preserving function on projective space descends from a complex linear operator. However, when we consider the unitary structure in Section 10.3 we find an imperfect but still useful converse — see Proposition 10.9.

At last, after several chapters of pretending that the state space of a quantum system is linear, we can finally be honest. The state space of each quantum system is a complex projective space. The reader may wish to review Section 1.2 at this point to see that while we were truthful there, we omitted to mention that unit vectors differing by a phase factor represent identical states. (In mathematics, as in life, "truthful" and "honest" are not synonyms.) In the next section, we apply our new insight to the spin state space of a spin-1/2 particle.

10.2 The Qubit

In this section we introduce the space of spin states of a *spin-1/2* particle, such as an electron. In quantum computation (the investigation of computers whose basic states are quantum, not deterministic), this space of states is called a *qubit*, pronounced "cue-bit." Just as a bit (a choice of 0 or 1) is the smallest unit of information in a deterministic computer, a qubit is the smallest unit of information in a quantum computer.

The usual presentation of a spin-1/2 particle starts with two physically distinguishable states. These states are usually labeled by *kets*, such as $|+\mathbf{z}\rangle$ and $|-\mathbf{z}\rangle$ (in physics texts) or $|1\rangle$ and $|0\rangle$ (in quantum computing texts). The name and asymmetrical notation connote the right half of the complex scalar product (also known as a bra*cket*) used in descriptions of quantum systems, $\langle \cdot | \cdot \rangle$. One posits that every quantum state can be written as a superposition of kets:

$$c_+ |+\mathbf{z}\rangle + c_- |-\mathbf{z}\rangle , \tag{10.2}$$

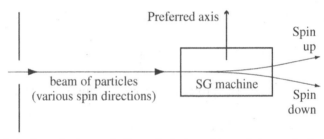

Figure 10.3. A schematic picture of a Stern–Gerlach machine and a beam of spin-1/2 particles.

where $c_\pm$ are complex numbers such that $|c_+|^2 + |c_-|^2 = 1$.

There is a piece of laboratory equipment that can take in a beam of electrons and put out two beams, one of electrons in the $|+\mathbf{z}\rangle$ state and one of electrons in the $|-\mathbf{z}\rangle$ state. This machine is called a *Stern–Gerlach machine*.[2] and depends on the reaction of a charged particle to a magnetic field that decreases along one axis. Such a magnetic field is sometimes called a *nonhomogeneous magnetic field*. Thus a Stern–Gerlach machine has a preferred axis; if one orients the machine so that the preferred axis is parallel to the z-axis, the machine will split a beam of spin-1/2 particles into a beam of spin-up (in the direction of the positive z-axis) particles and a beam of spin-down (in the direction of the negative z-axis) particles. See Figure 10.3. The condition on the coefficients in (10.2) comes from the physical interpretation of the $c_\pm$: if one puts a beam of particles in the state given by (10.2) through a Stern–Gerlach machine, the fraction of particles coming out in the $|+\mathbf{z}\rangle$ state is $|c_+|^2$, while the fraction of particles coming out in the $|-\mathbf{z}\rangle$ state is $|c_-|^2$. The fact that every particle comes out in one or the other state means that the sum of these two fractions should be 1. To put it another way, we must have $|c_+|^2 + |c_-|^2 = 1$, since the first summand is the probability that a particle will come out spin up, the second summand is the probability that the particle will come out spin down, and there are no other possible outcomes.

At this point it is useful to conduct a thought experiment. Consider a system for which the only possible measurement is by a Stern–Gerlach machine oriented along the z-axis. In other words, assume that once the probability for coming out of the machine spin up is known, every physically predictable feature of the state is known. Then the pair $(c_+, c_-) \in \mathbb{C}^2$ would contain more information than is necessary. Only $|c_+|^2$ and $|c_-|^2$ would have physical meaning, and because of the condition $|c_+|^2 + |c_-|^2 = 1$, even these two real numbers are dependent. Thus the phase space of this hypothetical system

---

[2]For more about the physics of Stern–Gerlach machines, see the Feynman Lectures [FLS, III-5].

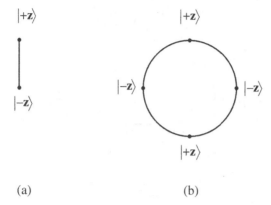

(a)   (b)

Figure 10.4. (a) A hypothetical phase space. (b) Four copies of the hypothetical phase space, glued together at the endpoints.

has one real dimension, and might be pictured as in Figure 10.4. We mention this hypothetical phase space because it is often pictured in computer science texts as a schematic drawing of a qubit. *Caveat emptor!* There is more to be known about the spin state of an electron than just its probability for emerging spin up from a z-axis Stern–Gerlach machine. For example, there are many different states corresponding to probability $1/2$ for a spin-up exit: for example, both

$$\frac{1}{\sqrt{2}}(|+\mathbf{z}\rangle + |-\mathbf{z}\rangle) \text{ and } \frac{1}{\sqrt{2}}(|+\mathbf{z}\rangle + i|-\mathbf{z}\rangle)$$

fit the bill, but these two states are physically distinguishable (Exercise 10.13). In fact there is a whole circle's worth of physically distinguishable points corresponding to this probability:

$$\frac{1}{\sqrt{2}}(|+\mathbf{z}\rangle + \lambda|-\mathbf{z}\rangle),$$

for any λ in the unit circle. Even in a quantum computation where the final step of the algorithm involves measuring whether a particle is spin up or spin down along the z-axis, there are intermediate steps involving interactions of more than one particle, and these interactions constitute a more complicated experiment whose outcome depends on more than just the probabilities for z-axis spin measurements. The drawing in Figure 10.4 can be misleading, since all but two points on the line stand for an infinite number of states of the qubit.

In order to define the correct state space for a qubit, one must determine the range of possible physical measurements. It turns out that one can predict the outcomes of experiments with Stern–Gerlach machines oriented any

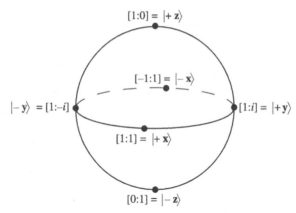

Figure 10.5. The qubit, a.k.a., the state space for a spin-1/2 particle, otherwise known as $\mathbb{P}(\mathbb{C}^2)$.

old way from the outcomes with machines oriented along the x-, y- and z-axes. In other words, the spin state of a spin-1/2 particle is determined by the probabilities associated to spin up vs. spin down along the three coordinate axes.

The natural model for a spin-1/2 particle, a model that incorporates all the possible spin experiments, is the projective space $\mathbb{P}(\mathbb{C}^2)$. We will wait for Section 10.3 to describe precisely how to predict experimental results from this model; in the meantime we hope the reader will be content with an appealing picture. We set $|{+}\mathbf{z}\rangle := [0:1]$ and $|{-}\mathbf{z}\rangle := [1:0]$. In other words, the north pole is the spin-up state for a Stern–Gerlach machine oriented along the z-axis (i.e., the z-spin-up state), while the south pole is the z-spin-down state. Next, set $|{+}\mathbf{x}\rangle := [1:1]$ and $|{-}\mathbf{x}\rangle := [1:-1]$. These are the x-spin-up and x-spin-down states, respectively, i.e., the up and down states for a Stern–Gerlach machine oriented along the x-axis. Finally, set $|{+}\mathbf{y}\rangle := [1:i]$ and $|{-}\mathbf{y}\rangle := [1:-i]$. These are the y-spin-up and y-spin-down states, respectively. See Figure 10.5.

In this model, the probability of emerging spin up (resp., down) from a Stern–Gerlach machine oriented along the z-axis is governed by the distance from the point $|{+}\mathbf{z}\rangle$ (resp., $|{-}\mathbf{z}\rangle$). For example, any point on the equator of the sphere labels a state of the spin-1/2 particle that has equal probability of being spin up or down along the z-axis. In particular, it is known experimentally that a particle coming out of a x-axis Stern–Gerlach machine is just as likely to be z-spin up as z-spin down after passing through a second Stern–Gerlach machine oriented along the z-axis. This experimental fact is encoded in the location of $|{+}\mathbf{x}\rangle$ and $|{-}\mathbf{x}\rangle$: on the equator, equidistant from the points $|{+}\mathbf{z}\rangle$ and $|{-}\mathbf{z}\rangle$.

We can reconcile the spherical picture with Figure 10.4(a) by noting that while the labeled points each refer to exactly one state of the qubit, each of the unlabeled points corresponds to a whole circle's worth of states, one circle of constant latitude on the sphere $\mathbb{P}(\mathbb{C}^2)$.

If $\mathbb{P}(\mathbb{C}^2)$ is indeed the right model for a qubit, how is it related to expressions such as (10.2)? What does the expression

$$c_0 \,|{-}\mathbf{z}\rangle + c_1 \,|{+}\mathbf{z}\rangle$$

mean? In the standard physics-style presentation, one assumes that two superpositions describe the same state if and only if they differ by overall multiplication by a phase factor. In other words, if $e^{i\alpha}$ is any *phase*, i.e., any complex number of modulus one, then the two superpositions

$$e^{i\alpha} c_+ \,|{+}\mathbf{z}\rangle + e^{i\alpha} c_- \,|{-}\mathbf{z}\rangle$$
$$c_+ \,|{+}\mathbf{z}\rangle + c_- \,|{-}\mathbf{z}\rangle$$

correspond to the same quantum state. However, if two superpositions are not related by a phase, then they stand for two different states of the particle. The nonuniqueness suggests an equivalence relation: define the symbol $\simeq$ by

$$\left(c_+ \,|{+}\mathbf{z}\rangle + c_- \,|{-}\mathbf{z}\rangle \right) \simeq \left(\tilde{c}_+ \,|{+}\mathbf{z}\rangle + \tilde{c}_- \,|{-}\mathbf{z}\rangle \right)$$

if and only if there is a complex number λ of modulus one such that

$$(\tilde{c}_-, \tilde{c}_+) = (\lambda c_-, \lambda c_+).$$

(Denoting the phase factor by "λ" instead of "$e^{i\alpha}$" is slightly cleaner notation.) The reader should verify that this is indeed an equivalence relation. Thus the mathematical state space of the qubit implicit in the standard presentation is the set of all possible pairs (c_+, c_-) such that $|c_+|^2 + |c_-|^2 = 1$ modulo the equivalence relation $\simeq$. Notice that the set of all satisfactory $c_\pm$'s is just the unit three-sphere S^3 inside $\mathbb{C}^2$. Because the equivalence comes from an action of the group $\mathbb{T}$ on this S^3, we call the state space

$$S^3/\mathbb{T}.$$

In fact, the space $S^3/\mathbb{T}$ suggested by the standard physics presentation is the same[3] as $\mathbb{P}(\mathbb{C}^2)$. To prove this, consider the function $h\colon S^3 \to \mathbb{P}(\mathbb{C}^2)$

---

[3] At this point the sophisticated reader will wonder what we mean by "the same." As we have seen, there are many different types of isomorphisms. To be precise, we should say that we will construct a *topological isomorphism*, i.e., an injective, surjective continuous function whose inverse is also continuous. We invite readers to show in Exercise 10.8 that the function H and its inverse H^{-1} are both continuous.

defined by

$$h(c_+, c_-) := [c_+ : c_-].$$

Notice that if $(c_+, c_-) \simeq (\tilde{c}_+, \tilde{c}_-)$, then $h(c_+, c_-) = h(\tilde{c}_+, \tilde{c}_-)$. So h descends to a function H on equivalence classes.

Let us show that the function H is injective. To this end, we suppose that $H(c_+, c_-) = H(b_+, b_-)$ and argue that $(c_+, c_-) \simeq (b_+, b_-)$. If $H(c_+, c_-) = H(b_+, b_-)$ then $[c_+ : c_-] = [b_+ : b_-]$. By the definition of $\mathbb{P}(\mathbb{C}^2)$, we know that there is a complex number λ such that $(c_+, c_-) = (\lambda b_+, \lambda b_-)$. Since $(c_+, c_-), (b_+, b_-) \in S^3$, we know that

$$|\lambda|^2 = |\lambda|^2 \left(|b_+|^2 + |b_-|^2 \right) = |c_+|^2 + |c_-|^2 = 1.$$

Since λ has modulus one, it follows that $(c_+, c_-) \simeq (b_+, b_-)$. So H is injective.

The function $H: S^3/\mathbb{T} \to \mathbb{P}(\mathbb{C}^2)$ is surjective as well. Suppose $[c_0 : c_1]$ is an arbitrary element of $\mathbb{P}(\mathbb{C}^2)$, i.e., that $(0, 0) \neq (c_0, c_1) \in \mathbb{C}^2$. Set

$$\lambda := \sqrt{|c_0|^2 + |c_1|^2} \neq 0.$$

Then $\lambda^{-1}(c_0, c_1) \in S^3$ and we have $f(\lambda^{-1}(c_0, c_1)) = [c_0 : c_1]$. The image of H is equal to the image of h, so H is surjective.

Since the function $H: S^3/\mathbb{T} \to \mathbb{P}(\mathbb{C}^2)$ is well defined, injective and surjective, the sets $S^3/\mathbb{T}$ and $\mathbb{P}(\mathbb{C}^2)$ are indeed equivalent. With the function H in our bag of tools, we are free to consider the qubit either way: as the complex projective space $\mathbb{P}(\mathbb{C}^2)$ or as superpositions $c_+ |+\mathbf{z}\rangle + c_- |-\mathbf{z}\rangle$ modulo phase factors. We will take advantage of this flexibility in the sections that follow, often assuming without loss of generality that the entries in a point $[c_0 : c_1]$ satisfy $|c_0|^2 + |c_1|^2 = 1$.

The reader familiar with the presentation of the state space of a spin-1/2 particle as $S^3/\mathbb{T}$ (i.e., the set of normalized pairs of complex numbers modulo a phase factor) may wonder why we even bother to introduce $\mathbb{P}(\mathbb{C}^2)$. One reason is that complex projective spaces are familiar to many mathematicians; in the interest of interdisciplinary communication, it is useful to know that the state space of a spin-1/2 particle (and other spin particles, as we will see in Section 10.4) are complex projective spaces. Another reason is that in order to apply the powerful machinery of representation theory (including eigenvalues and superposition), there must be a linear space somewhere in the background; by considering a projective space, we make the role of the linear space explicit. Finally, as we discuss in the next section, the effects of the complex scalar product on a linear space linger usefully in the projective space.

10.3   Projective Hilbert Spaces

It is natural to ask which operations descend from V to $\mathbb{P}(V)$. Is $\mathbb{P}(V)$ a complex vector space? Usually not. If V has a complex scalar product, does $\mathbb{P}(V)$ have a complex scalar product? No. But, as we will see in this section, a complex scalar product on V does endow $\mathbb{P}(V)$ with a useful notion of orthogonality. Furthermore, using the complex scalar product on V we can measure angles in $\mathbb{P}(V)$. At the end of the section we apply this new technology to the qubit $\mathbb{P}(\mathbb{C}^2)$.

Physics books on quantum mechanics are full of expressions such as

$$|+\mathbf{y}\rangle = \frac{1}{\sqrt{2}} |+\mathbf{z}\rangle + \frac{i}{\sqrt{2}} |-\mathbf{z}\rangle .$$

If the kets label individual states, i.e., points in projective space, and if addition makes no sense in projective space, what could this addition mean? The answer lies with the unitary structure (i.e., the complex scalar product) on V and how it descends to $\mathbb{P}(V)$. If V models a quantum mechanical system, then there is a complex scalar product $\langle \cdot, \cdot \rangle$ on V. Naively speaking, the complex scalar product does not descend to an operation on $\mathbb{P}(V)$. For example, if $v, w \in V \setminus \{0\}$ and $\langle v, w \rangle \neq 0$ we have $v \sim 2v$ but $\langle v, w \rangle \neq 2 \langle v, w \rangle = \langle 2v, w \rangle$. So the bracket is not well defined on equivalence classes. Still, one important consequence of the bracket survives the equivalence: orthogonality.

The notion of orthogonality descends to projective space.

Definition 10.3 *Suppose V is a complex scalar product space. Two elements* $[v], [w] \in \mathbb{P}(V)$ *are* orthogonal *if $\langle v, w \rangle = 0$.*

Note that this definition does not depend on the choice of v and w inside their equivalence classes. If $\langle v, w \rangle = 0$, then for any $\tilde{v} \sim v$ and $\tilde{w} \sim w$ we have nonzero complex numbers c_v and c_w such that

$$\langle \tilde{v}, \tilde{w} \rangle = \langle c_v v, c_w w \rangle = c_v^* c_w \langle v, w \rangle = 0.$$

So it does make sense to say that two equivalence classes, i.e., two points of projective space, are orthogonal.

Now we can define an orthogonal basis of a projective space.

Definition 10.4 *Suppose V is a complex scalar product space and $\mathbb{P}(V)$ is its projectivization. An* orthogonal basis *of $\mathbb{P}(V)$ is a subset $B \subset \mathbb{P}(V)$ whose members*

1. are mutually orthogonal, *i.e., if b_1 and b_2 are two distinct elements of B then b_1 and b_2 are orthogonal;*

2. span V, *i.e.,* $B^\perp = \emptyset$, *i.e., no element of* $\mathbb{P}(V)$ *is orthogonal to every element of B.*

Physically, a basis for a quantum mechanical system is a list of mutually exclusive states, a list long enough to capture all physically distinguishable properties of the system. Mutual exclusivity is the physical meaning of mutual orthogonality: if $|\langle b_1, b_2 \rangle| = 0$, then the probability that a particle in state $|b_1\rangle$ will be measured in state $|b_2\rangle$ is zero. The spanning requirement of the definition ensures that the list of states in the basis will be long enough. One way to find a basis for a physical system is to consider a particular measurement (say, spin up along the z-axis) and make a list of *pure states* for that measurement, i.e., states that are certain to yield a given value when measured. (States that are not pure are called *mixed states* for the measurement.) For the list of pure states to span, the measurement should be fine enough. Trouble arises when one or more values of the measurement have *multiplicities*, i.e., when there is more than one pure state corresponding to that value of the energy. For example, the measurement of electron energy in the hydrogen atom has *multiplicities*, i.e., several different states can have the same energy. One cannot distinguish between two different states of, for example, a p-orbital (i.e., a three-dimensional orbital corresponding to the spherical harmonics of degree one) by measuring energy. However, for any quantum system, one can always find measurements without multiplicities.

For example, consider a particle of spin 1/2. We can build a basis of the corresponding projective space by considering spin along the z-axis. There are only two certain spin states, up and down. These are mutually exclusive: if a particle is spin up, then it will not exit spin down from a z-axis Stern–Gerlach machine, and vice versa. But is this set of states large enough? Do either of these states have multiplicities? In other words, is there some measurement that can distinguish between two pure spin-up particles, or between two pure spin-down particles? The answer is no. As far as experiments have been done, any two z-spin-up (resp., spin-down) spin-1/2 particles are absolutely identical. So the list

$$\{\text{spin up, spin down}\}$$

is a basis the quantum model of a spin-1/2 particle.

Let us express this situation mathematically: the set $\{[1:0], [0:1]\}$ (a.k.a. $\{|-\mathbf{z}\rangle, |+\mathbf{z}\rangle\}$, a.k.a. $\{|0\rangle, |1\rangle\}$) is an orthogonal basis of $\mathbb{P}(\mathbb{C}^2)$. First, because the equivalence class [1:0] contains the point $(1, 0) \in \mathbb{C}^2$ and [0:1] contains the point $(0, 1)$, the calculation $\langle (1, 0), (0, 1) \rangle = 0$ shows that [1:0] and [0:1] are mutually orthogonal. Second, every point of $\mathbb{P}(\mathbb{C}^2)$ is of the form $[c_0:c_1]$ for some nonzero $(c_0, c_1) \in \mathbb{C}^2$. If $c_0 \neq 0$, then $[c_0:c_1]$ is not orthogonal to [1:0], since

$$\langle (1, 0), (c_0, c_1) \rangle = c_0 \neq 0.$$

But if $c_0 = 0$ then, by the definition of projective space, $c_1 \neq 0$ and hence, by a similar argument, $[c_0:c_1]$ is not orthogonal to [0:1]. So $\{[1:0], [0:1]\}$ spans $\mathbb{P}(\mathbb{C}^2)$. Thus $\{[1:0], [0:1]\}$ satisfies the criteria of Definition 10.4.

Orthogonality in $\mathbb{P}(\mathbb{C}^2)$ is quite different from Euclidean orthogonality in three-space. In other words, although the projective space $\mathbb{P}(\mathbb{C}^2)$ can be thought of as the sphere S^2, as indicated in Figure 10.5, the two points [1:0] and [0:1], which are orthogonal as elements of the projective space, correspond to two points on the sphere that are antipodal, not orthogonal, in the Euclidean sense.

Still, the right angles in Euclidean space do have meaning in our model. The three standard axes in the $\mathbb{R}^3$ in which the sphere sits correspond to three different orthogonal bases of $\mathbb{P}(\mathbb{C}^2)$. Along the x-axis we have the two points $[1: \pm 1]$, corresponding to the two states $|\pm\mathbf{x}\rangle$. Along the y-axis we have the two points $[1: \pm i]$, corresponding to the states $|\pm\mathbf{y}\rangle$. Each of these pairs of states forms an orthogonal basis for $\mathbb{P}(\mathbb{C}^2)$. The fact that the x-axis is at right angles to the y-axis shows up in the fact that a particle in state $|+\mathbf{x}\rangle$ has probability $1/2$ of emerging y-spin up from a Stern–Gerlach machine oriented along the y-axis.

Furthermore, every state of a spin-1/2 system is the pure spin-up or -down state for a Stern–Gerlach machine along some axis. We will not prove this assertion, just as we did not prove that [1:1] corresponds to the positive x-axis. But we can think of the sphere $\mathbb{P}(\mathbb{C}^2)$ as sitting inside the physical three-space. See Figure 10.6. Then each point $[c_0:c_1]$ on the sphere determines an axis, as well as a choice of positive direction along that axis. Particles exiting a Stern–Gerlach machine oriented along that axis will be either spin up, i.e., in the state $[c_0:c_1]$ or spin down, i.e., in the orthogonal state $[-c_1^*:c_0^*]$. In Exercise 10.7 we encourage the reader to show that $[-c_1^*:c_0^*]$ is the antipodal point to $[c_0:c_1]$. It follows that any pair of antipodal states (states on the same straight line through the origin) are mutually exclusive.

A good model must allow the user to express the outcome of any experiment (at least in theory). The only possible physical measurements of a spin-1/2 particle boil down to finding the probability that a particle in any given state will exit spin up from any given Stern–Gerlach machine. In other words, we can orient our Stern–Gerlach machine any way we like, shoot a

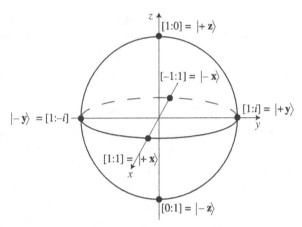

Figure 10.6. A correspondence between directions in $\mathbb{R}^3$ and the state space $\mathbb{P}(\mathbb{C}^2)$.

beam of particles in any known state through the machine, and count the fraction exiting spin up (or, equivalently, the fraction exiting spin down). For example, we might use one Stern–Gerlach machine oriented along the z-axis to create a beam of z-spin-up particles and then send them through a y-oriented Stern–Gerlach machine. Here is the calculation that predicts the fraction of particles exiting y-spin up from the second Stern–Gerlach machine: take any point (y_0, y_1) in the three-sphere S^3 inside $\mathbb{C}^2$ such that $[y_0{:}y_1] = |+\mathbf{y}\rangle$. Likewise, take any point (z_0, z_1) in the three-sphere S^3 inside $\mathbb{C}^2$ such that $[z_0{:}z_1] = |+\mathbf{z}\rangle$. The fraction of particles exiting y-spin up will be $|\langle(y_0, y_1), (z_0, z_1)\rangle|^2$. Note that this expression does not depend on our choices of (y_0, y_1) and (z_0, z_1): different choices would have differed by a phase factor, but because the phase factor has modulus one, it would not affect the final answer. We choose $(y_0, y_1) = \frac{1}{\sqrt{2}}(1, i)$ and $(z_0, z_1) = (0, 1)$ to find the probability

$$\left| \left\langle \frac{1}{\sqrt{2}}(1, i), (0, 1) \right\rangle \right|^2 = \left| \frac{i}{\sqrt{2}} \right|^2 = \frac{1}{2}.$$

Note the importance of the normalization factor $1/\sqrt{2}$: We could not have used $(1, i)$ because it does not lie in the sphere S^3, and it would have given a different answer. We can use this method to calculate any experimental outcomes. See for example Exercise 10.10.

To emphasize the difference between the upstairs bracket (on V) and the downstairs bracket (on $\mathbb{P}(V)$), we define special notation for the downstairs bracket.

Definition 10.5 *Suppose V is a complex scalar product space with complex scalar product* $\langle \cdot, \cdot \rangle$. *We define the* absolute bracket *on the projective space* $\mathbb{P}(V)$ *by*

$$\left| \langle [v] | [w] \rangle \right| := \left| \left\langle \frac{v}{\|v\|}, \frac{w}{\|w\|} \right\rangle \right| = \frac{|\langle v, w \rangle|}{\|v\| \, \|w\|}.$$

We have chosen the notation to match the physicists' convention. The reader should check that the absolute bracket is well defined, i.e., that its value does not depend on the choice of v and w in their respective equivalence classes.

The angle between two arbitrary points $[c_0{:}c_1]$ and $[b_0{:}b_1]$ on the sphere in Figure 10.6 determines the probability that a particle originally in the state $[b_0{:}b_1]$ will end up in the $[c_0{:}c_1]$ state upon exiting a Stern–Gerlach machine oriented along the axis corresponding to $[c_0{:}c_1]$. This angle can be calculated from the bracket.

Proposition 10.2 *The angle* θ *between two vectors A and B in the unit sphere in* $\mathbb{R}^3$ *satisfies*

$$\left| \langle [a_0, a_1] | [b_0, b_1] \rangle \right|^2 = \frac{1 + \cos \theta}{2},$$

where $[a_0{:}a_1]$ *and* $[b_0{:}b_1]$ *are the corresponding points in* $\mathbb{P}(\mathbb{C}^2)$ *(via stereographic projection).*

In particular, two points $[a_0{:}a_1]$ and $[b_0{:}b_1]$ satisfy

$$\left| \langle [a_0, a_1] | [b_0, b_1] \rangle \right|^2 = \frac{1}{2}$$

if and only if $\cos \theta = 0$, i.e., if and only if the two corresponding points on the unit sphere in $\mathbb{R}^3$ are orthogonal as vectors in $\mathbb{R}^3$.

Proof. The proof is by calculation. Without loss of generality, we may assume that $|a_0|^2 + |a_1|^2 = |b_0|^2 + |b_1|^2 = 1$. By Exercise 10.6 we have explicit formulas for the two corresponding points, and $\cos \theta$ is equal to their inner product (a.k.a. dot product, a.k.a. real scalar product),

$$\left(2\Re(a_0^* a_1), 2\Im(a_0^* a_1), |a_1|^2 - |a_0|^2 \right) \cdot \left(2\Re(b_0^* b_1), 2\Im(b_0^* b_1), |b_1|^2 - |b_0|^2 \right)$$
$$= 4\Re(a_0^* a_1)\Re(b_0^* b_1) + 4\Im(a_0^* a_1)\Im(b_0^* b_1) + \left(|a_1|^2 - |a_0|^2 \right)\left(|b_1|^2 - |b_0|^2 \right)$$
$$= 4\Re(a_0^* a_1 b_0 b_1^*) + |a_1|^2 |b_1|^2 + |a_0|^2 |b_0|^2 - |a_0|^2 |b_1|^2 - |a_1|^2 |b_0|^2$$
$$= 2\left| a_0^* b_0 + a_1^* b_1 \right|^2 - \left(|a_0|^2 + |a_1|^2 \right)\left(|b_0|^2 + |b_1|^2 \right)$$
$$= 2\left| \langle [a_0, a_1] | [b_0, b_1] \rangle \right|^2 - 1.$$

We conclude that $\cos \theta = 2 \left| \langle [a_0, a_1] | [b_0, b_1] \rangle \right|^2 - 1$, from which the proposition follows easily.   □

As far as experiments have been done, the state of a spin-1/2 particle is completely determined by its probabilities of exiting x-, y- and z-spin up from Stern–Gerlach machines oriented along the coordinate axes. This fact is consistent with the mathematical model for a qubit, as the following proposition shows.

Proposition 10.3 *The point on the sphere* $S^2 \subset \mathbb{R}^3$ *corresponding to* $[c_0:c_1] \in \mathbb{P}(\mathbb{C}^2)$ *via stereographic projection is*

$$
\begin{pmatrix}
2 \left| \langle [1{:}1] | [c_0, c_1] \rangle \right|^2 - 1 \\
2 \left| \langle [1, i] | [c_0, c_1] \rangle \right|^2 - 1 \\
2 \left| \langle [0, 1] | [c_0, c_1] \rangle \right|^2 - 1
\end{pmatrix}.
$$

Conversely, the three absolute brackets

$$
\left| \langle [0, 1] | [c_0, c_1] \rangle \right|, \quad \left| \langle [1, 1] | [c_0, c_1] \rangle \right|, \quad \left| \langle [1{:}i] | [c_0, c_1] \rangle \right|
$$

determine the point $[c_0:c_1]$.

Once one visualizes the sphere and circles of constant angle from a given point, it is intuitively clear that the three brackets should determine the point. If you cannot conjure the sphere accurately in your imagination, try an orange and a pen.

Proof. The coordinates of a point on the unit sphere in $\mathbb{R}^3$ are

$$
(\cos \theta_x, \cos \theta_y, \cos \theta_z),
$$

where θ_x denotes the angle between the radius from the origin to the point in question and the positive x-axis, while the angles θ_y and θ_z are the angles with the positive y- and z-axes, respectively. See Figure 10.7. Because $[1:1]$ lies on the positive x-axis, $[1:i]$ lies on the positive y-axis and $[0:1]$ lies on the positive z-axis, it follows from Proposition 10.2 that

$$
\cos \theta_x = 2 \left| \langle [1{:}1] | [c_0{:}c_1] \rangle \right|^2 - 1;
$$
$$
\cos \theta_y = 2 \left| \langle [1{:}i] | [c_0{:}c_1] \rangle \right|^2 - 1;
$$
$$
\cos \theta_z = 2 \left| \langle [0{:}1] | [c_0{:}c_1] \rangle \right|^2 - 1.
$$

Because stereographic projection is an injective function, the point $[c_0:c_1]$ is completely determined by the three values of the angles, which in turn are completely determined by the three absolute brackets.   □

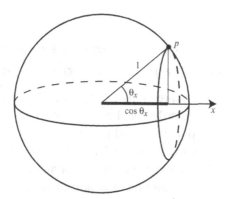

Figure 10.7. The x-coordinate of the point p is $\cos\theta_x$.

Orthogonal bases can help us understand superpositions of kets. For example, the superposition

$$c_- \, |-\mathbf{z}\rangle + c_+ \, |+\mathbf{z}\rangle \, .$$

is physicists' notation for the element $[c_-:c_+]$ in $\mathbb{P}(\mathbb{C}^2)$. To make sense of superpositions, let us first consider expressions such as

$$\langle +\mathbf{z}| + \mathbf{y}\rangle \, ,$$

which are common in physics textbooks. Can we make sense of such quantities without the absolute value? For example, some might guess that the bracket

$$\langle +\mathbf{z}|[c_0:c_1]\rangle$$

takes a value, namely c_1. Others might guess that the value is determined only up to multiplication by a phase factor. The truth is in between. It turns out that the pair

$$\Big(\langle -\mathbf{z}|[c_0:c_1]\rangle \, , \, \langle +\mathbf{z}|[c_0:c_1]\rangle\Big) \tag{10.3}$$

is determined up to a phase factor. In other words, the magnitudes of $\langle +\mathbf{z}|[c_0:c_1]\rangle$ and $\langle -\mathbf{z}|[c_0:c_1]\rangle$, as well as the phase difference between them, are physically meaningful quantities. In other words, the pair in (10.3) is a point in $\mathbb{P}(\mathbb{C}^2)$, namely

$$\big[\langle -\mathbf{z}|[c_0:c_1]\rangle : \langle +\mathbf{z}|[c_0:c_1]\rangle\big] = [c_0:c_1].$$

The superposition notation makes the following calculations natural: up to an overall phase (i.e., multiplying both equations by the same phase factor), we have

$$\langle -\mathbf{z}| \left(c_- \, |-\mathbf{z}\rangle + c_+ \, |+\mathbf{z}\rangle \right) = c_- \, \langle -\mathbf{z}| - \mathbf{z}\rangle + c_+ \, \langle -\mathbf{z}| + \mathbf{z}\rangle = c_-,$$
$$\langle +\mathbf{z}| \left(c_- \, |-\mathbf{z}\rangle + c_+ \, |+\mathbf{z}\rangle \right) = c_- \, \langle +\mathbf{z}| - \mathbf{z}\rangle + c_+ \, \langle +\mathbf{z}| + \mathbf{z}\rangle = c_+.$$

To put it another way, the expression $c_- \, |-\mathbf{z}\rangle + c_+ \, |+\mathbf{z}\rangle$ is a list of normalized coefficients for a ket in the orthogonal basis $\{|-\mathbf{z}\rangle, |+\mathbf{z}\rangle\}$. These normalized coefficients are unique up to an overall phase.

More generally, in any quantum mechanical system, any superposition of mutually orthogonal kets can be interpreted as an expansion in an orthogonal basis. However, superpositions of nonorthogonal kets are meaningless. Indeed, all of the ket superpositions given in the standard physics references involve only mutually orthogonal kets.

In this section we have studied the shadow downstairs (in projective space) of the complex scalar product upstairs (in the linear space). We have found that although the scalar product itself does not descend, we can use it to define angles and orthogonality. Up to a phase factor, we can expand kets in orthogonal bases. We will use this *projective unitary structure* to define projective unitary representations and physical symmetries.

10.4   Projective Unitary Irreducible Representations and Spin

In this section we define irreducible projective representations and find the irreducible projective representations of $SO(3)$. These turn out to correspond to the different kinds of spin elementary particles can have, namely, 0, $1/2$, 1, $3/2, \ldots$.

We start by defining the projective unitary representations. Recall the unitary group $\mathcal{U}(V)$ of a complex scalar product space V from Definition 4.2. The following definition is an analog of Definition 4.11.

Definition 10.6 *Suppose V is a complex scalar product space. The* projective unitary group of V *is*

$$\mathbb{P}\mathcal{U}(V) := \mathcal{U}(V)/\sim,$$

where $\mathcal{U}(V)$ is the group of unitary operators from $V \to V$ and $T_1 \sim T_2$ if and only if $T_1 T_2^{-1}$ is a scalar multiple of the identity.

Proposition 10.4 *Suppose V is a complex scalar product space. Then the projective unitary group $\mathbb{P}\mathcal{U}(V)$ is indeed a group, with the multiplication of equivalence classes that descends from the group multiplication on $\mathcal{U}(V)$.*

Proof. First we must check that the multiplication on $\mathcal{U}(V)$ descends to the equivalence classes. That is, we must check that if $T_1 T_2^{-1}$ and $T_3 T_4^{-1}$ are both scalar multiples of the identity then so is $T_1 T_3 (T_2 T_4)^{-1}$. Recalling that scalar multiples of the identity commute with every element of $\mathcal{U}(V)$, we find that

$$T_1 T_3 (T_2 T_4)^{-1} = T_1 (T_3 T_4^{-1}) T_2^{-1} = T_1 T_2^{-1} T_3 T_4^{-1},$$

which is the product of two scalar multiples of the identity and hence is a scalar multiple of the identity. So multiplication is well defined on $\mathbb{P}\mathcal{U}(V)$. The identity element in $\mathbb{P}\mathcal{U}(V)$ is the set of scalar multiples of the identity in $\mathcal{U}(V)$. Finally, the group axioms (listed in Definition 4.1 follow easily from the fact that $\mathcal{U}(V)$ is a group. $\qquad\square$

Definition 10.7 *Suppose G is a group and V is a complex scalar product space. Then the triple (G, V, ρ) is called a* projective unitary representation *if and only if ρ is a group homomorphism from G to $\mathbb{P}\mathcal{U}(V)$.*

Sometimes, to stress the distinction between unitary group representations as defined in Chapter 4 and projective unitary representations, we will call the former *linear unitary representations*. Any (linear) unitary representation descends to a projective unitary representation. More specifically, suppose G is a group, suppose V is complex scalar product space and suppose $\rho: G \to \mathcal{U}(V)$ is a (linear) unitary representation. Then we can define a projective unitary representation $\tilde{\rho}: G \to \mathbb{P}(V)$ by

$$\tilde{\rho}(g) := [\rho(g)] \in \mathbb{P}\mathcal{U}(V).$$

Not every projective representation arises in such a simple way. For example, set $G := SO(3)$ and set $V := \mathbb{C}^2$. Recall the group homomorphism $\Phi: SU(2) \to SO(3)$ defined in Section 4.3. Recall that this group homomorphism is two-to-one: if $\Phi(U) = \Phi(\tilde{U}) \in SO(3)$, then $U = \pm\tilde{U} \in SU(2)$, so $[U] = [\tilde{U}] \in \mathbb{P}\mathcal{U}(\mathbb{C}^2)$. Hence for any element $g \in SO(3)$, we can set, without ambiguity,

$$\rho_{\frac{1}{2}}(g) := [U] \in \mathbb{P}\mathcal{U}(\mathbb{C}^2),$$

where $U \in SU(2)$ satisfies $\Phi(U) = g$. Note that $\rho_{\frac{1}{2}}: SO(3) \to \mathbb{P}\mathcal{U}(\mathbb{C}^2)$. We must show that $\rho_{\frac{1}{2}}$ is a group homomorphism. Fix any $g_1, g_2 \in SO(3)$ and let $U_1, U_2 \in SU(2)$ be such that $\Phi(U_1) = g_1$ and $\Phi(U_2) = g_2$. Then $\Phi(U_1 U_2) = g_1 g_2$, so

$$\rho_{\frac{1}{2}}(g_1 g_2) = [U_1 U_2] = [U_1][U_2] = \rho_{\frac{1}{2}}(g_1) \circ \rho_{\frac{1}{2}}(g_2).$$

So $\rho_{\frac{1}{2}}$ is a projective unitary representation of $SO(3)$. In fact, $\rho_{\frac{1}{2}}$ is a *bona fide* projective Lie group representation, i.e,. it is a differentiable function, as we will show in Proposition 10.5. However, $\rho_{\frac{1}{2}}$ does not descend from any linear unitary representation of $SU(2)$ (Exercise 10.20).

The representation $\rho_{\frac{1}{2}}$ is called the *spin-1/2 representation*. It arises from the rotation of three-dimensional physical space and its effect on the qubit. In other words, experiments show that if two observers differ by a rotation g, then their observations of states of the qubit differ by a projective unitary transformation $[U]$ such that $\Phi(U) = g$. For example, consider a rotation $\mathbf{X}_\theta$ of angle θ around the x-axis. By Exercise 4.38, we have

$$\mathbf{X}_\theta = \Phi \begin{pmatrix} e^{-i\theta/2} & 0 \\ 0 & e^{i\theta/2} \end{pmatrix}.$$

Hence

$$\rho_{\frac{1}{2}}(\mathbf{X}_\theta) = \left[\begin{pmatrix} e^{-i\theta/2} & 0 \\ 0 & e^{i\theta/2} \end{pmatrix} \right].$$

Physically, as an observer rotates around the x-axis, the corresponding equivalence class in $\mathbb{P}(\mathbb{C}^2)$ rotates at one-half the speed of the observer. Rotating a vector at half speed would get us into trouble, for we need $\rho_{\frac{1}{2}}(\mathbf{X}_0) = \rho_{\frac{1}{2}}(\mathbf{X}_{2\pi})$. But our state is not a vector; it is an equivalence class of vectors. Note that $[c_0 : c_1] = [-c_0 : -c_1]$. So

$$\rho_{\frac{1}{2}}(\mathbf{X}_0) = \left[\begin{pmatrix} 1 & 0 \\ 0 & 1 \end{pmatrix} \right] = \left[\begin{pmatrix} -1 & 0 \\ 0 & -1 \end{pmatrix} \right] = \rho_{\frac{1}{2}}(\mathbf{X}_{2\pi}).$$

The existence of spin-1/2 particles is evidence that the projective-space model is correct. For a description of the relevant experiments with Stern–Gerlach machines, see the Feynman Lectures [FLS, Vol. III, Chapter 6].

Notice that the representation $\rho_{\frac{1}{2}}$ is reminiscent of a push-forward representation (see Section 5.6). It is inherently problematic to push forward along a two-to-one function; however, because of the projective equivalence, the push-forward turns out to be well defined in this case. We can use this trick to define a whole family of projective representations of $SO(3)$. These representations arise in the study of spin angular momentum.

Not all particles are spin 1/2 like electrons and neutrinos. Some, such as pions, are spin-0 particles, some, such as photons, are spin-1 particles. We can speak of the spin of aggregates of particles too: the silver atoms used by Stern and Gerlach in their original experiments were spin 1/2 ([To, Section 1.1]); a tennis ball is essentially spin 0. Note that the word "spin" here does not refer to any kind of turning in three-space, except for the hypothetical turning of a hypothetical observer. You can certainly put topspin on a tennis ball, but that tennis ball is still a spin-0 particle in the quantum-mechanical sense. The topspin is an example of *orbital spin*. The word "spin" is used in quantum mechanics because the mathematics reminded the discoverers of the mathematics of angular momentum; one might even go so far as to say that "spin" is a synonym for "representation of $SO(3)$".

All the irreducible linear Lie group representations of $SU(2)$ correspond to spin representations of particles, i.e., to *irreducible projective representations*. The definition is quite natural.

Definition 10.8 *Suppose G is a group, V is a complex scalar product space and $\rho: G \to \mathbb{PU}(V)$ is a projective unitary representation. We say that ρ is* irreducible *if the only subspace W of V such that $[W]$ is invariant under ρ is V itself.*

In other words, if ρ is irreducible and a subspace $W \neq 0$ of V satisfies $\rho(g)([w]) \in [W]$ for all $g \in G$, then $W = V$.

From the irreducible (linear) representations of the Lie group $SU(2)$ we can construct a family of irreducible projective representations of the Lie group $SO(3)$. Recall the (linear) representations $(SU(2), \mathcal{P}^n, R_n)$ (for $n = 0, 1, \dots$) from Section 4.6. The projective unitary representation $\rho_{\frac{n}{2}}$ defined in Proposition 10.5 corresponds to a particle of spin $n/2$.

Proposition 10.5 *Suppose n is a nonnegative integer. Define a function $\rho_{\frac{n}{2}}:$ $SO(3) \to \mathbb{PU}(\mathcal{P}^n)$ by*

$$\rho_{\frac{n}{2}}(g) := [R_n(U)],$$

where $U \in SU(2)$ satisfies $\Phi(U) = g$. Then $\rho_{\frac{n}{2}}$ is an irreducible projective unitary Lie group representation of $SO(3)$.

Our proof uses some differential geometry from Appendix B. One can replace the theory by a concrete calculation of the derivative of Φ.

Proof. We saw in Section 10.4 that when $n = 2$ the function $\rho_{\frac{n}{2}}$ is a well-defined homomorphism despite the ambiguity in choice of U. The same argument works for arbitrary n, since $[R_n(-U)] = [(-1)^n R_n(U)] = [R_n(U)]$ for any $U \in SU(2)$ by Exercise 4.35.

Next we show that $\rho_{\frac{n}{2}}$ is a Lie group representation, i.e., that it is a differentiable function from $SO(3)$ to $\mathbb{PU}(\mathcal{P}^n)$. To this end, consider Figure 10.8. By Proposition 4.5 we know that Φ is surjective. So given an arbitrary element $A \in SO(3)$, there is an element $g \in SU(2)$ such that $\Phi(g) = A$. By Proposition B.1, we know that Φ is a local diffeomorphism (Definition B.2). Hence there is a neighborhood N of g such that $\Phi|_N$ has a differentiable inverse. By

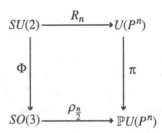

Figure 10.8. A commutative diagram for the proof of Proposition 10.5.

hypothesis ρ is a Lie group representation, and hence it is differentiable. Finally, from Theorem B.3 we know that π is a differentiable function. Hence the function

$$\left.\rho_{\frac{n}{2}}\right|_N = \pi \circ \rho \circ \left(\left.\Phi\right|_N\right)^{-1}$$

is differentiable. So $\rho_{\frac{n}{2}}$ is differentiable at A. But A was arbitrary, so $\rho_{\frac{n}{2}}$ is differentiable. Since $\rho_{\frac{n}{2}}$ is also a group homomorphism, it must be a Lie group homomorphism.

Finally, we must show that $\rho_{\frac{n}{2}}$ is irreducible. Suppose $W \neq 0$ is an invariant subspace of $\mathcal{P}^n$ invariant under $\rho_{\frac{n}{2}}$. Then, for any $w \neq 0 \in W$ and any $g \in SU(2)$ we have

$$[R_n(g)w] = [R_n(g)][w] \in [W].$$

Hence $R_n(g)w \in W$, so W is invariant under R_n. But R_n is an irreducible representation. So $W = V$. We have shown that $\rho_{\frac{n}{2}}$ is irreducible.   □

It is customary to think of the projective vector space of the spin-$n/2$ representation as $\mathbb{P}(\mathbb{C}^{n+1})$ instead of $\mathcal{P}^n$. In a sense, this is a distinction without a difference, as the two vector spaces are isomorphic, and it is possible to choose an isomorphism that preserves the complex scalar product. See Exercise 10.21.

It is useful to know that these spin representations have no multiplicities. Recall from Proposition 10.1 that multiplicities of eigenvalues on linear spaces correspond to dimensions of linear subspaces of fixed points of the projectivization. For example, the points $\{[1 : 0], [0 : 1]\}$ are the only fixed points of $\rho_{\frac{1}{2}}(\mathbf{X}_\theta)$ (for $\theta \notin \pi\mathbb{Z}$). They form a basis of $\mathbb{P}(\mathbb{C}^2)$. Similarly, in any spin representation the eigenstates of rotations around any one axis have isolated fixed points that form a basis for the state space.

Particles of integer spin (e.g., spin 1 or spin 0) are called *bosons*. Particles with half-integer spin (e.g., 1/2, or 3/2) are *fermions*. The fact that wave functions differing by a phase factor label the same physical state of the particle makes fermions possible. Bosons and fermions behave very differently. For example, the Pauli exclusion principle applies only to fermions. On the other hand, a curious phenomenon in photon emission is due to the bosonic nature of the photon: the probability that an atom will emit a photon in a particular state increases if there are already photons in that particular state. See the Feynman Lectures [FLS, III.15].

It is natural to wonder whether we have missed any irreducible projective unitary representations of $SO(3)$. Are there any others besides those that come from irreducible linear representations? The answer is no.

Proposition 10.6 *The irreducible projective unitary representations of the Lie group $SO(3)$ are in one-to-one correspondence with the irreducible (linear) unitary representations of the Lie group $SU(2)$.*

The proof, which requires a knowledge of differential geometry beyond the prerequisites of the text, is in Appendix B.

The results of this section are another confirmation of the philosophy spelled out in Section 6.2. We expect that the irreducible representations of the symmetry group determined by equivalent observers should correspond to the elementary systems. In fact, the experimentally observed spin properties of elementary particles correspond to irreducible projective unitary representations of the Lie group $SO(3)$. Once again, we see that representation theory makes a testable physical prediction.

10.5 Physical Symmetries

In Section 10.4 we studied projective unitary representations, important because they are symmetries of quantum systems. It is natural to wonder whether projective unitary symmetries are the only symmetries of quantum systems. In this section, we will show that complex conjugation, while not projective unitary, is a *physical symmetry*, i.e., it preserves all the physically relevant quantities. The good news is that complex conjugation is essentially the only physical symmetry we missed. More precisely, each physical symmetry is either projective unitary or it is the composition of a projective unitary symmetry with complex conjugation. This result (Proposition 10.10) is known as *Wigner's theorem on quantum mechanical symmetries*. The original proof can be found in the appendix to Chapter 20 in Wigner's book [Wi].

In this section we will freely interchange row and column vectors. To be precise, the expression

$$\left[\begin{pmatrix} z_0 \\ z_1 \end{pmatrix} \right]$$

denotes the element $[z_0 : z_1] \in \mathbb{P}(\mathbb{C}^2)$. For example, we have

$$\left[\begin{pmatrix} 1 & 0 \\ 0 & e^{i\alpha} \end{pmatrix} \begin{pmatrix} z_0 \\ z_1 \end{pmatrix} \right] = \left[\begin{pmatrix} z_0 \\ e^{i\alpha} z_1 \end{pmatrix} \right] = [z_0 : e^{i\alpha} z_1].$$

A symmetry of a quantum system, also known as a *physical symmetry*, is a function from $\mathbb{P}(V)$ to $\mathbb{P}(V)$ that preserves the absolute bracket $|\langle \cdot | \cdot \rangle|$.

Definition 10.9 *Suppose V is a complex vector space with a complex scalar product $\langle \cdot, \cdot \rangle$. A function $\mathcal{S} \colon \mathbb{P}(V) \to \mathbb{P}(V)$ is a physical symmetry of $\mathbb{P}(V)$ if for every $[v]$, $[w]$ in $\mathbb{P}(V)$ we have*

$$\left| \langle [v] | [w] \rangle \right|^2 = \left| \langle \mathcal{S}([v]) | \mathcal{S}([w]) \rangle \right|^2.$$

It follows easily from the definition that the composition of two physical symmetries is a physical symmetry and that every physical symmetry is injective (see Exercise 10.24).

As a first example, consider the state space of the qubit, $\mathbb{P}(\mathbb{C}^2)$. Let α be any real number. Then the function

$$Z_\alpha \colon \mathbb{P}(\mathbb{C}^2) \to \mathbb{P}(\mathbb{C}^2)$$
$$[z_0 : z_1] \mapsto [z_0 : e^{i\alpha} z_1]$$

is well defined and, for any $[z_0 : z_1]$ and $[\tilde{z}_0 : \tilde{z}_1]$ we have (assuming without loss of generality that $|z_0|^2 + |z_1|^2 = |\tilde{z}_0|^2 + |\tilde{z}_1|^2 = 1$)

$$\left| \langle [z_0 : z_1] | [\tilde{z}_0 : \tilde{z}_1] \rangle \right|^2 = \left| z_0^* \tilde{z}_0 + z_1^* \tilde{z}_1 \right|^2 = \left| z_0^* \tilde{z}_0 + (e^{i\alpha} z_1)^* (e^{i\alpha} \tilde{z}_1) \right|^2$$
$$= \left| \langle Z_\alpha([z_0 : z_1]) | Z_\alpha([\tilde{z}_0 : \tilde{z}_1]) \rangle \right|^2.$$

The function Z_α preserves the absolute value of the bracket and hence Z_α is a physical symmetry of the state space. This transformation corresponds to rotating the sphere in Figure 10.6 through an angle of α around the vertical axis (Exercise 10.16). The function Z_α descends from a linear transformation on $\mathbb{C}^2$, namely,

$$Z_\alpha([z_0 : z_1]) = \left[\begin{pmatrix} 1 & 0 \\ 0 & e^{i\alpha} \end{pmatrix} \begin{pmatrix} z_0 \\ z_1 \end{pmatrix} \right]$$

for any $(z_0, z_1) \in \mathbb{C}^2$.

Complex conjugation is another physical symmetry of the qubit.[4] We will find the following nomenclature useful.

Definition 10.10 *Suppose n is a natural number. The function*

$$\tau : \mathbb{C}^n \to \mathbb{C}^n$$

$$\begin{pmatrix} v_1 \\ \vdots \\ v_n \end{pmatrix} \mapsto \begin{pmatrix} v_1^* \\ \vdots \\ v_n^* \end{pmatrix}$$

is called the conjugation function *on $\mathbb{C}^n$.*

The function τ descends to equivalence classes in $\mathbb{P}(\mathbb{C}^n)$, so in particular we can write

$$\tau : \mathbb{P}(\mathbb{C}^2) \to \mathbb{P}(\mathbb{C}^2)$$

$$[z_0 : z_1] \mapsto [z_0^* : z_1^*].$$

For any $[z_0 : z_1]$ and $[\tilde{z}_0 : \tilde{z}_1]$ we have (assuming without loss of generality that $|z_0|^2 + |z_1|^2 = |\tilde{z}_0|^2 + |\tilde{z}_1|^2 = 1$)

$$\left| \langle [z_0 : z_1] | [\tilde{z}_0 : \tilde{z}_1] \rangle \right|^2 = \left| z_0^* \tilde{z}_0 + z_1^* \tilde{z}_1 \right|^2 = \left| z_0 \tilde{z}_0^* + z_1 \tilde{z}_1^* \right|^2$$

$$= \left| \langle [z_0^* : z_1^*] | [\tilde{z}_0^* : \tilde{z}_1^*] \rangle \right|^2 = \left| \langle \tau([z_0 : z_1]) | \tau([\tilde{z}_0 : \tilde{z}_1]) \rangle \right|^2.$$

So the function τ preserves the absolute value of the bracket and hence complex conjugation is a physical symmetry of the state space. This transformation corresponds to reflecting the sphere in Figure 10.6 in the xz-plane (Exercise 10.16). Complex conjugation does not descend from a (complex) linear transformation; however, we have

$$\tau([z_0, z_1]) = \left[\begin{pmatrix} 1 & 0 \\ 0 & 1 \end{pmatrix} \begin{pmatrix} z_0^* \\ z_1^* \end{pmatrix} \right]$$

for any $(z_0, z_1) \in \mathbb{C}^2$.

The first result in this section is a useful tool in uniqueness arguments.

---

[4]Students of complex geometry should note that complex conjugation is not an automorphism of $\mathbb{P}(\mathbb{C}^2)$, which has a natural complex structure inherited from $\mathbb{C}^2$.

Proposition 10.7 *Suppose S is a physical symmetry of the qubit such that*

$$S([1:1]) = [1:1],$$
$$S([1:i]) = [1:i],$$
$$S([0:1]) = [0:1].$$

Then S is the identity, i.e., for every $[c_0 : c_1] \in \mathbb{P}(\mathbb{C}^2)$ we have

$$S([c_0 : c_1]) = [c_0 : c_1].$$

Proof. Consider an arbitrary point $[c_0 : c_1] \in \mathbb{P}(\mathbb{C}^2)$. Because S is a physical symmetry and S fixes $[1:1]$, we have

$$\left|\langle[1:1]|S([c_0:c_1])\rangle\right| = \left|\langle S([1:1])|S([c_0:c_1])\rangle\right| = \left|\langle[1:1]|[c_0:c_1]\rangle\right|.$$

Similarly, we have

$$\left|\langle[1:i]|S([c_0:c_1])\rangle\right| = \left|\langle[1:1]|[c_0:c_1]\rangle\right|,$$
$$\left|\langle[1:i]|S([c_0:c_1])\rangle\right| = \left|\langle[1:1]|[c_0:c_1]\rangle\right|.$$

But by Proposition 10.2, these three brackets determine a point in $\mathbb{P}(\mathbb{C}^2)$. Hence $S([c_0 : c_1]) = [c_0 : c_1]$. $\qquad\square$

The next proposition classifies the physical symmetries of the qubit. As promised in the introduction to this section, these symmetries consist of the projective unitary symmetries (rotations) and compositions of projective unitary symmetries with complex conjugation (reflections). It follows easily from Proposition 10.1 that for any unitary operator $T \in \mathcal{U}(\mathbb{C}^2)$ both $[v] \mapsto [Tv]$ and $[v] \mapsto [T(\Re v - i\Im v)]$ are well-defined physical symmetries of $\mathbb{P}(\mathbb{C}^2)$. In fact, every physical symmetry of the qubit is of this form.

Proposition 10.8 *Suppose $S: \mathbb{P}(\mathbb{C}^2) \to \mathbb{P}(\mathbb{C}^2)$ preserves $|\langle\cdot,\cdot\rangle|$. Then there is an element T of $\mathcal{U}(\mathbb{C}^2)$, i.e, there is a two-by-two unitary matrix T and a function $\kappa: \mathbb{C}^2 \to \mathbb{C}^2$, equal either to the identity or the conjugation function, such that*

$$S[v] = [T\kappa(v)]$$

for any $v \in \mathbb{C}^2$. The function κ is unique, and the unitary matrix T is unique up to multiplication by a complex number of modulus one.

It follows from this proposition that every physical symmetry of the qubit is surjective.

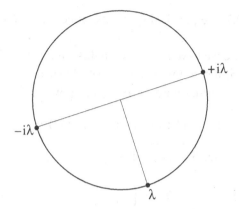

Figure 10.9. If $\lambda, \mu \in \mathbb{T}$ and $\mu^*\lambda \in \mathbb{T}$ is pure imaginary, then $\mu = \pm i\lambda$.

Proof. First we show that any physical symmetry fixing the north pole $[0 : 1]$ is of the desired form. Then we extend the result to arbitrary physical symmetries.

Suppose $\mathcal{S}$ is a physical symmetry of $\mathbb{P}(\mathbb{C}^2)$ such that $\mathcal{S}([0 : 1]) = [0 : 1]$. Then by the injectivity of $\mathcal{S}$ (Exercise 10.24) we have $\mathcal{S}([1 : 1]) \neq [0 : 1]$. So there is a complex number λ such that $\mathcal{S}([1 : 1]) = [1 : \lambda]$. Because $\mathcal{S}$ is a physical symmetry we have

$$\frac{|\lambda|^2}{1 + |\lambda|^2} = \left|\langle [0 : 1], [1 : \lambda]\rangle\right|^2 = \left|\langle [0 : 1], [1 : 1]\rangle\right|^2 = \frac{1}{2},$$

and hence $|\lambda| = 1$. A similar argument shows that we can write $\mathcal{S}([1 : i]) = [1 : \mu]$ with $|\mu| = 1$. Then we must also have

$$\frac{|1 + \mu^*\lambda|^2}{4} = \left|\langle [1 : \mu], [1 : \lambda]\rangle\right|^2 = \left|\langle [1 : i], [1 : 1]\rangle\right|^2 = \frac{1}{2},$$

and hence $\mu^*\lambda$ must be pure imaginary. It follows that $\mu = \pm i\lambda$, i.e., we have $\mathcal{S}([1 : i]) = [1 : \pm i\lambda]$. See Figure 10.9. We take two cases.

In the first case, $\mu = i\lambda$, consider the unitary 2×2 matrix

$$T := \begin{pmatrix} 1 & 0 \\ 0 & \lambda \end{pmatrix}$$

and define a function $\tilde{\mathcal{S}} := [T^{-1}] \circ \mathcal{S}$. Note that

$$\tilde{\mathcal{S}}([0 : 1]) = [T^{-1}] \circ \mathcal{S}([0 : 1]) = [T^{-1}]([0 : 1]) = [0 : \lambda^*] = [0 : 1]$$

$$\tilde{\mathcal{S}}([1 : 1]) = [T^{-1}] \circ \mathcal{S}([1 : 1]) = [T^{-1}]([1 : \lambda]) = [1 : \lambda^*\lambda] = [1 : 1]$$

$$\tilde{\mathcal{S}}([1 : i]) = [T^{-1}] \circ \mathcal{S}([1 : i]) = [T^{-1}]([1 : i\lambda]) = [1 : \lambda^* i\lambda] = [1 : i].$$

By Proposition 10.7, the physical symmetry function $\tilde{S}$ must be the identity. Hence $S = [T]$.

In the second case we have $\mu = -i\lambda$. In this case define $\tilde{S}$ by $\tilde{S}[z_0 : z_1] := S[z_0^* : z_1^*]$. Then $\tilde{S}$, the composition of two physical symmetries, is itself a physical symmetry. Furthermore, $\tilde{S}$ fixes $[0 : 1]$, while

$$\tilde{S}([1 : 1]) = [1 : \lambda] \quad \text{and} \quad \tilde{S}([1 : i]) = [1 : i\lambda].$$

Hence, by the first case, $\tilde{S} = [T]$, where

$$T := \begin{pmatrix} 1 & 0 \\ 0 & \lambda \end{pmatrix}.$$

It follows that $S[v] = [T\kappa(v)]$ for all $v \in V$, where κ denotes the conjugation function.

The last task in the proof of the first statement is to generalize to an arbitrary physical symmetry S. Set $[c_0 : c_1] := S([0 : 1])$, where we assume without loss of generality that $|c_0|^2 + |c_1|^2 = 1$. Consider the unitary matrix

$$U := \begin{pmatrix} c_1 & -c_0 \\ c_0^* & c_1^* \end{pmatrix}.$$

Then $[U] \circ S([0 : 1]) = [0 : 1]$, so we can apply the first part of the proof to find a unitary linear transformation $\tilde{T}$ such that $[U] \circ S(v) = [\tilde{T}\kappa(v)]$ for all $v \in \mathbb{C}^2$, where κ is either the identity function or the conjugation function. Set $T := U\tilde{T}$. Then T is unitary and we have $S(v) = [T\kappa(v)]$, as required.

Finally, we must show that κ is unique and T is unique up to multiplication by a scalar of modulus one. Suppose T_1, κ_1 and T_2, κ_2 both satisfy the requirements of the proposition. We must show that $\kappa_1 = \kappa_2$ and there is a real number θ such that $T_1 = e^{i\theta} T_2$. We know that for any element $v \in \mathbb{C}^2$ we have $[T_1\kappa_1(v)] = [T_2\kappa_2(v)]$. Applying the physical symmetry $[T_1^{-1}]$ to both sides we find that

$$[\kappa_1(v)] = [T_1^{-1} T_2 \kappa_2(v)].$$

Now if $v \in \mathbb{R}^2 \subset \mathbb{C}^2$, then $\kappa_1(v) = \kappa_2(v) = v$. So for every $v \in \mathbb{R}^2$ we have

$$[v] = [T_1^{-1} T_2 v]$$

and hence there is a complex number c such that $T_1^{-1} T_2 = cI$. Because both T_1 and T_2 are unitary, we have

$$|c|^2 = |\det(cI)| = \left| \det(T_1^{-1} T_2) \right| = 1,$$

so $|c| = 1$. So the matrices T_1 and T_2 are equal, up to multiplication by a complex number of modulus one. Thus for any $v \in \mathbb{C}^2$ (not necessarily real) we have $[v] = [\kappa_1^{-1}\kappa_2(v)]$, which implies that $\kappa_1 = \kappa_2$. We have shown that given a physical symmetry $\mathcal{S}$, the unitary operator T is unique up to multiplication by a scalar of modulus one, and κ is unique. □

To extend this result to projective space of arbitrary finite dimension we will need the technical proposition below. Since addition does not descend to projective space, it makes no sense to talk of linear maps from one projective space to another. Yet something of linearity survives in projective space: subspaces, as we saw in Proposition 10.1. The next step toward our classification is to show that physical symmetries preserve finite-dimensional linear subspaces and their dimensions.

Proposition 10.9 *Suppose U and V are complex scalar product spaces and $\mathcal{S}: \mathbb{P}(U) \to \mathbb{P}(V)$ preserves the absolute value of the bracket. Suppose U is finite-dimensional. There is a linear subspace $V_{\mathcal{S}}$ of V such that $[V_{\mathcal{S}}]$ is the image of $[U]$ under $\mathcal{S}$. Furthermore, $\dim V_{\mathcal{S}} = \dim U$.*

Proof. Set $n := (\dim U) - 1$, so $\dim U = n + 1$. We proceed by induction on n. The case $n = 0$ is trivial: set $V_{\mathcal{S}} := \mathcal{S}([u])$ for any u in the one-dimensional space U.

The case $n = 1$ will help us with the inductive step. Choose an orthonormal basis $\{u_0, u_1\}$ of U. For $j = 0, 1$, define $v_j \in V$ to be a vector of length one such that $[v_j] = \mathcal{S}([u_j])$. Let $V_{\mathcal{S}}$ denote the subspace of V spanned by $\{v_0, v_1\}$. Note that $\{v_0, v_1\}$ is an orthonormal basis of $V_{\mathcal{S}}$. We will show that $[V_{\mathcal{S}}]$ is the image of $[U]$ under $\mathcal{S}$.

First we show that the image of $[U]$ under $\mathcal{S}$ lies inside $[V_{\mathcal{S}}]$. Let u be an arbitrary element of U and let v be a vector of length one ($\|v\| = 1$) such that $[v] = \mathcal{S}([u])$. Since $V_{\mathcal{S}}$ is finite-dimensional, the orthogonal projection $\Pi_\perp := \Pi_{V_{\mathcal{S}}^\perp}$ onto the subspace $V_{\mathcal{S}}^\perp$ perpendicular to $V_{\mathcal{S}}$ exists, by Proposition 3.5. Since $\{u_0, u_1\}$ and $\{v_0, v_1\}$ are orthonormal bases, we have

$$\begin{aligned}
|\langle \Pi_\perp v, \Pi_\perp v \rangle|^2 &= 1 - |\langle v_0, v \rangle|^2 - |\langle v_1, v \rangle|^2 \\
&= 1 - |\langle \mathcal{S}([u_0])|\mathcal{S}([u])\rangle|^2 - |\langle \mathcal{S}([u_1])|\mathcal{S}([u])\rangle|^2 \\
&= 1 - |\langle [u_0]|[u]\rangle|^2 - |\langle [u_1]|[u]\rangle|^2 \\
&= 0.
\end{aligned}$$

Hence $\Pi_\perp v = 0$, so $\mathcal{S}([u]) \in [V_{\mathcal{S}}]$. Since u was an arbitrary element of U, it follows that the image of U under $\mathcal{S}$ is a subset of $[V_{\mathcal{S}}]$.

Next we show that $[V_S]$ lies inside the image of $[U]$ under S. Since V_S is two-dimensional, there is a function $f \colon [V_S] \to \mathbb{P}(\mathbb{C}^2)$, defined by

$$f \colon [c_0 v_0 + c_1 v_1] \mapsto [c_0 : c_1].$$

Similarly, there is a function $g \colon \mathbb{P}(\mathbb{C}^2) \to [U]$ defined by

$$g \colon [c_0 : c_1] \mapsto [c_0 u_0 + c_1 u_1].$$

Since $\{u_0, u_1\}$ and $\{v_0, v_1\}$ are orthonormal bases, both these functions are injective and preserve absolute values of brackets. Therefore the function $f \circ S \circ g \colon \mathbb{P}(\mathbb{C}^2) \to \mathbb{P}(\mathbb{C}^2)$ preserves absolute values of brackets. Since the image of $S \circ g$ lies in the domain of f, the domain of $f \circ S \circ g$ is all of $\mathbb{P}(\mathbb{C}^2)$. Hence $f \circ S \circ g$ is a physical symmetry of the qubit. By Proposition 10.8, the physical symmetry $f \circ S \circ g$ is surjective. Hence S must be surjective onto the domain of f, namely, $[V_S]$. In particular, $[V_S]$ lies inside the image of $[U]$ under S.

Putting our two results together, we conclude that $[V_S]$ is the image of U under S. This proves the proposition in the special case $\dim U = 2$.

Finally, we must prove the inductive step. Suppose $n \geq 2$, suppose $\dim U = n + 1$, and suppose that the statement is known to be true for subspaces of dimension n and fewer. Fix any $u_0 \in U$ such that $\|u_0\| = 1$. Then the subspace $[u_0]^\perp$ of U has dimension n. By the inductive hypothesis, there is a subspace V_0 of V such that $[V_0]$ is the image of $[u_0]^\perp$ under S and $\dim V_0 = n$. Choose $v_0 \in V$ such that $[v_0] = S([u_0])$. Set

$$V_S := V_0 \oplus [v_0].$$

By definition, this V_S is a linear subspace of V. Since S preserves brackets, v is orthogonal to V_0. Hence

$$\dim V_S = n + 1 = \dim U.$$

Let us show that for any $u \in U$ we have $S([u]) \in [V_S]$. If u is a scalar multiple of u_0, then the proof is trivial: $S([u]) = S([u_0]) = [v_0] \in V_S$. Similarly, if $u \in [u_0]^\perp$, then the proof is trivial. So assume u is neither a scalar multiple of u_0 nor an element of $[u_0]^\perp$. Then we can write $u = x_0 + x_1$, where $0 \neq x_0 \in [u_0]$ and $0 \neq x_1 \in [u_0]^\perp$. Let X denote the subspace of U spanned by $\{x_0, x_1\}$. Note that because X is two-dimensional, we know that the image of X under S is a two-dimensional subspace of V. Furthermore, this two-dimensional image space is spanned by the two distinct lines $S([x_0]) = [v_0]$

and $\mathcal{S}([x_1]) \subset V_0$. Hence the image of X is a subspace of of $V_\mathcal{S} = V_0 \oplus [v_0]$. In particular, since $u \in X$ we find $\mathcal{S}([u]) \in [V_\mathcal{S}]$. So the image of $[U]$ under $\mathcal{S}$ is a subset of $[V_\mathcal{S}]$.

To finish the inductive step, let us show that for any $v \in V_\mathcal{S}$ there is a $u \in U$ such that $\mathcal{S}([u]) = v$. If either $v \in [v_0]$ or $v \in V_0$, then the conclusion holds, either because $\mathcal{S}([u_0]) = [v]$ or by induction. So assume $v \notin [v_0]$ and $v \notin V_0$. Then we can write $v = w_0 + w_1$, where $0 \neq w_0 \in [v_0]$ and $0 \neq w_1 \in V_0$. Now $\mathcal{S}([u_0]) = [w_0]$ and, by induction, there is a $u_1 \in [u_0]^\perp$ such that $\mathcal{S}([u_1]) = [w_1]$. Consider the two-dimensional subspace of U spanned by the set $\{u_0, u_1\}$ and the image of that set under $\mathcal{S}$. The image is of the form $[W]$ for some two-dimensional subspace W of V. Since both $w_0, w_1 \in W$, we have $v \in W$. Hence by the special case $n = 2$, there must be an element u in the span of $\{u_0, u_1\}$ satisfying $\mathcal{S}([u]) = [v]$. Note that $[u] \in [U]$. Thus $[V_\mathcal{S}]$ is a subset of the image of $[U]$ under $\mathcal{S}$.

Thus by construction dim $V_\mathcal{S} = \dim U$, and we have shown that $[V_\mathcal{S}]$ is the image of $[U]$ under $\mathcal{S}$. This completes the inductive step and the proof. $\square$

With Proposition 10.8 and the technical result Proposition 10.9 in hand, we are ready to classify the physical symmetries of complex projective spaces of arbitrary finite dimension.

Proposition 10.10 *Suppose n is a natural number and $\langle \cdot, \cdot \rangle$ is the standard complex scalar product on $\mathbb{C}^n$. Suppose $\mathcal{S} \colon \mathbb{P}(\mathbb{C}^n) \to \mathbb{P}(\mathbb{C}^n)$ is a physical symmetry. Then there is a unitary operator $T \colon \mathbb{C}^n \to \mathbb{C}^n$ and a function κ, equal to either the identity or the conjugation function, such that*

$$\mathcal{S}[v] = [T\kappa(v)]$$

for any $v \in \mathbb{C}^n$. The function κ is unique, and the unitary operator T is unique up to scalar multiplication by a complex number of modulus one.

Proof. We proceed by induction on n. For the base case ($n = 1$), consider $\mathbb{P}(\mathbb{C})$. By Exercise 10.1, $\mathbb{P}(\mathbb{C})$ consists of a single point. So $\mathcal{S}$ must be the identity function, in which case the desired unitary transformation of V is the identity linear operator (or any modulus-one complex multiple of the identity operator) and the function κ is the identity.

The next case ($n = 2$) is the content of Proposition 10.8.

For the inductive step, fix a natural number $n \geq 2$. Suppose that the proposition is known for all $\mathbb{C}^k$ with $k \leq n$. Suppose $\mathcal{S} \colon \mathbb{P}(\mathbb{C}^{n+1}) \to \mathbb{P}(\mathbb{C}^{n+1})$ is a physical symmetry. We must find a function κ and a unitary transformation T satisfying the conclusion of the proposition and show uniqueness up

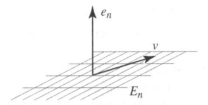

Figure 10.10. The vector e_n is orthogonal to the subspace E_n in the complex scalar product space $\mathbb{C}^n$.

to scalar multiplication of T. We first consider a special case and then deduce the general case.

Define

$$
e_n := \begin{pmatrix} 0 \\ \vdots \\ 0 \\ 1 \end{pmatrix} \in \mathbb{C}^{n+1}
$$

$$
E_n := \left\{ \begin{pmatrix} v_0 \\ \vdots \\ v_{n-1} \\ 0 \end{pmatrix} \in \mathbb{C}^{n+1} : v := (v_0, \ldots, v_{n-1}) \in \mathbb{C}^n \right\}.
$$

Note that $E_n = e_n^\perp$, as illustrated in Figure 10.10.

Consider the special case where $\mathcal{S}([e_n]) = [e_n]$ and $\mathcal{S}([v]) = [v]$ for every $v \in E_n$. We will show that S has the form given in Equation 10.4 below. Fix any element $v \in E_n$ such that $[v] \neq [v^*]$ and consider the two-dimensional subspace V of $\mathbb{C}^{n+1}$ spanned by v and e_n. By Proposition 10.9, the image of this subspace under S is a two-dimensional subspace. Because of our special assumptions on S, this subspace must contain both v and e_n. Hence the image of V under S is V. By Proposition 10.8, the restriction of S to the set V descends from a unitary operator (possibly preceded by complex conjugation). In other words, there is a function κ_v, equal either to the identity function or the conjugation function, and a 2×2 unitary matrix $M_v : \mathbb{C}^2 \to \mathbb{C}^2$ (determined up to a scalar multiple) such that for any $(c_0, c_1) \in \mathbb{C}^2$ we have $S([c_0 v + c_1 e_n]) = [\tilde{c}_0 v + \tilde{c}_1 e_n]$, where $\tilde{c}_0$ and $\tilde{c}_1$ are defined by

$$
\begin{pmatrix} \tilde{c}_0 \\ \tilde{c}_1 \end{pmatrix} := M_v \kappa_v \begin{pmatrix} c_0 \\ c_1 \end{pmatrix}.
$$

Since the function S fixes the points $[v]$ and $[e_n]$, the matrix M_v must be diagonal; because M_v is unitary and unique up to constant multiplication, there must be a real number $\theta_v \in [0, 2\pi)$ such that

$$\forall [c_0 : c_1] \in \mathbb{P}(\mathbb{C}^2), \quad \mathcal{S}([c_0 v + c_1 e_n]) = [\kappa_v(c_0)v + e^{i\theta_v}\kappa_v(c_1)e_n].$$

We will show that every κ_v must be the identity, and that we can make one choice of θ that works for all choices of $v \neq 0$. This will establish the conclusion of the proposition in the special case of a function $\mathcal{S}$ fixing $[e_n]$ and every element of $[E_n]$.

Consider any two linearly independent elements $v, \tilde{v} \in E_n$. Then, for any nonzero scalars a and $\tilde{a}$ we have

$$\mathcal{S}([v + ae_n]) = [v + e^{i\theta_v}\kappa_v(a)e_n],$$
$$\mathcal{S}([\tilde{v} + \tilde{a}e_n]) = [v + e^{i\theta_{\tilde{v}}}\kappa_{\tilde{v}}(\tilde{a})e_n].$$

Now we let W denote the two-dimensional subspace of $\mathbb{C}^{n+1}$ spanned by $\{v + ae_n, \tilde{v} + \tilde{a}e_n\}$. It is useful to consider the vector

$$w_{a,\tilde{a}} := \tilde{a}v - a\tilde{v} = \tilde{a}(v + ae_n) - a(\tilde{v} + \tilde{a}e_n) \in E_n \cap W.$$

Because $\mathcal{S}$ preserves linear subspaces and dimensions (by Proposition 10.9), the image X of W under $\mathcal{S}$ is two-dimensional and is spanned by the lines $\mathcal{S}([v + ae_n])$ and $\mathcal{S}([\tilde{v} + \tilde{a}e_n])$. Hence, the line $\mathcal{S}([w_{a,\tilde{a}}]) = [w_{a,\tilde{a}}]$, which lies in E_n, must also lie in the subspace X. Since $a \neq 0$, the subspace X does not lie entirely in E_n; hence the two-dimensional subspace X must intersect the n-dimensional subspace E_n in a one-dimensional subspace. The intersection must be the subspace $[w_{a,\tilde{a}}]$; on the other hand, we can calculate the intersection explicitly from the basis of X. Since v and $\tilde{v}$ are linearly independent, we can construct a nonzero element of $E_n \cap X$:

$$X \ni e^{i\theta_{\tilde{v}}}\kappa_{\tilde{v}}(\tilde{a})(v + e^{i\theta_v}\kappa_v(a)e_n) - e^{i\theta_v}\kappa_v(a)(\tilde{v} + e^{i\theta_{\tilde{v}}}\kappa_{\tilde{v}}(\tilde{a})e_n)$$
$$= e^{i\theta_{\tilde{v}}}\kappa_{\tilde{v}}(\tilde{a})v - e^{i\theta_v}\kappa_v(a)\tilde{v} \in E_n.$$

Since this vector and the vector $w_{a,\tilde{a}}$ both lie in the one-dimensional set $E_n \cap X$ we find that

$$[e^{i\theta_{\tilde{v}}}\kappa_{\tilde{v}}(\tilde{a})v - e^{i\theta_v}\kappa_v(a)\tilde{v}] = [w_{a,\tilde{a}}] = [\tilde{a}v - a\tilde{v}].$$

Since v and $\tilde{v}$ are linearly independent and $a, \tilde{a}$ are arbitrary nonzero scalars, it follows that for any nonzero $a, \tilde{a} \in \mathbb{C}$ we have

$$[\tilde{a} : -a] = [e^{i\theta_{\tilde{v}}}\kappa_{\tilde{v}}(\tilde{a}) : -e^{i\theta_v}\kappa_v(a)].$$

For any real number a, $\tilde{a}$ we have $\kappa_{\tilde{v}}(\tilde{a}) = \tilde{a}$ and $\kappa_v(a) = a$, so we conclude that $e^{i\theta_v} = e^{i\theta_{\tilde{v}}}$. Because $\theta_v, \theta_{\tilde{v}} \in [0, 2\pi)$, it follows that $\theta_v = \theta_{\tilde{v}}$. Furthermore, if we set $a := 1$ and $\tilde{a} := i$ we find that

$$\kappa_{\tilde{v}}(i) = \frac{\kappa_{\tilde{v}}(i)}{\kappa_v(1)} = \frac{\tilde{a}}{a} = i.$$

Hence $\kappa_{\tilde{v}}$ is the identity; similarly, we can show that κ_v is the identity. Thus there is one value of θ such that for all nonzero $v \in E_n$ and all $[c_0, c_1] \in \mathbb{P}(\mathbb{C}^2)$ we have

$$S([c_0 v + c_1 e_n]) = [c_0 v + c_1 e^{i\theta}].$$

In other words, for any $z \in \mathbb{C}^{n+1}$ we have

$$
S[z] = \left[\begin{pmatrix} 1 & 0 & 0 & \cdots & 0 \\ 0 & 1 & 0 & \cdots & 0 \\ \vdots & & \ddots & & \vdots \\ 0 & \cdots & 0 & 1 & 0 \\ 0 & \cdots & 0 & 0 & e^{i\theta} \end{pmatrix} z \right]. \tag{10.4}
$$

Note that the matrix in this formula is unitary. This completes the proof in the special case where S fixed the point $[e_n]$ and every element of $[E_n]$.

Now we must prove the general case. Let v_0 denote a length-one vector in $S([e_n])$. Because v_0 is of unit length, it is possible to construct a unitary transformation T_0 on $\mathbb{C}^{n+1}$ such that $T_0 v_0 = e_n$. Then $[T_0] \circ S$ is a physical symmetry fixing $[e_n]$. Hence $[T_0] \circ S$ also takes points in $[E_n]$ to (possibly different) points in $[E_n]$. By induction, there is a unitary operator $M_n : E_n \to E_n$ and a function κ_n, equal to the identity or to conjugation, such that for every $[v] \in E_n$ we have

$$[T_0] \circ S([v]) = [M_n \kappa_n(v)].$$

Let $T_1 : \mathbb{C}^{n+1} \to \mathbb{C}^{n+1}$ denote the unitary transformation that agrees with M_n on E_n and that takes e_n to itself. Then the physical symmetry

$$\kappa_n \circ [T_1^{-1}] \circ [T_0] \circ S$$

satisfies the hypotheses of the special case above. So there is a unitary operator $T_2 : \mathbb{C}^{n+1} \to \mathbb{C}^{n+1}$ such that $\kappa_n \circ [T_1^{-1}] \circ [T_0] \circ S([v]) = [T_2 v]$ for all $v \in \mathbb{C}^{n+1}$. But then $S([v]) = [T_0^{-1} T_1 T_2 \kappa(v)]$ for all $v \in \mathbb{C}^{n+1}$. In other words, the first conclusion of the theorem is satisfied for $T := T_0^{-1} T_1 T_2$ and $\kappa = \kappa_n$.

The proof that κ and T are unique up to multiplication of T by a scalar of modulus one is exactly the same as in the proof of Proposition 10.8.   □

In this section we have classified the physical symmetries of a finite-dimensional quantum system. Half of these symmetries are projective unitary transformations; the other half are projective unitary transformations preceded by complex conjugation. This result means that by studying projective unitary transformations and complex conjugation, one can understand all physical symmetries.

10.6 Exercises

Exercise 10.1 (Used in Proposition 10.10) *Let* 0 *denote the zero-dimensional vector space. Is* $\mathbb{P}(0)$ *a vector space? Is* $\mathbb{P}(\mathbb{C})$ *a vector space? What is the cardinality of these two sets, i.e., how many points do they have?*

Exercise 10.2 (For students of differential geometry) *Show that for any natural number n, the set* $\mathbb{P}(\mathbb{C}^{n+1})$ *is a real manifold of dimension 2n.*

Exercise 10.3 *Suppose V is a complex vector space and [A] and [B] are projective linear operators on* $\mathbb{P}(V)$. *Show that* $[A] \circ [B] = [AB]$.

Exercise 10.4 *Consider the function* $f : \mathbb{C}^n \to \mathbb{P}(\mathbb{C}^{n+1})$ *defined by*

$$f(c_1, \ldots, c_n) := [1 : c_1 : \cdots : c_n].$$

Find a natural injective function from $\mathbb{P}(\mathbb{C}^n)$ *onto*

$$\mathbb{P}(\mathbb{C}^{n+1}) \setminus \{\text{Image } f\},$$

i.e., onto the points at infinity.

Exercise 10.5 *Show that Equation 10.1 corresponds to the picture in Figure 10.2. In other words, show that for any* (x, y), *the points* $(1,0,0)$, $(x, y, 0)$ *and* $F(x, y)$ *are collinear and* $\|F(x, y)\| = 1$. *Show that F is injective and that its image is* $S^2 \setminus \{(0, 0, 1)\}$.

Exercise 10.6 (Used in Proposition 10.2) *Show that stereographic projection (the correspondence between* $\mathbb{P}(\mathbb{C}^2)$ *and the unit sphere in* $\mathbb{R}^3$ *defined in Section 10.1) is given by*

$$[c_0 : c_1] \mapsto \left(\frac{2\Re(c_0^* c_1)}{|c_1|^2 + |c_0|^2}, \frac{2\Im(c_0^* c_1)}{|c_1|^2 + |c_0|^2}, \frac{|c_1|^2 - |c_0|^2}{|c_1|^2 + |c_0|^2} \right).$$

Show that this function is injective.

Exercise 10.7 (Used in Sections 10.3 and 11.2) *Suppose $[c_0:c_1]$ is a point on $\mathbb{P}(\mathbb{C}^2)$. Let p denote the corresponding point on the two-sphere in $\mathbb{R}^3$. Show that the antipodal point to p (i.e., the point that lies on the opposite end of the diameter containing p and the center of the sphere) corresponds to $[-c_0^*:c_0^*]$. Show that*

$$\{[c_0:c_1], [-c_1^*:c_0^*]\}$$

is an orthonormal basis of $\mathbb{P}(\mathbb{C}^2)$.

Exercise 10.8 (For students of topology) *Consider the topology on $S^3/\mathbb{T}$ inherited from the Euclidean topology on S^3 and the topology on $\mathbb{P}(\mathbb{C}^2)$ inherited from the norm topology on $\mathbb{C}^2$. Show that the function F defined in Section 10.2 and its inverse F^{-1} are both continuous functions with respect to these topologies.*

Exercise 10.9 *Suppose V is a complex scalar product space and $v, w \in V$. Show that $[v] = [w] \in \mathbb{P}(V)$ if and only if $\left|\langle [v]|[w]\rangle\right|^2 = 1$.*

Exercise 10.10 *Consider a particle in the state $[2 + i:1] \in \mathbb{P}(\mathbb{C}^2)$. Find its probability of exiting z-spin up from a Stern–Gerlach machine oriented along the z-axis. Find its probability of exiting y-spin up from a Stern–Gerlach machine oriented along the y-axis. Find its probability of exiting x-spin up from a Stern–Gerlach machine oriented along the x-axis.*

Find the same three probabilities for an arbitrary point $[c_0, c_1] \in \mathbb{P}(\mathbb{C}^2)$.

Exercise 10.11 *Consider the ket $|\phi\rangle := \frac{1}{\sqrt{2}}(|+\mathbf{z}\rangle + |+\mathbf{x}\rangle)$. Evaluate:*

$$\langle +\mathbf{x}|\phi\rangle, \qquad \langle +\mathbf{y}|\phi\rangle, \qquad \langle +\mathbf{z}|\phi\rangle.$$

Exercise 10.12 *Consider the kets of a spin-1/2 system. Any ket $c_+|+\mathbf{z}\rangle + c_-|-\mathbf{z}\rangle$ can be expressed in terms of the x-axis basis. That is, there are complex numbers b_+ and b_- such that $c_+|+\mathbf{z}\rangle + c_-|-\mathbf{z}\rangle$ and $b_+|+\mathbf{x}\rangle + b_-|-\mathbf{x}\rangle$ designate the same state. Is the function from $\mathbb{P}(\mathbb{C}^2)$ to $\mathbb{P}(\mathbb{C}^2)$ taking $[c_+:c_-]$ to $[b_+:b_-]$ a projective linear transformation? (Compare Exercise 2.16.)*

Exercise 10.13 *Design an experiment with a Stern–Gerlach machine to distinguish these two states:*

$$\frac{1}{\sqrt{2}}|+\mathbf{z}\rangle + \frac{1}{\sqrt{2}}|-\mathbf{z}\rangle$$

$$\frac{1}{\sqrt{2}}|+\mathbf{z}\rangle + \frac{i}{\sqrt{2}}|-\mathbf{z}\rangle.$$

Exercise 10.14 *What is wrong with the following argument?*
Since states are equivalent up to multiplication by a phase factor, $|+\mathbf{z}\rangle = i\,|+\mathbf{z}\rangle$. Hence

$$\frac{1}{\sqrt{2}}\,|+\mathbf{z}\rangle + \frac{1}{\sqrt{2}}\,|-\mathbf{z}\rangle = \frac{i}{\sqrt{2}}\,|+\mathbf{z}\rangle + \frac{1}{\sqrt{2}}\,|-\mathbf{z}\rangle .$$

Hence $[\frac{1}{\sqrt{2}}:\frac{1}{\sqrt{2}}] = [\frac{i}{\sqrt{2}}:\frac{1}{\sqrt{2}}]$. It follows that $[1:1] = [i:1]$, so $1 = i$.

Exercise 10.15 *Show that for any element*

$$[c_0:c_1] = c_0\,|-\mathbf{z}\rangle + c_1\,|+\mathbf{z}\rangle$$

of $\mathbb{P}(\mathbb{C}^2)$ we have

$$\left[\langle -\mathbf{x}|[c_0:c_1]\rangle : \langle +\mathbf{x}|[c_0:c_1]\rangle\right] = [c_0 + c_1 : c_0 - c_1]$$
$$\left[\langle -\mathbf{y}|[c_0:c_1]\rangle : \langle +\mathbf{z}|[c_0:c_1]\rangle\right] = [c_0 + ic_1 : c_0 - ic_1].$$

Exercise 10.16 *Show that the function*

$$\mathbb{P}(\mathbb{C}^2) \to \mathbb{P}(\mathbb{C}^2),\ [z_0:z_1] \mapsto [z_0:e^{i\theta}z_1]$$

corresponds to rotating the sphere in Figure 10.6 through an angle of θ around the z-axis. Show that the function

$$\mathbb{P}(\mathbb{C}^2) \to \mathbb{P}(\mathbb{C}^2),\ [z_0:z_1] \mapsto [z_0^*:z_1^*]$$

corresponds to reflecting the sphere in Figure 10.6 in the xz-plane.
Find an explicit formula for the physical symmetry of $\mathbb{P}(\mathbb{C}^2)$ that corresponds to a rotation through an angle of θ around the y-axis.

Exercise 10.17 *Suppose p_1, p_2 and p_3 are points on the unit sphere in $\mathbb{R}^3$ that do not lie on one common great circle. (A great circle is the intersection of the unit sphere with a plane through the origin in $\mathbb{R}^3$.) Show that every point p on the sphere is uniquely defined by its distances from the three points p_1, p_2, p_3. Interpret great circles in $\mathbb{P}(\mathbb{C}^2)$ physically; i.e., give a definition in terms of experiments and probabilities.*

Exercise 10.18 (For students of topology) *Show that there is a topology on $\mathbb{P}(\mathbb{C}^{n+1})$ whose basic open sets are of the form*

$$\left\{[z] \in \mathbb{P}(\mathbb{C}^{n+1}) : 1 - \left|\langle [z]|[z_0]\rangle\right| < \epsilon\right\},$$

where $[z_0] \in \mathbb{P}(\mathbb{C}^{n+1})$ and $\epsilon > 0$. Show that any $[z]$ in $\mathbb{P}(\mathbb{C}^{n+1})$ can be approximated in this topology by elements of the form $[y + ae_n]$, where $[y] \neq [y^*]$ and $a \neq 0$. Show that any physical symmetry $F : \mathbb{P}(\mathbb{C}^{n+1}) \to \mathbb{P}(\mathbb{C}^{n+1})$ is continuous in this topology.

Exercise 10.19 *Is $\mathbb{P}\mathcal{U}(V) = \mathbb{P}(\mathcal{U}(V))$?*

Exercise 10.20 *Show that the projective unitary representation $\rho_{\frac{1}{2}}$ of $SO(3)$ does not descend from any linear unitary representation of $SO(3)$.*

Exercise 10.21 *Find an isomorphism of complex scalar product spaces between $\mathbb{C}^{n+1}$ (with the standard scalar product) and $\mathcal{P}^n$ (with the scalar product space defined in Proposition 4.7).*

Exercise 10.22 *Find a group isomorphism between $SO(3)$ and a subgroup of the physical symmetries of the qubit. Use Proposition 10.1 to find a nontrivial group homomorphism from $SU(2)$ into the group of physical symmetries of the qubit. Finally, express the group homomorphism $\Phi : SU(2) \to SO(3)$ from Section 4.3 in terms of these functions.*

Exercise 10.23 *Find out what the Hopf fibration of the sphere is and relate it to the contents of this chapter.*

Exercise 10.24 (Injectivity used in Proposition 10.8) *Show that the composition of two physical symmetries is a physical symmetry. Show that every physical symmetry is injective and surjective.*

Exercise 10.25 *Show that the group of physical symmetries of the qubit is isomorphic to the group $O(3)$.*

Exercise 10.26 (For students of topology) *Suppose G is a connected Lie group, V is a finite-dimensional complex scalar product space and*

$$\rho : G \to \text{Physical symmetries of } \mathbb{P}(V).$$

is a Lie group homomorphism. Show that ρ is a projective unitary representation.

11
Independent Events and Tensor Products

Die Quantenmechanik ist sehr achtung-gebietend. Aber eine innere Stimme sagt mir, daß das doch nicht der wahre Jakob ist. Die Theorie liefert viel, aber dem Geheimnis des Alten bringt sie uns kaum näher. Jedenfalls bin ich überzeugt, daß *der* nicht würfelt. Wellen im $3n$-dimensionalen Raum, deren Geschwindigkeit durch potentielle Energie (z. B. Gummibänder) reguliert wird.... Ich plage mich damit herum, die Bewegungsgleichungen von als Singularitäten aufgefaßten materiellen Punkten aus den Differentialgleichungen der allgemeinen Relativität abzuleiten.

> — Albert Einstein, in a letter to Max Born [BBE, 4.12.26]

Quantum mechanics is certainly imposing. But an inner voice tells me that it is not yet the real thing. The theory says a lot, but does not really bring us any closer to the secret of the 'old one.' I, at any rate, am convinced that *He* is not playing at dice. Waves in $3n$-dimensional space, whose velocity is regulated by potential energy (for example, rubber bands).... I am working very hard at deducing the equations of motion of material points regarded as singularities, given the differential equation of general relativity.

> — Albert Einstein, in a letter to Max Born,
> translated by Irene Born [BBE′, 4 December 1926]

In this chapter we investigate an appropriate mathematical framework for treating quantum systems with several particles, or with a single particle of several attributes. In classical mechanics, the phase space of a system of many particles is the Cartesian sum of the phase spaces of the individual particles. The situation in quantum mechanics is different. Even if there are no forces between the particles, there are still some states of the system where measurements of one particle can affect measurements of another particle. These are called *entangled states*. The empirical verification of the existence of entangled states (most famously in the Einstein–Podolsky–Rosen paradox, discussed in Section 11.3) implies that the Cartesian sum is not the right mathematical tool. Instead, the phase space of the system of many particles should be the tensor product of the phase spaces of the individual particles, as we will see in Section 11.1. In Section 11.2 we discuss the quantum mechanics of partial measurements. Section 11.3 introduces physical entanglement and its simplest mathematical counterpart. Finally, in Section 11.4, we apply these insights to the hydrogen atom in order to incorporate the spin of the electron into our model. The reward is a much-desired factor of two.

11.1 Independent Measurements

In this section we show that a tensor product of state spaces is the proper mathematical formulation for a system with independent measurements. Each measurement has a corresponding complex scalar product space whose projectivization is the state space for that measurement; the projectivization of the tensor product of the complex scalar product spaces is the state space for all of the measurements together. We work with one example, leaving generalization to the reader.

Consider a quantum system consisting of the spin state of two particles, one of spin 1/2 (a fermion) and one of spin 1 (a boson). How do we model the two of them together? The (experimentally justified) prescription in quantum mechanics is to find a basis of states and build a projective space out of them, as we discussed briefly in Section 10.3. In this section we will show that the set

$$\left\{ \, |{+}{+}\rangle \, , |{+}\, 0\rangle \, , |{+}{-}\rangle \, , |{-}{+}\rangle \, , |{-}\, 0\rangle \, , |{-}{-}\rangle \right\}$$

is a basis, where we introduce the notation $|ab\rangle$ to denote the state of the system where the spin-1/2 particle is in state a (where $a = +$ or $a = -$) and the spin-1 particle is in state b (where $b = +$ or $b = 0$ or $b = -$). Of course, we must specify axes along which to measure the spins; we will measure spins in the direction of the positive z-axis for both particles. We will use this basis to describe the spin state space of a system comprised of one fermion and one boson.

First we verify that the states in the set are mutually exclusive. For instance, the two states $|++\rangle$ and $|-+\rangle$ have mutually exclusive states for the spin-1/2 particle, and hence must be mutually exclusive. Similarly, in any pair either the spin-1/2 or the spin-1 particle states are mutually exclusive.

Next we must verify that the list is long enough. If we measure the z-spins of both particles, we must find one of the six listed states. Also, none of these states have multiplicities because the spin states of the two individual particles have no multiplicities. Hence the set of six ordered pairs above is a basis for the quantum system consisting of one spin-1/2 and one spin-1 particle.

Notice how the independence of the two particles came into the analysis: because the state of one particle is not restricted by the state of the other particle, all six of the listed states are indeed possible. More fundamentally, the expression of the states as pairs joined by "and" is possible only because measuring the state of the spin-1 particle does not affect future measurements of the spin-1/2 particle, and vice versa. In the typical physics-style presentation of this material, this idea might be stated in the form $\hat{J}_{\frac{1}{2}} \hat{J}_1 = \hat{J}_1 \hat{J}_{\frac{1}{2}}$, where the $\hat{J}$'s are the operators corresponding to spin around the z-axis. In other words, we can measure the z-spin states of the two particles simultaneously. In contrast, it is impossible to measure both the x- and z-spins of a single particle simultaneously; nor is it possible to measure both the position and the momentum of a dynamical particle simultaneously, by Heisenberg's uncertainty principle. The independence of our two measurements is crucial.

To model the two-particle system mathematically we need to find a mathematical projective space whose basis corresponds to the list of six states. We want more than just dimensions to match; we want the physical representations on the individual particle phase spaces to combine naturally to give the physical representations on the combined phase space. The space that works is

$$\mathbb{P}(\mathbb{C}^2 \otimes \mathbb{C}^3).$$

Recall from Section 10.4 that if an observer undergoes a rotation of g (with $g \in SO(3)$, the spin-1/2 state space $\mathbb{P}(\mathbb{C}^2)$ transforms via the linear operator $\rho_{\frac{1}{2}}(g)$, while the spin-1 state space $\mathbb{P}(\mathbb{C}^3)$ transforms via $\rho_1(g)$. Hence the corresponding transformation of a vector $v \otimes w$ in $\mathbb{C}^2 \otimes \mathbb{C}^3$ is

$$\left[\left(\rho_{\frac{1}{2}}(g)v \right) \otimes \left(\rho_1(g)w \right) \right].$$

Note that because $\rho_1(g)$ is well defined and $\rho_{\frac{1}{2}}(g)v$ is well defined up to a scalar multiple, the displayed expression is a well-defined element of $\mathbb{P}(\mathbb{C}^2 \otimes \mathbb{C}^3)$. This is precisely the tensor product representation.

More concretely, a basis for the state space of the spin-1/2 particle is $\{|+\rangle, |-\rangle\}$, while a basis for the state space of the spin-1 particle is $\{|+\rangle, |0\rangle, |-\rangle\}$. The state space for the system of two particles is the tensor product, for which the kets

$$|+\rangle \otimes |+\rangle = |++\rangle, \quad |+\rangle \otimes |0\rangle = |+0\rangle, \quad |+\rangle \otimes |-\rangle = |+-\rangle$$

$$|-\rangle \otimes |+\rangle = |-+\rangle, \quad |-\rangle \otimes |0\rangle = |-0\rangle, \quad |-\rangle \otimes |-\rangle = |--\rangle$$

form a basis. For an example of the group action, recall that a physical rotation through an angle θ around the spin axis corresponds to the actions

$$\left(c_+ |+\rangle + c_- |-\rangle\right) \mapsto \left(e^{i\theta/2}c_+ |+\rangle + e^{-i\theta/2}c_- |-\rangle\right)$$

and

$$\left(a_+ |+\rangle + a_0 |0\rangle + a_- |-\rangle\right) \mapsto \left(e^{i\theta}a_+ |+\rangle + a_0 |0\rangle + e^{-i\theta}a_- |-\rangle\right).$$

It follows that the action of such a rotation of the physical space on the state space of the two-particle system takes a state

$$c_{++} |++\rangle + c_{+0} |+0\rangle + c_{+-} |+-\rangle + c_{-+} |-+\rangle + c_{-0} |-0\rangle + c_{--} |--\rangle$$

to the state

$$e^{3i\theta/2}c_{++} |++\rangle + e^{i\theta/2}c_{+0} |+0\rangle + e^{-i\theta/2}c_{+-} |+-\rangle + e^{i\theta/2}c_{-+} |-+\rangle$$
$$+ e^{-i\theta/2}c_{-0} |-0\rangle + e^{-3i\theta/2}c_{--} |--\rangle.$$

The construction in this section generalizes. Any time there are two (or more) independent quantum-mechanical measurements, a tensor product is appropriate. We will see another example in Section 11.4, where we consider the independent measurements of position and spin of an electron.

11.2 Partial Measurement

When we discussed the physics of one quantum particle, we did not explicitly confront one of the deepest mysteries of quantum mechanics, sometimes called the *collapse of the wave function*. Recall that a spin-1/2 particle exiting a z-axis Stern–Gerlach machine must be in either the state $|+\mathbf{z}\rangle$ or $|-\mathbf{z}\rangle$.

If the entering particle was in a mixed state (relative to the z-spin measurement), then the act of measurement changes the state of the particle. No one understands how this happens, but it is an essential feature of the quantum mechanical model. For example, this phenomenon contributes to Heisenberg's uncertainty principle, whose most famous implication is that one cannot measure both the position and the momentum of a particle exactly. The point is that a position measurement changes the state of the particle in a way that erases information about the momentum, and vice versa.

In the case of a spin measurement on a single particle, the final states are all pure states, without multiplicities. In this case there is only one state corresponding to each possible result of the measurement. But what if the measurement has multiplicities? In other words, what if we make only a partial measurement? Then there are several states corresponding to one particular result of the measurement; which state is the final state for the measured particle?

To answer this question, we must first introduce the quantum mechanical model for measurement. First we discuss measurement on finite-dimensional phase spaces, to avoid mathematical complications. Then we say a few words about the infinite-dimensional case.

One assumption of the model is that each measurable quantity A (also known as an *observable*) of a finite-dimensional quantum system $\mathbb{P}(V)$ determines a decomposition of the vector space V into orthogonal subspaces, with a measurement value corresponding to each subspace. In other words, there is a set $\{W_j : j = 1, \dots, n\}$ of mutually orthogonal subspaces, where $n \in \mathbb{N}$, such that

$$\bigoplus_{j=1}^{n} W_j = V,$$

and a set of numbers $\{\lambda_j : j = 1, \dots, n\}$ such that if $w \in W_j$, then measuring the state $[w]$ is sure to yield the value λ_j. Typically, the information about the orthogonal subspaces and measurement values is encoded in a linear operator $\hat{A}$ on the vector space V. The λ_j's are the eigenvalues, and the W_j's are the corresponding eigenspaces. This information completely determines the operator $\hat{A}$ corresponding to the observable A. Since the eigenvalues are real and the eigenspaces are orthogonal to one another, the operator $\hat{A}$ is Hermitian-symmetric with respect to the standard complex scalar product on V, as the reader may check in Exercise 11.1. Conversely, because every Hermitian-symmetric linear operator on a finite-dimensional vector space can be diagonalized (by the Spectral Theorem for Hermitian-symmetric matrices,

Exercise 11.2), every such operator corresponds to an observable of the quantum system.

The second assumption of the quantum mechanical model is that we can calculate the probabilities of various outcomes of the measurement A on an arbitrary state $[v]$ from the W_j's and the λ_j's. Specifically, the probability of an outcome of λ_j for the measurement A on the state $[v]$ is

$$\frac{|\langle \Pi_{W_j} v, v \rangle|}{\|v\|^2}.$$

Recall from Proposition 3.5 that given any finite-dimensional linear subspace W of a scalar product space V, there is an orthogonal projection Π_W on V whose kernel is $W^\perp$. Note that the expression giving the probability does not depend on the choice of vector in the equivalence class $[v]$.

We can argue for the plausibility of this second assumption by working out an example. Consider a spin-1/2 particle in the state

$$[(c_+, c_-)] = c_+ \, |+\mathbf{z}\rangle + c_- \, |-\mathbf{z}\rangle,$$

where we assume without loss of generality that $|c_+|^2 + |c_-|^2 = 1$. We expect that the probability for finding such a particle to be spin up after a measurement of z-axis spin should be $|c_+|^2$. In this case the subspace W corresponding to a spin-up measurement is $\mathbb{C} \oplus \{0\} \subset \mathbb{C}^2$. Thus our second assumption implies that the probability is

$$\frac{|\langle (c_+, 0), (c_+, c_-) \rangle|}{|c_+|^2 + |c_-^2|} = |c_+|^2,$$

as expected.

To specify the final state of a measured particle, we need one more tool, orthogonal projection in projective space. We would like to consider the "projectivization" of Π_W, but since Π_W is not necessarily invertible, we cannot apply Proposition 10.1. To evade this technical difficulty we restrict the domain of Π_W. Recall the notation for set subtraction: $A \setminus B := \{x \in A : x \notin B\}$.

Definition 11.1 *Suppose V is a complex scalar product space and W is a linear subspace of V. Then we define an operator $\Pi_{[W]} \colon \mathbb{P}(V) \setminus \mathbb{P}(W^\perp) \to \mathbb{P}(V)$ such that for all $[v] \in \mathbb{P}(V) \setminus \mathbb{P}(W^\perp)$, we have*

$$\Pi_{[W]}[v] := [\Pi_W v].$$

We call $\Pi_{[W]}$ the orthogonal projection onto $[W]$.

Finally we have the tools to state the answer to our original question: what is the final state of a measured particle? Consider a measurement A on a finite-dimensional vector space V, possibly with multiplicities. Suppose that a particle enters the measuring device in a state $[v]$ and the measurement yields the result λ. Let W_λ denote the subspace of states whose measurement is sure to yield λ. Note that $[v] \notin [W_\lambda^\perp]$ because there is a nonzero chance of the measurement result λ. Hence $[v]$ is in the domain of $\Pi_{[W_\lambda]}$; the third assumption of the model is that the state of the particle on exit is $\Pi_{[W_{\lambda_j}]}[v]$.

Consider, for example, the measurement of the spin of a spin-1/2 particle via a Stern–Gerlach machine oriented along an arbitrary axis. Let $[c_0 : c_1]$ be the point in projective space corresponding to the positive axis of the Stern–Gerlach machine. Then a spin-up measurement corresponds to the one-dimensional subspace W_{up} of $\mathbb{C}^2$ spanned by (c_0, c_1), while a spin-down measurement corresponds (by Exercise 10.7) to the one-dimensional subspace W_{down} spanned by $(-c_1^*, c_0^*)$. Note that both $[W_{\text{up}}]$ and $[W_{\text{down}}]$ consist of single points, $[c_0 : c_1]$ and $[-c_1^* : c_0^*]$, respectively. So any particle that exits the machine spin up will be in the pure spin-up state, namely $[c_0 : c_1]$, while any particle exiting spin down will be in the pure spin-down state, $[-c_1^* : c_0^*]$. The same phenomenon occurs whenever the measurement has no multiplicities: the end result of a measurement is a particle in the single pure state corresponding to the result of the measurement.

For the next example, consider an arbitrary state of the two-particle system from Section 11.1:

$$c_{++} \,|{+}{+}\rangle + c_{+0} \,|{+}0\rangle + c_{+-} \,|{+}{-}\rangle + c_{-+} \,|{-}{+}\rangle + c_{-0} \,|{-}0\rangle + c_{--} \,|{-}{-}\rangle,$$

where $[c_{++} : c_{+0} : c_{+-} : c_{-+} : c_{-0} : c_{--}] \in \mathbb{P}(\mathbb{C}^6)$. Suppose we measure the z-axis spin of the fermion (i.e., the spin-1/2 particle) and find it to be spin up. To find the final state of the the system, we must first identify the subspace of $\mathbb{C}^6$ corresponding to this result: it is $\mathbb{C} \times \mathbb{C} \times \mathbb{C} \times \{0\} \times \{0\} \times \{0\}$. Then we project orthogonally onto this subspace. Hence the final state is

$$c_{++} \,|{+}{+}\rangle + c_{+0} \,|{+}0\rangle + c_{+-} \,|{+}{-}\rangle.$$

(Note that we have not taken the trouble to normalize the coefficients, preferring to think of them as a point in $\mathbb{P}(\mathbb{C} \times \mathbb{C} \times \mathbb{C} \times \{0\} \times \{0\} \times \{0\})$.) Similarly, the final state of a particle exiting spin down is

$$c_{-+} \,|{-}{+}\rangle + c_{-0} \,|{-}0\rangle + c_{--} \,|{-}{-}\rangle.$$

The situation for infinite-dimensional quantum-mechanical systems is similar, but the mathematics is more subtle. The operators that carry the information about observables are known as *self-adjoint* operators. We saw several examples of such operators (some missing a factor of i) in Chapter 8: **H**, the $\mathbf{R_q}$'s, the $\mathbf{L_q}$'s, etc. There is a Spectral Theorem for operators on infinite-dimensional scalar product spaces, but even the correct generalization of Hermitian symmetry is a technical challenge. Readers interested in the mathematical details ("continuous spectrum," "spectral projections," "dense subspaces" and more) should consult a book on functional analysis, such as Reed and Simon [RS].

Note that in this section we have introduced three assumptions of the quantum mechanical model. We recall them here and add a fourth.

1. Each observable A of a finite-dimensional quantum system $\mathbb{P}(V)$ determines a decomposition of the vector space V into orthogonal subspaces, with a measurement value corresponding to each subspace.

2. The probability of an outcome of λ for the measurement A on the state $[v]$ is
$$\frac{|\langle v, \Pi v \rangle|}{\|v\|^2},$$
where Π is the orthogonal projection onto the subspace W_λ of states which are sure to yield the result λ for the measurement A. (Note that if v is a unit vector then the expression simplifies to $|\langle v, \Pi v \rangle|^2$.

3. After a particle in a state $[v]$ is subjected to the measurement A and yields a value λ, the particle will be in the state $[\Pi v]$, where Π is the orthogonal projection onto the subspace W_λ, as in Assumption 2

4. Given any two distinct states of a quantum system, there is a measurement that distinguishes them.

This last assumption is natural: if there is no measurement that distinguishes two states, then there is no meaningful physical difference between the two states, i.e., they are not distinct. The quantum mechanical assumptions about measurement introduced in this section will allow us to discuss *entanglement* in the next section.

11.3   Entanglement and Quantum Computing

One of the strange phenomena of quantum mechanics, with no counterpart in classical mechanics, is *entanglement*. Consider the state

$$\frac{1}{\sqrt{2}} \, |++\rangle + \frac{1}{\sqrt{2}} \, |--\rangle \tag{11.1}$$

of the quantum system introduced in Section 11.1. This is a perfectly legitimate quantum state. However, a pair of particles in this state behave counter-
intuitively — that is, counter to the intuition of the author and many others,
including Albert Einstein. Imagine measuring the z-axis spin of the fermion
(i.e., the spin-1/2 particle). If the result is spin up, then the pair of particles
must be in the state $|++\rangle$. It follows that the boson (i.e., the spin-1 particle)
is also spin up, even though we never measured it directly. In other words,
the states of the two particles are *entangled*: it is possible to get information about one by measuring the other. Entanglement is possible even if the
particles exert no force on one another.

This bizarre prediction, known as the *Einstein–Podolsky–Rosen paradox*,
has been verified many times in the laboratory. The most famous version
involves two electrons manipulated into a mixed state with combined spin
of 0. The electrons are separated in space before the spin of one (and only
one) electron is measured, say, in a Stern–Gerlach machine. If that electron
is found to be spin up, then by conservation of spin angular momentum,
the other electron must be spin down, and vice versa. This holds true even
if the ratio of the distance between the measurements to the time between
the measurements is greater than the speed of light. See the discussion in
Townsend [To, Sections 5.4 and 5.5] and the references therein.

Note that entanglement occurs independently of any classical interaction
of the particles. In other words, entanglement occurs for free particles as well
as for particles exerting forces on one another. To put it yet another way, the
possibility of entanglement arises from the quantum mechanical state space
itself, not from any differential equation or differential operator used to describe the evolution of the system.

Not every state of the system is entangled. Consider, for example, the state

$$\frac{1}{\sqrt{2}}\,|++\rangle + \frac{i}{\sqrt{2}}\,|-+\rangle.$$

A pair of particles in this state has the property that the boson is sure to be
found spin up, while the fermion is equally likely to be spin up as spin down.
Measuring the spin of the fermion gives us no new information about the
boson — for example, if the fermion is found to be spin up, then the system
is in the $|++\rangle$ state, so we know that the boson is spin up, but we knew that
before. And measuring the boson's state yields no information at all, since it
is sure to be spin up.

Of course, it is not enough to consider z-axis spin measurements alone.
Perhaps a spin measurement around another axis would show entanglement.

In fact, no measurement would show entanglement, as we shall prove once we have found a convenient mathematical description of entanglement in Proposition 11.1.

We need mathematical notation for measuring one or more particles in a multiparticle system; more generally, if a quantum system V is a combination of several quantum systems $V_1, \ldots, V_n$, i.e., if

$$V := \bigotimes_{k=1}^{n} V_k,$$

we need to express mathematically the notion of a measurement limited to some of the several quantum systems. Because both the probabilities of various measurement outcomes and the state of the measured system after measurement can be calculated in terms of projections, our criterion is stated in terms of projections. Readers accustomed to thinking of an observable as a Hermitian operator should note that a Hermitian operator on a finite-dimensional complex scalar product space is determined by its eigenspaces and eigenvalues. Hence actual measurements correspond to eigenspaces or, equivalently, projections onto eigenspaces.

For any $K := \{k_1, \ldots, k_m\} \subset \{1, \ldots, K\}$, we define

$$V_K := \bigotimes_{j=1}^{m} V_{k_j}.$$

It will be useful to have notation for the complement of K; we let $\hat{K}$ denote the unique set such that

$$K \cup \hat{K} = \{1, \ldots, K\}, \quad K \cap \hat{K} = \emptyset.$$

For example, if $n = 5$ and $K = \{1, 4\}$ then $\hat{K} = \{2, 3, 5\}$. By Exercise 11.3 we know that the vector space V is unitarily isomorphic to the vector space $V_K \otimes V_{\hat{K}}$. Now suppose Π is a projection operator on V_K and let I denote the identity operator on $V_{\hat{K}}$. We will use the notation $\widetilde{\Pi}$ for the linear operator on V corresponding to the operator

$$\Pi \otimes I : V_K \otimes V_{\hat{K}} \to V_K \otimes V_{\hat{K}}. \tag{11.2}$$

With this notation in hand, we are now ready to define *entangled* and *unentangled* states.

Definition 11.2 *Suppose $V_1, \ldots, V_n$ are complex vector spaces, and consider their tensor product:*

$$V := \bigotimes_{k=1}^{n} V_k.$$

A state $[x] \in \mathbb{P}(V)$ is entangled *if and only if there is a set $K \subset \{1, \ldots, n\}$ and nonzero orthogonal projections $\Pi_1 \colon V_K \to V_K$ and $\Pi_2 \colon V_{\hat{K}} \to V_{\hat{K}}$, with $\tilde{\Pi}_2 x \neq 0$, such that*

$$\frac{|\langle \tilde{\Pi}_1 x, x \rangle|}{\|x\|^2} \neq \frac{|\langle \tilde{\Pi}_1 \tilde{\Pi}_2 x, \tilde{\Pi}_2 x \rangle|}{\|\tilde{\Pi}_2 x\|^2}. \tag{11.3}$$

A state that is not entangled is unentangled.

This definition matches the common use of "entangled" in the literature. If there are projections Π_1 and Π_2 making Inequality 11.3 true, then we can find measurements A_1 and A_2 whose results on a system in state x depend on the order of the measurement. The image of Π_1 (resp., Π_2) is, for some possible value λ_1 (resp., λ_2), the set of states for which the measurement A_1 (resp., A_2) is sure to yield the value λ_1 (resp., λ_2).

Note that the definition of entanglement depends on a choice of factors for the tensor product. That is, a state $[v] \in V$ might be entangled with respect to one factorization $V = V_1 \otimes \cdots \otimes V_n$ but not with respect to a different factorization $V = W_1 \otimes \cdots \otimes W_m$. This distinction is especially important in the proof of Proposition 11.1.

Now recall the definition of an elementary tensor (Definition 2.15). It turns out that elementary tensors always correspond to unentangled states, and vice versa.

Proposition 11.1 *Suppose $V_0, \ldots, V_n$ are complex vector spaces, and consider a nonzero element of their tensor product:*

$$x \in V := \bigotimes_{k=0}^{n} V_k.$$

The state $[x] \in \mathbb{P}(V)$ is entangled if and only if x is not an elementary tensor.

In the proof we will use various tensor products and entanglement with respect to these various tensor products. Where we do not specify the tensor product we mean entanglement with respect to $V_0 \otimes \cdots \otimes V_n$.

Proof. First, consider an elementary tensor, i.e., a vector of the form $x := v_0 \otimes \cdots \otimes v_n$. We will show that the state $[x]$ is not entangled. Suppose K is a subset of $\{0, \ldots, n\}$ and that $\Pi_1 : V_k \to V_k$ and $\Pi_2 : V_{\hat{K}} \to V_{\hat{K}}$ are arbitrary nonzero orthogonal projections. We will write

$$v_K := \bigotimes_{k \in K} v_k \quad \text{and} \quad v_{\hat{K}} := \bigotimes_{j \notin K} v_j.$$

Recall from Section 5.3 that the natural complex scalar product on a tensor product is obtained by multiplying the individual complex scalar products of the factors. Letting $\langle \cdot, \cdot \rangle_1$ and $\langle \cdot, \cdot \rangle_2$ denote the complex scalar products on the factors V_K and $V_{\hat{K}}$, respectively, we find

$$\langle \tilde{\Pi}_1 x, x \rangle = \langle \Pi_1 v_K, v_K \rangle_1 \langle v_{\hat{K}}, v_{\hat{K}} \rangle_2$$
$$\|x\|^2 = \langle v_K, v_K \rangle_1 \langle v_{\hat{K}}, v_{\hat{K}} \rangle_2$$
$$\langle \tilde{\Pi}_1 \tilde{\Pi}_2 x, \tilde{\Pi}_2 x \rangle = \langle \Pi_1 v_K, v_K \rangle_1 \langle \Pi_2 v_{\hat{K}}, \Pi_2 v_{\hat{K}} \rangle_2$$
$$\|\tilde{\Pi}_2 x\|^2 = K \langle v_K, v_K \rangle_1 \langle \Pi_2 v_{\hat{K}}, \Pi_2 v_{\hat{K}} \rangle_2.$$

These equations imply that

$$\frac{\langle \tilde{\Pi}_1 x, x \rangle}{\|x\|^2} = \frac{\langle \Pi_1 v_K, v_K \rangle_1}{\langle v_K, v_K \rangle_1} = \frac{\langle \tilde{\Pi}_1 \tilde{\Pi}_2 x, \tilde{\Pi}_2 x \rangle}{\|\tilde{\Pi}_2 x\|^2}.$$

Hence the state x is not entangled.

Next we suppose that a state $[x]$ is not entangled and show that it is elementary. We proceed by induction on n, the number of factors in the tensor product. The base case is trivial: if $n = 1$, then any $x \in V_1$ is elementary.

We may assume, without loss of generality, that every one of the vector spaces V_k is finite dimensional, due to the following argument. Because x is an element of the tensor product, it must be a finite sum of elementary tensors:

$$x = \sum_{k=1}^{m} \left(\bigotimes_{j=0}^{n} x_{kj} \right), \tag{11.4}$$

where each $x_{kj} \in V_k$. Without loss of generality, we may replace each V_k by the span of the set $\{x_{kj} \in V_k : k = 1, \ldots, m\}$. These spans are all finite-dimensional.

It remains to prove the inductive step. Suppose that n is a natural number and suppose that the proposition holds true for tensor products with n factors. Suppose

$$V = \bigotimes_{k=0}^{n} V_k = V_0 \otimes V_K,$$

where $K := \{1, \ldots, n\}$. Suppose $[x] \in \mathbb{P}(V)$ is not entangled. We must show that x is an elementary tensor.

We will exploit the vector space isomorphism between the scalar product space $V_0 \otimes V_K$ and the scalar product space $\text{Hom}(V_0^*, V_K)$, introduced in Proposition 5.14 and Exercise 5.22. (Note that $(V^*)^* = V$ by Exercise 2.15.) Instead of working directly with $x \neq 0$, we will work with the corresponding linear transformation $X \neq 0$. We will show that $X : V_0^* \to V_K$ has rank one and that its image is generated by an elementary element of the tensor product $V_1 \otimes \cdots \otimes V_n$. Then we will deduce that x itself is elementary in the tensor product $V_0 \otimes V_1 \otimes \cdots \otimes V_n$.

We would like to show that X itself has rank one. Because $X \neq 0 \in \text{Hom}(V_0^*, V_K)$ there is a rank-one projection $\Pi_K : V_K \to V_K$ such that $\Pi_K X \neq 0 \in \text{Hom}(V_0^*, V_K)$. Hence, recalling the adjoint transformation $X^* \in \text{Hom}(V_K, V_0^*)$ from Definition 3.9 and the complex scalar product on $\text{Hom}(V_K, V_0^*)$ from Section 5.5, we have

$$0 \neq \|\Pi_K X\| = \text{Tr}(X^* \Pi_K^* \Pi_K X) = \text{Tr}(X^* \Pi_K X),$$

and the rank of $X^* \Pi_K X$ must be at least one. On the other hand, Π_K has rank one, so $X^* \Pi_K X$ has rank at most one. We conclude that

$$\text{rank}(X^* \Pi_K X) = 1.$$

Let $Q : V_0^* \to V_0^*$ be the orthogonal projection onto the kernel of $X^* \Pi_K X$. Define a corresponding orthogonal projection $P : V_0 \to V_0$ by

$$P := \tau^{-1} \circ Q \circ \tau,$$

where τ is the natural function from V_0 to V_0^* defined in Equation 5.3. By Exercise 11.5, P is an orthogonal projection and

$$(Q\alpha)v = \alpha(Pv)$$

for every $\alpha \in V_0^*$ and every $v \in V$.

Since $[x]$ is unentangled we have

$$\frac{|\langle \tilde{P}x, x \rangle|}{\|x\|^2} = \frac{|\langle \tilde{P}\tilde{\Pi}_K x, x \rangle|}{\|\tilde{\Pi}_K x\|^2}.$$

We use Exercise 11.6 to rewrite this equation in terms of the complex scalar product space $\text{Hom}(V_0^*, V_K)$. We obtain

$$\frac{|\text{Tr}(X^* X Q)|}{\text{Tr}(X^* X)} = \frac{|\text{Tr}(X^* \Pi_K X Q)|}{\text{Tr}(X^* \Pi_K X)} = 0,$$

since the image of Q is the kernel of $X^* \Pi_K X$. Hence by Proposition 2.10

$$\|XQ\| = \mathrm{Tr}(QX^*XQ) = \mathrm{Tr}(X^*XQ^2) = \mathrm{Tr}(X^*XQ) = 0.$$

So the kernel of X contains the image of Q. But the image of Q is the kernel of $X^* \Pi X$, which has dimension $(\dim V_0) - 1$. Hence the rank of X is at most one. Because $X \neq 0$, we conclude that the rank of X exactly one.

Because the rank of X is one, Exercise 5.14 implies that x is elementary in the tensor product $V_0 \otimes V_K$. In other words, there are vectors $x_0 \in V_0$ and $x_K \in V_K$ such that $x = x_0 \otimes x_K$. It remains to show that x_K is elementary.

By the inductive hypothesis, it suffices to show that $[x_K]$ is not entangled with respect to the tensor product $V_K = V_1 \otimes \cdots \otimes V_n$. Consider any subset $J \subset \{1, \ldots, n\}$. Let $P_1 : V_J \to V_J$ and $P_2 : V_j \to V_j$ be arbitrary orthogonal projections. Let $\hat{P}_1, \hat{P}_2 : V_K \to V_K$ denote the corresponding orthogonal projections on V_K, while $\tilde{P}_1$ and $\tilde{P}_2$ denote, as usual, the corresponding orthogonal projections on V. We have

$$\frac{|\langle \hat{P}_1 x_K, x_K \rangle|}{\|x_K\|^2} = \frac{\langle x_0, x_0 \rangle \langle \hat{P}_1 x_K, x_K \rangle}{\|x_0\|^2 \|x_K\|^2} = \frac{\langle \tilde{P}_1 x, x \rangle}{\|x\|^2},$$

where the third equality follows because X and $P_1 X$ are rank one and $w \perp \ker(X)$; similarly, if $\hat{P}_2 x_K \neq 0$ we have

$$\frac{|\langle \hat{P}_1 \hat{P}_2 x_K, \hat{P}_2 x_K \rangle|}{\|\hat{P}_2 x_K\|^2} = \frac{|\langle \tilde{P}_1 \tilde{P}_2 x, \tilde{P}_2 x \rangle|}{\|\tilde{P}_2 x\|^2}.$$

Because $[x]$ is not entangled, these equations imply that $[x_K]$ is not entangled. By the inductive hypothesis, the vector $x_K \in V_K$ must be elementary in $V_K = V_1 \otimes V_n$. Hence there are vectors $x_j \in V_j$, for $j = 1, \ldots, n$ such that

$$x = x_0 \otimes x_K = x_0 \otimes x_1 \otimes \cdots \otimes x_n.$$

In other words, x is an elementary vector in $V = V_0 \otimes V_1 \otimes \cdots \otimes V_n$. $\quad\square$

Let us illustrate this proposition by showing that the example state from Formula 11.1 at the beginning of the section is indeed unentangled. Because we can write

$$\frac{1}{\sqrt{2}} |{++}\rangle + \frac{i}{\sqrt{2}} |{-+}\rangle = \frac{1}{\sqrt{2}} |{+}\rangle \otimes |{+}\rangle + \frac{i}{\sqrt{2}} |{-}\rangle \otimes |{+}\rangle$$

$$= \frac{1}{\sqrt{2}} \Big(|{+}\rangle + i |{-}\rangle \Big) \otimes |{+}\rangle,$$

Proposition 11.1 ensures that this state is unentangled. Recall that our one calculation on this state (comparing measurement of the z-axis spins of the two particles) indicated only that the state might not be entangled. Proposition 11.1 removes all doubt.

Quantum computation exploits entanglement. The simplest kind of quantum computer is an *n-qubit register*, i.e., a system of n electrons. Each electron is a spin-1/2 particle so, by the analysis we did in Section 10.2, the state space is

$$\mathbb{P}\left(\bigotimes_{j=1}^{n} \mathbb{C}^2\right) = \mathbb{P}(\mathbb{C}^2 \otimes \cdots \otimes \mathbb{C}^2),$$

where there are n factors in the tensor product on the right-hand side. The subset of unentangled states is

$$\mathbb{P}(\mathbb{C}^2) \times \cdots \times \mathbb{P}(\mathbb{C}^2).$$

Note that this subset is not a subspace.

To see how entanglement can be used in quantum computation, consider Shor's algorithm for factoring a product N of two prime numbers. At the heart of the algorithm is a periodic function $f : \mathbb{Z}/N \rightarrow \mathbb{Z}/N$ whose period one must calculate in order to find the two prime factors of N. The phase space for computation is a pair of registers of size L, where $2^{L-1} < N \leq 2^L$. In other words, the state space for the quantum computer is

$$\mathbb{P}\left(\left(\bigotimes_{j=0}^{L-1} \mathbb{C}^2\right) \otimes \left(\bigotimes_{j=0}^{L-1} \mathbb{C}^2\right)\right). \tag{11.5}$$

We can use binary expansions to define a convenient basis for this state space. For any integer k between 0 and $2^L - 1$, we let $|k\rangle$ denote the element

$$\bigotimes_{j=0}^{L-1} |b_j\rangle,$$

where the binary expression for k is $b_{L-1}b_{L-2}\cdots b_1 b_0$. For example, if $L = 3$ and $k = 6$, we would have $|6\rangle = |1\rangle \otimes |1\rangle \otimes |0\rangle$. Then the set

$$\left\{|k\rangle \otimes |\ell\rangle : k, \ell \in \mathbb{Z}, 0 \leq k, \ell < 2^L\right\}$$

is a basis for the state space 11.5. A crucial step of the algorithm encodes the function f into an entangled state of the system, namely,

$$\frac{1}{2^L} \sum_{k=0}^{2^L-1} |k\rangle \otimes |f(k)\rangle,$$

where we abuse notation slightly by letting $f(k)$ denote the element of the equivalence class $f(k)$ that lies between 0 and $2^L - 1$. Without entanglement, Shor's algorithm would not be possible. For more details of Shor's algorithm and a more comprehensive introduction to quantum computing, see [Bare, Section 6].

In this section we have presented a mathematical foundation for entanglement of quantum systems. This foundation lies behind most modern discussions of quantum computing, as well as the Einstein–Podolsky–Rosen paradox.

11.4 The State Space of a Mobile Spin-1/2 Particle

The electron in a hydrogen atom moves, but it also has spin. In previous chapters we have shown how to model the dynamics and how to model the spin, but how do we model them both together? In this section we will use a tensor product to find the factor of two we need to make our model of the mobile electron correspond to the experimental results.

Because position measurements and spin measurements commute (i.e., measuring position and then spin yields the same result as measuring spin first and then position), the state space of a mobile electron (or any mobile spin-1/2 particle) is

$$\mathbb{P}(L^2(\mathbb{R}^3) \otimes \mathbb{C}^2).$$

Elements of $L^2(\mathbb{R}^3) \otimes \mathbb{C}^2$ are $\mathbb{C}^2$-valued functions on $\mathbb{R}^3$. Once we have chosen a basis for $\mathbb{C}^2$, every element of the tensor product $L^2(\mathbb{R}^3) \otimes \mathbb{C}^2$ has the form

$$\begin{pmatrix} f_0(x, y, z) \\ f_1(x, y, z) \end{pmatrix} = f_0(x, y, z) \begin{pmatrix} 1 \\ 0 \end{pmatrix} + f_1(x, y, z) \begin{pmatrix} 0 \\ 1 \end{pmatrix},$$

where $f_0, f_1 \in L^2(\mathbb{R}^3)$. The complex scalar product on this space is

$$\left\langle \begin{pmatrix} f_0(x, y, z) \\ f_1(x, y, z) \end{pmatrix}, \begin{pmatrix} g_0(x, y, z) \\ g_1(x, y, z) \end{pmatrix} \right\rangle = \int_{\mathbb{R}^3} (f_0 g_0^* + f_1 g_1^*), \tag{11.6}$$

as the reader can verify in Exercise 11.9.

Let us check that in at least one case, position and spin measurements commute. Suppose we build a machine that can measure whether the electron lies in the unit cube U in $\mathbb{R}^3$ (see Figure 1.1). Also suppose we can measure the z-axis spin regardless of position (with some technology far more advanced than a Stern–Gerlach machine, which is located at a particular position). Suppose that the state of the particle is given by $(f_0, f_1)^T$, where we assume without loss of generality that

$$\int_{\mathbb{R}^3} \left(|f_0|^2 + |f_1|^2 \right) = 1.$$

Then the probability that we first find the particle in the unit cube U and afterwards find it to be spin up is

$$\left| (1 \quad 0) \left(\int_U \begin{pmatrix} f_0 \\ f_1 \end{pmatrix} \right) \right|^2 = \left| (1 \quad 0) \begin{pmatrix} \int_U f_0 \\ \int_U f_1 \end{pmatrix} \right|^2 = \left| \int_U f_0 \right|^2$$

$$= \left| \int_U \left((1 \quad 0) \begin{pmatrix} f_0 \\ f_1 \end{pmatrix} \right) \right|^2,$$

which is precisely the probability that we find the particle to be spin up and afterwards find it to be in the unit cube U. The other three cases (in the cube spin down, out of the cube spin up, out of the cube spin down) can be verified in a similar manner. We leave it to the reader to generalize to all position measurements and spin measurements in Exercise 11.8.

Our next task is to identify the projective representation of $SO(3)$ on the state space. This representation is determined by the representations on the factors, but the projection must be handled carefully. The spin-1/2 projective representation of $SO(3)$ on $\mathbb{P}(\mathbb{C}^2)$ descends from the linear representation $\rho_{\frac{1}{2}}$ on $\mathbb{C}^2$. The natural representation of $SO(3)$ on $L^2(\mathbb{R}^3)$ (Section 4.4) descends to a projective representation of $SO(3)$ on $\mathbb{P}(L^2(\mathbb{R}^3))$. To put these together, we pull the natural representation of $SO(3)$ on $L^2(\mathbb{R}^3)$ back to a representation of $SU(2)$ under the two-to-one group homomorphism Φ (Section 4.3). Let us call the resulting representation σ. Then, by the natural tensor product of representations (Section 5.3) we have a representation

$$\sigma \otimes \rho_{\frac{1}{2}} : SU(2) \to \mathcal{GL} \left(L^2(\mathbb{R}^3) \otimes \mathbb{C}^2 \right).$$

Let I denote the identity matrix in $SU(2)$. Then $-I \in SU(2)$ and, for any $f \in L^2(\mathbb{R}^3)$ and any $c \in \mathbb{C}^2$, we have

$$\left((\sigma \otimes \rho_{\frac{1}{2}})(-I) \right)(f \otimes c) = \left(\sigma(-I)f \right) \otimes \left(\rho_{\frac{1}{2}}(-I)c \right) = f \otimes (-c) = -f \otimes c.$$

Hence on the projective space we have

$$\left([\sigma \otimes \rho_{\frac{1}{2}}](-I) \right)[f \otimes c] = [f \otimes c],$$

so the linear representation of $SU(2)$ on $L^2(\mathbb{R}^3) \otimes \mathbb{C}^2$ descends to a projective representation of $SO(3)$ on $\mathbb{P}(L^2(\mathbb{R}^3) \otimes \mathbb{C}^2)$.

Physically natural states of the system correspond to invariant subspaces of $\mathbb{P}(L^2(\mathbb{R}^3) \otimes \mathbb{C}^2)$. Each electronic shell with principal quantum number n corresponds to a subrepresentation V of $L^2(\mathbb{R}^3)$ of dimension n^2, as we saw in Section 8.6. That same electronic shell will correspond (in the tensor product model we are considering in this section) to the subspace $V \otimes \mathbb{C}^2$ of $L^2(\mathbb{R}^3) \otimes \mathbb{C}^2$. Note that

$$\dim(V \otimes \mathbb{C}^2) = 2 \dim V = 2n^2.$$

Similarly, within any one electronic shell, the set of orbitals with azimuthal quantum number ℓ corresponds to a subspace V of $L^2(\mathbb{R}^3)$ of dimension $2\ell + 1$, as we saw in Section 7.3. Hence in this new model such a set of orbitals corresponds to the set $V \otimes \mathbb{C}^2$, which has dimension $2(2\ell + 1)$.

Thus the new model, incorporating the spin state of the electron, predicts the right number of electrons. What is more, one can use this state space to model the *spin-orbit coupling*, a relativistic effect, with an operator that uses both differentiation in $L^2(\mathbb{R}^3)$ and 2×2 *Pauli matrices* acting on $\mathbb{C}^2$. The resulting equation is called the *Pauli equation* (see [BeS, Sections 12, 13]). However, even without further investigation, the tensor product model introduced in this section correctly predicts the experimental observations of Sections 1.3 and 1.4.

11.5   Conclusion

Here ends our story of the hydrogen atom. The author hopes that this story will encourage readers, as they go their separate ways, to continue to make the effort to connect ideas from different disciplines. Crossing boundaries is difficult, important, rewarding work. Languages and goals differ in subtle, unmarked ways. Yet the underlying phenomena and major ideas are often similar. In this age of specialization, we need to clarify similarities and build bridges. You can contribute. Go to it!

11.6   Exercises

Exercise 11.1 (Used in Section 11.2) *Suppose* V *is a finite-dimensional complex scalar product space and* $\hat{A}: V \rightarrow V$ *is a linear transformation* *whose eigenvalues*

$$\{\lambda_j : j = 1, \ldots, n\}$$

are all real and whose eigenspaces are mutually orthogonal. Suppose further that the eigenspaces $\{W_j : j = 1, \ldots, n\}$ span V, i.e.,

$$V = \bigoplus_{j=1}^{n} W_j.$$

Then $\hat{A}$ is Hermitian-symmetric. (Recall Definition 3.10).

Exercise 11.2 (Used in Section 11.2) *Suppose M is a Hermitian-symmetric, finite-dimensional matrix (as defined in Exercise 3.25). Show that there exists a real diagonal matrix D and a unitary matrix B (see Definition 3.5) such that*

$$M = B^{-1}DB.$$

This is the Spectral Theorem for Hermitian-symmetric matrices. (Hint: Use induction on the number of distinct eigenvalues of M.)

Exercise 11.3 *Suppose that $n \in \mathbb{N}$. Show that, for any permutation σ of n elements and any vector spaces $V_1, \ldots, V_n$ we have*

$$\bigotimes_{k=1}^{n} V_k \simeq \bigotimes_{k=1}^{n} V_{\sigma(k)}$$

as complex scalar product spaces.

Exercise 11.4 *Can you exploit the Einstein–Podolsky–Rosen paradox to send information faster than the speed of light?*

Exercise 11.5 (Used in Exercise 11.6 and Proposition 11.1) *Let V be a finite-dimensional complex scalar product space and suppose V^* is its dual space. Suppose $Q : V^* \to V^*$ is an orthogonal projection. Define a function $P : V \to V$ by*

$$P := \tau^{-1} \circ Q \circ \tau,$$

where τ denotes the natural function from V to V^*. Show that P is an orthogonal projection and, for every $\alpha \in V^*$ and every $v \in V$ we have

$$(Q\alpha)v = \alpha(Pv).$$

Exercise 11.6 (Used in Proposition 11.1) *Suppose V and W are finite-dimensional vector spaces and let $T: V \otimes W \to Hom(V^*, W)$ denote the isomorphism from the proof of Proposition 5.14. Suppose $\Pi: W \to W$ is an orthogonal projection and consider $x \in V \otimes W$. Set $X := T(x)$. Show that*

$$T(\tilde{\Pi}x) = \Pi X \in \text{Hom}(V^*, W),$$

where $\tilde{\Pi}$ is defined in Formula 11.2.

Next suppose $Q: V^* \to V^*$ is a orthogonal projection. Define P as in Exercise 11.5. Show that

$$T(\tilde{P}x) = XQ \in \text{Hom}(V^*, W).$$

Exercise 11.7 *For each nonnegative integer ℓ, decompose the representation of $SU(2)$ on $\mathcal{P}^\ell \otimes \mathbb{C}^2$ into a Cartesian sum of its irreducible components. Conclude that this representation is reducible. Is there a meaningful physical consequence or interpretation of this reducibility?*

Exercise 11.8 *Show that if H is any operator on $L^2(\mathbb{R}^3)$ and p is any direction in $\mathbb{R}^3$, then measurement of H and measurement of the spin in the p-direction commute on the state space of a mobile particle with spin 1/2.*

Exercise 11.9 *Show that the complex scalar product on the tensor product $L^2(\mathbb{R}^3) \otimes \mathbb{C}^2$, defined in terms of the complex scalar products on $L^2(\mathbb{R}^3)$ and $\mathbb{C}^2$ from Equation 5.2, agrees with the complex scalar product given in Equation 11.6.*

Appendix A
Spherical Harmonics

The goal of this appendix is to prove that the restrictions of harmonic polynomials of degree ℓ to the sphere do in fact correspond to the spherical harmonics of degree ℓ. Recall that in Section 1.6 we used solutions to the Legendre equation (Equation 1.11) to define the spherical harmonics. In this appendix we construct bona fide solutions $\mathbf{P}_{\ell,m}$ to the Legendre equation; then we show that each of the span of the spherical harmonics of degree ℓ is precisely the set of restrictions of harmonic polynomials of degree ℓ to the sphere.

Physicists and chemists know the Legendre functions well. One very useful explicit expression for these functions is given in terms of derivatives of a polynomials.

Definition A.1 *Let ℓ be a nonnegative integer and let m be an integer satisfying $0 \leq m \leq \ell$. Define the ℓ, m Legendre function by*

$$\mathbf{P}_{\ell,m}(t) := \frac{(-1)^m}{\ell! 2^\ell}(1 - t^2)^{m/2}\partial_t^{\ell+m}(t^2 - 1)^\ell.$$

For each ℓ, the function $\mathbf{P}_{\ell,0}$ is called the Legendre polynomial *of degree ℓ.*

Note that the so-called Legendre polynomial is in fact a polynomial of degree ℓ, as it is the ℓth derivative of a polynomial of degree 2ℓ. Legendre functions with $m \neq 0$ are often called *associated Legendre functions*.

Recall the Legendre equation (Equation 1.11):

$$(1 - t^2)P''(t) - 2t\,P'(t) + \left(\ell(\ell + 1) - \frac{m^2}{(1 - t^2)}\right)P(t) = 0. \qquad \text{(A.1)}$$

Proposition A.1 *The Legendre functions of Definition A.1 satisfy the Legendre equation.*

There are many ways to prove this proposition. Our proof is straightforward, elementary and rather ugly. For a more elegant proof via the "Rodrigues formula," see [WW, Chapter XV] or [DyM, Section 4.12].

Proof. First we will show that the Legendre polynomial of degree ℓ satisfies the Legendre equation with $m = 0$. Then we will deduce that for any $m = 1, \ldots, \ell$, the Legendre function $\mathbf{P}_{\ell,m}$ satisfies the Legendre equation.

When $m = 0$ the Legendre equation reduces to

$$(1 - t^2)P''(t) - 2t\,P'(t) + \ell(\ell + 1)P(t) = 0. \qquad \text{(A.2)}$$

We use the binomial expansion to find the coefficients of the Legendre polynomial of degree ℓ. For convenience, we multiply through by $2^\ell \ell!$:

$$
\begin{aligned}
(2^\ell \ell!)\mathbf{P}_{\ell,0}(t) &= \partial_t^\ell (t^2 - 1)^\ell \\
&= \partial_t^\ell \sum_{k=0}^{\ell} \binom{\ell}{k}(-1)^{\ell-k}t^{2k} \\
&= \sum_{k=(\ell+\epsilon)/2}^{\ell} \binom{\ell}{k}(-1)^{\ell-k}\frac{(2k)!}{(2k-\ell)!}t^{2k-\ell}.
\end{aligned}
$$

Differentiating once we find

$$(2^\ell \ell!)\mathbf{P}'_{\ell,0}(t) = \sum_{k=1+(\ell-\epsilon)/2}^{\ell} \binom{\ell}{k}(-1)^{\ell-k}\frac{(2k)!}{(2k-\ell-1)!}t^{2k-\ell-1}$$

and, differentiating again,

$$(2^\ell \ell!)\mathbf{P}''_{\ell,0}(t) = \sum_{k=1+(\ell+\epsilon)/2}^{\ell} \binom{\ell}{k}(-1)^{\ell-k}\frac{(2k)!}{(2k-\ell-2)!}t^{2k-\ell-2},$$

where $\epsilon = 0$ if ℓ is even and $\epsilon = 1$ if ℓ is odd. Hence to show that

$$(1 - t^2)\mathbf{P}''_{\ell,0}(t) - 2t\mathbf{P}'_{\ell,0}(t) + \ell(\ell + 1)\mathbf{P}_{\ell,0}(t) = 0,$$

it suffices to show the vanishing of the following expression:

$$\sum_{k=1+(\ell+\epsilon)/2}^{\ell} \binom{\ell}{k} \frac{(-1)^{\ell-k}(2k)!}{(2k-\ell-2)!} t^{2k-\ell-2}$$

$$- \sum_{k=1+(\ell+\epsilon)/2}^{\ell} \binom{\ell}{k} \frac{(-1)^{\ell-k}(2k)!}{(2k-\ell-2)!} t^{2k-\ell}$$

$$- 2 \sum_{k=1+(\ell-\epsilon)/2}^{\ell} \binom{\ell}{k} \frac{(-1)^{\ell-k}(2k)!}{(2k-\ell-1)!} t^{2k-\ell}$$

$$+ \ell(\ell+1) \sum_{k=(\ell+\epsilon)/2}^{\ell} \binom{\ell}{k} \frac{(-1)^{\ell-k}(2k)!}{(2k-\ell)!} t^{2k-\ell}.$$

We will show that the coefficient of each power of t is zero. The coefficient of t^ℓ is

$$- \frac{(2\ell)!}{(\ell-2)!} - 2\frac{(2\ell)!}{(\ell-1)!} + \ell(\ell+1)\frac{(2\ell)!}{\ell!}$$

$$= \frac{2\ell}{(\ell-1)!} (-(\ell-1) - 2 + (\ell+1)) = 0.$$

The coefficients of $t^{\ell-1}$ through t^2 take the form (with an appropriate choice of k, and ignoring an overall factor of $(-1)^{\ell-k}$):

$$- \binom{\ell}{k} \frac{(2k+2)!}{(2k-\ell)!} - \binom{\ell}{k}\frac{(2k)!}{(2k-\ell-2)!} - 2\binom{\ell}{k}\frac{(2k)!}{(2k-\ell-1)!}$$

$$+ \ell(\ell+1)\binom{\ell}{k}\frac{(2k)!}{(2k-\ell)!}$$

$$= \binom{\ell}{k}\frac{(2k)!}{(2k-\ell)!}\Big(-2(\ell-k)(2k+1) - (2k-\ell-1)(2k-\ell)$$

$$- 2(2k-\ell) + \ell(\ell+1)\Big) = 0.$$

There is one more term: t^1 if ℓ is odd and t^0 if ℓ is even. We will leave the even case to the reader. If ℓ is odd, then the coefficient of t^1 is

$$\binom{\ell}{(\ell+3)/2}(-1)^{\frac{\ell+1}{2}}(\ell+3)! - 2\binom{\ell}{(\ell+1)/2}(-1)^{(\ell-1)/2}(\ell+1)!$$

$$+ \ell(\ell+1)\binom{\ell}{(\ell+1)/2}(-1)^{(\ell-1)/2}(\ell+1)!$$

$$= \binom{\ell}{(\ell+1)/2}(\ell+1)! \left((\ell+2)(\ell+1) + 2 - \ell(\ell+1)\right) = 0.$$

The calculation for the case of even ℓ is similar. So we have shown that the Legendre polynomial $\mathbf{P}_{\ell,0}$ of degree ℓ satisfies the Legendre equation with $m = 0$.

Next we fix an integer m with $1 \leq m \leq \ell$ and show that $\mathbf{P}_{\ell,m}$ satisfies the Legendre equation (Equation A.1). Since the function $\mathbf{P}_{\ell,0}$ satisfies Equation A.2, we have

$$(1 - t^2)\mathbf{P}''_{\ell,0}(t) - 2t\mathbf{P}'_{\ell,0}(t) + \ell(\ell + 1)\mathbf{P}_{\ell,0}(t) = 0.$$

Differentiating m times with respect to t, we find that

$$\left((1-t^2)\partial_t^{m+2} - 2(m+1)t\partial_t^{m+1} + (\ell(\ell+1) - m(m+1))\partial_t^m\right)\mathbf{P}_{\ell,0}(t) = 0. \quad \text{(A.3)}$$

Define $c := (-1)^m/(\ell! 2^\ell)$. From Definition A.1 we know that $c\,\partial_t^m \mathbf{P}_{\ell,0}(t) = (1 - t^2)^{(-\frac{m}{2})}\mathbf{P}_{\ell,m}(t)$. Differentiating this expression twice in a row we obtain

$$c\,\partial_t^{m+1}\mathbf{P}_{\ell,0}(t) = (1 - t^2)^{(-\frac{m}{2})}\left(\frac{mt}{1 - t^2}\mathbf{P}_{\ell,m}(t) + \mathbf{P}'_{\ell,m}(t)\right)$$

$$c\,\partial_t^{m+2}\mathbf{P}_{\ell,0}(t) = (1 - t^2)^{(-\frac{m}{2})}\left(\frac{m}{1 - t^2} + \frac{(m + 2)t^2}{(1 - t^2)^2}\right)\mathbf{P}_{\ell,m}(t)$$

$$+ (1 - t^2)^{(-\frac{m}{2})}\left(\frac{2mt}{1 - t^2}\mathbf{P}'_{\ell,m}(t) + \mathbf{P}''_{\ell,m}(t)\right).$$

Here we have used the fact (easily verified by induction) that for any sufficiently differentiable function $f(t)$ we have

$$\partial_t^m (1 - t^2)f(t) = (1 - t^2)\partial_t^m f + 2mt\partial_t^{m-1} f(t) + m(m - 1)\partial_t^{m-2} f(t).$$

Plugging these expressions into Equation A.3, multiplying by $c(1 - t^2)^{(m/2)}$ and simplifying we find that

$$0 = (1 - t^2)\mathbf{P}''_{\ell,m}(t) - 2t\mathbf{P}'_{\ell,m}(t) + \left(\ell(\ell + 1) - \frac{m}{1 - t^2}\right)\mathbf{P}_{\ell,m}(t).$$

In other words, the function $\mathbf{P}_{\ell,m}$ satisfies the Legendre equation (Equation A.1). $\square$

It is natural to wonder whether there are any other solutions to Legendre's equation. Since the equation is linear (in P, P' and P''), there should be two solutions for each value of m^2. For $m^2 \neq 0$ there are indeed two solutions: $\mathbf{P}_{\ell,\pm m}$. The case $m^2 = 0$ is discussed in detail in Simmons' undergraduate text on ordinary differential equations [Sim, Sections 28, 29 and 44]. The point is that a solution corresponds to a continuous function on the sphere only if it is bounded near $t = \pm 1$ and only one of the solutions to the Legendre equation is bounded near $t = \pm 1$.

Now we are ready to define the spherical harmonic functions. In Section 1.6 we gave examples for $\ell = 0, 1, 2$; here is the general definition.

Definition A.2 *Let ℓ be a nonnegative integer and let m be an integer satisfying $-\ell \leq m \leq \ell$. Define the ℓ, m spherical harmonic function $Y_{\ell,m} : [0, \pi] \oplus (-\pi, \pi] \to \mathbb{C}$ by*

$$Y_{\ell,m}(\theta, \phi) := c_{\ell,m} \mathbf{P}_{\ell,|m|}(\cos \theta) e^{im\phi},$$

where the the constant $c_{\ell,m}$ takes the value

$$\sqrt{\frac{(\ell - m)!(2\ell + 1)}{(\ell + m)!4\pi}}.$$

For each ℓ, linear combinations of the vectors

$$\left\{ Y_{\ell,m} : m = -\ell, \ldots, \ell \right\}$$

are spherical harmonics of degree ℓ.

In fact, every spherical harmonic function is the restriction to the sphere S^2 in $\mathbb{R}^3$ of a harmonic polynomial on $\mathbb{R}^3$. Recall the vector space $\mathcal{Y}^\ell$ of restrictions of harmonic polynomials of degree ℓ in three variables to the sphere S^2 (Definition 2.6).

Proposition A.2 *Suppose ℓ is a nonnegative integer. Then the span of the set $\{Y_{\ell,-\ell}, \ldots, Y_{\ell,\ell}\}$ is $\mathcal{Y}^\ell$.*

Proof. First we will show that the set $\left\{ Y_{\ell,m} : m = -\ell, \ldots, \ell \right\}$ is linearly independent. Next we will show it is a subset of $\mathcal{Y}^\ell$. The proof ends with a dimension count.

To show linear independence, consider an arbitrary linear combination equalling zero:

$$0 = \sum_{m=-\ell}^{\ell} C_m \mathbf{P}_{\ell,|m|}(\cos \theta) e^{im\phi}.$$

We must show that each $C_m = 0$. By Exercise 2.2, the set

$$\left\{ e^{im(\cdot)} : m = -\ell, \ldots, \ell \right\},$$

where $e^{im(\cdot)} \colon [0, \pi] \to \mathbb{C}, x \to e^{imx}$, is linearly independent, so we can conclude that for each m we have $C_m \mathbf{P}_{\ell,m} = 0$. Hence we will be done with the proof of linear independence if we can show that for each m, the function $\mathbf{P}_{\ell,m}(\cos\theta)$ is not the zero function. Now $(-1)^m/\ell!2^\ell$ is a nonzero constant, and $(1 - \cos^2(\pi/2))^{m/2} \neq 0$, so it suffices to show that $\partial_t^{\ell+m}(t^2 - 1)^\ell$ is not the zero polynomial. But $(t^2 - 1)^\ell$ is a polynomial of degree 2ℓ in t, so its first 2ℓ derivatives are nonzero. Since $m \leq \ell$, it follows that $\mathbf{P}_{\ell,m}$ is not the zero function. We have shown the required linear independence.

A longer argument is required to show that $\{Y_{\ell,m} : m = -\ell, \ldots, \ell\} \subset \mathcal{Y}^\ell$. We begin by showing that for any nonnegative integer k the expression $\partial_t^k(t^2 - 1)^\ell$ is a polynomial in the variables $\alpha := 1 - t^2$ and $\beta := t$. According to the chain rule for partial derivatives we have $\partial_t = 2\beta\partial_\alpha + \partial_\beta$, so applying ∂_t to any polynomial in α and β yields a polynomial in α and β. Consider $(t^2 - 1)^\ell = \alpha^\ell$, which is a polynomial in α and β. Hence, by induction on k, we can conclude that $\partial_t^k(t^2 - 1)^\ell$ is a polynomial in α and β.

Another induction on k shows that for any nonnegative integer k the expression $\partial_t^k(t^2 - 1)^\ell$ is a homogeneous polynomial of degree $2\ell - k$ in the variables $\sqrt{\alpha}$ and β. The key to the inductive step is that $(t^2 - 1)^\ell = \sqrt{\alpha}^{2\ell}$, a polynomial of degree 2ℓ, while applying $\partial_t = 2\beta\partial_\alpha + \partial_\beta$ lowers the degree (in $\sqrt{\alpha}$ and β) by one.

The point is that if $t = \cos\theta$, then we have

$$r^2\alpha = r^2(1 - t^2) = r^2 \sin^2\theta = x^2 + y^2$$
$$r\beta = r\cos\theta = z.$$

So a polynomial of degree d in $\sqrt{\alpha}$ and β that is also a polynomial in α will be homogeneous of degree d in x, y and z. Setting $k = \ell + m$ and applying the results of our inductions above, we find that

$$r^{\ell-m}\partial_t^{\ell+m}(t^2 - 1)^\ell$$

is a polynomial of degree $\ell - m$ in x, y, and z. Also,

$$r^m(1 - t^2)^{m/2}e^{\pm im\phi} = r^m \sin^m\theta(\cos\phi \pm i \sin\phi)^m = (x \pm iy)^m,$$

which is a homogeneous polynomial in x and y of degree m when $m \geq 0$. Note that $\theta \in [0, \pi]$, so $\sin\theta \geq 0$. Hence the function

$$r^\ell(1 - t^2)^{m/2}\partial_t^{\ell+m}(t^2 - 1)^\ell e^{im\phi}$$

is a polynomial of degree ℓ in x, y and z; by inspection, it is homogeneous. We know from Equation 1.12 and Proposition A.1 that if we evaluate this function at $t = \cos\theta$ we obtain a harmonic function. Restricting this homogeneous polynomial to the sphere we obtain $\mathbf{P}_{\ell,m}$. Hence $\mathbf{P}_{\ell,m} \in \mathcal{Y}^\ell$ for $m = 0, 1, \ldots, \ell$.

Next we show that the harmonic function from Equation 1.12 is a polynomial in x, y and z of degree ℓ even when $m < 0$. To see this, note that (by yet another induction, this time on $-m$ and left to the reader), for any nonnegative integer ℓ and any integer m with $-\ell \le m < 0$ there is a polynomial q of two variables such that $q(\alpha, \beta)$ has degree $\ell + m$ in $\sqrt{\alpha}$ and β and

$$\partial_t^{\ell+m}(t^2 - 1)^\ell = \alpha^{-m}q(\alpha, \beta).$$

Note that $r^{\ell+m}q(\alpha, \beta)$ is a polynomial of degree $\ell + m$ in x, y and z. Hence, for $m < 0$ we have

$$r^\ell(1 - t^2)^{m/2}\partial_t^{\ell+m}(t^2 - 1)^\ell e^{im\phi} = r^\ell\alpha^{-m/2}(e^{-i\phi})^{-m}q(\alpha, \beta)$$
$$= (x - iy)^{-m}r^{\ell+m}q(\alpha, \beta),$$

which is a polynomial of degree $-m + \ell + m = \ell$ in x, y and z. This polynomial is clearly homogeneous, and by Equation 1.12 it is harmonic. Restricting this homogeneous polynomial to the sphere we obtain $\mathbf{P}_{\ell,m}$. Hence $\mathbf{P}_{\ell,m} \in \mathcal{Y}^\ell$ for $m = -\ell, \ldots, -1$. Thus we have shown that each function $\mathbf{P}_{\ell,m}$ is the restriction to the sphere S^2 of a harmonic polynomial of degree ℓ on $\mathbb{R}^3$. In other words, $\{Y_{\ell,m} : m = -\ell, \ldots, \ell\} \subset \mathcal{Y}^\ell$.

Finally, since the $Y_{\ell,m}$'s are linearly independent, they span a $(2\ell + 1)$-dimensional subset of $\mathcal{Y}^\ell$. But we know by Proposition 7.1 that $\mathcal{Y}^\ell$ has dimension at most $(2\ell + 1)$. Hence $\mathcal{Y}^\ell$ is equal to the span of the $Y_{\ell,m}$'s. □

The following proposition justifies the reliance on spherical harmonics in spherically symmetric problems involving the Laplacian. To state it succinctly, we introduce the vector space $C_2 \subset L^2(\mathbb{R}^3)$ of continuous functions whose first and second partial derivatives are all continuous.

Proposition A.3 *Suppose D is a differential operator of the form*

$$D = \nabla^2 + u(r),$$

where u is a real-valued function of r. Then the vector space

$$K := \{f \in L^2(\mathbb{R}^3): f \in C_2 \text{ and } Df = 0\}$$

of solutions to the differential equation $Df = 0$ is spanned by solutions of the form $\alpha \otimes Y_{\ell,m}$, where $\alpha \in \mathcal{I}$, ℓ is a nonnegative integer and m is an integer such that $|m| \le \ell$.

The technical conditions on f are quite reasonable: if a physical situation has a discontinuity, we might look for solutions with discontinuities in the function f and its derivatives. In this case, we might have to consider, e.g., piecewise-defined combinations of smooth solutions to the differential equation. These solutions might not be linear combinations of spherical harmonics.

Proof. Let V denote the set of solutions in $L^2(\mathbb{R}^3)$ obtained by multiplying a spherical harmonic by a spherically symmetric function:

$$V := \left\{ \alpha \otimes Y_{\ell,m} \in \mathcal{I} \otimes \mathcal{Y} \colon \alpha \in C_2 \text{ and } D(\alpha \otimes Y_{\ell,m}) = 0 \right\}.$$

It suffices to show that $K \cap (V^\perp) = 0$. So suppose that $f \in K \cap (V^\perp)$, i.e., suppose that f and its first and second partial derivatives are continuous, that $Df = 0$ and that f is orthogonal to every solution obtained by separation of variables. We will show that $f = 0$.

By Fubini's Theorem (Theorem 3.1), the function $\|f\|_{S^2}$ defined by

$$\|f\|_{S^2} \colon r \mapsto \sqrt{\int_{S^2} |f(r, \theta, \phi)|^2 \sin \theta \, d\theta d\phi}$$

lies in $\mathcal{I}$ because $f \in L^2(\mathbb{R}^3)$. Now for any nonnegative integer ℓ and any integer m with $|m| \le \ell$, the function $Y_{\ell,m} f$ is measurable and

$$\int_{\mathbb{R}^3} |Y_{\ell,m}(\theta, \phi) f(r, \theta, \phi)|^2 r^2 \, dr \, \sin \theta \, d\theta d\phi < \infty$$

because $Y_{\ell,m}$ is bounded and $f \in L^2(\mathbb{R}^3)$. Again by Fubini's Theorem,

$$\alpha_{\ell,m}(r) := \int_{S^2} Y_{\ell,m}(\theta, \phi) f(r, \theta, \phi) \sin \theta \, d\theta d\phi = \langle Y_{\ell,m}, f(r, \cdot, \cdot) \rangle_{S^2}$$

defines a measurable function $\alpha_{\ell,m}$ on $\mathbb{R}^{\ge 0}$. Note that by the Schwarz Inequality (Proposition 3.6) on $L^2(S^2)$ we have

$$|\alpha_{\ell,m}|^2 = \left| \int_{S^2} Y_{\ell,m}(\theta, \phi) f(\cdot, \theta, \phi) \sin \theta \, d\theta d\phi \right|^2 \le \|Y_{\ell,m}\|_{S^2}^2 \|f\|_{S^2}^2.$$

Since $\|Y_{\ell,m}\|^2$ does not depend on r and $\|f\|_{S^2} \in \mathcal{I}$, it follows that $\alpha_{\ell,m} \in \mathcal{I}$.

Next we introduce some convenient notation. By Exercise 1.12 we know that $\nabla^2 = \nabla_r^2 + \nabla_{\theta,\phi}^2$, where we set

$$\nabla_r^2 := \partial_r^2 + \frac{2}{r} \partial_r$$

$$\nabla_{\theta,\phi}^2 := \frac{1}{r^2} \partial_\theta^2 + \frac{\cos \theta}{r^2 \sin \theta} \partial_\theta + \frac{1}{r^2 \sin^2 \theta} \partial_\phi^2.$$

Note that $\nabla^2_{\theta,\phi}$ is Hermitian-symmetric on $L^2(S^2)$ by Exercise 3.26.

Since $Df = 0$ we have $(\nabla^2_r + u)f = -\nabla^2_{\theta,\phi}f$. Hence for $r \in (0, \infty)$ we have

$$
\begin{aligned}
(\nabla^2_r + u)\alpha_{\ell,m}(r) &= \left\langle Y_{\ell,m}, (\nabla^2_r + u)f(r, \cdot, \cdot)\right\rangle_{S^2} \\
&= -\left\langle Y_{\ell,m}, \nabla^2_{\theta,\phi}f(r, \cdot, \cdot)\right\rangle_{S^2} \\
&= -\left\langle \nabla^2_{\theta,\phi}Y_{\ell,m}, f(r, \cdot, \cdot)\right\rangle \\
&= \ell(\ell+1)\left\langle Y_{\ell,m}, f(r, \cdot, \cdot)\right\rangle \\
&= \ell(\ell+1)\alpha_{\ell,m}(r).
\end{aligned}
$$

Here the first equality follows from the fact that $f \in C_2$. The technical continuity condition on f and its first and second partial derivatives allows us to exchange the derivative and the integral sign (disguised as a complex scalar product). See, for example, [Bart, Theorem 31.7]. The third equality follows from the Hermitian symmetry of $\nabla^2_{\theta,\phi}$. It follows that $\alpha_{\ell,m}Y_{\ell,m}$ is an element of the kernel of $D = \nabla^2 + u$, as we can verify:

$$
\begin{aligned}
(\nabla^2 + u)\alpha_{\ell,m}(r)Y_{\ell,m}(\theta, \phi) &= \left((\nabla^2_r + u)\alpha_{\ell,m}(r)\right)Y_{\ell,m}(\theta, \phi) \\
&\quad + \alpha_{\ell,m}(r)\nabla^2_{\theta,\phi}Y_{\ell,m}(\theta, \phi) \\
&= \ell(\ell+1)\alpha_{\ell,m}(r)Y_{\ell,m}(\theta, \phi) \\
&\quad - \alpha_{\ell,m}(r)\ell(\ell+1)Y_{\ell,m}(\theta, \phi) \\
&= 0.
\end{aligned}
$$

Hence $\alpha_{\ell,m} \otimes Y_{\ell,m} \in V$. Next we examine the norm of $\alpha_{\ell,m} = 0$ and recall that $f \in V^\perp$ by hypothesis:

$$
\begin{aligned}
\left\| \alpha_{\ell,m} \right\|^2_\mathcal{I} &= \left\langle \alpha_{\ell,m}, \left\langle Y_{\ell,m}, f\right\rangle_{S^2}\right\rangle = \int_0^\infty \int_{S^2} \alpha^*_{\ell,m}Y^*_{\ell,m}f \\
&= \left\langle \alpha_{\ell,m}Y_{\ell,m}, f\right\rangle_{\mathbb{R}^3} \\
&= 0.
\end{aligned}
$$

Hence $\alpha_{\ell,m} = 0$. But this implies that for any $h \otimes Y \in (\mathcal{I} \otimes \mathcal{Y})$ we have

$$
\langle h \otimes Y, f\rangle_{\mathbb{R}^3} = \int_0^\infty \langle Y, f(r, \cdot, \cdot)\rangle_{S^2}\, r^2\, dr = \left\langle h, \alpha_{\ell,m}\right\rangle_{\mathbb{R}^{\geq 0}} = 0.
$$

But, by Proposition 7.5, $\mathcal{I} \otimes \mathcal{Y}$ spans $L^2(\mathbb{R}^3)$. Hence $f = 0$.   □

Note that the application of Fubini's Theorem here mirrors the argument in Proposition 7.7. Also note that this proposition could easily be generalized to differential operators of the form

$$\nabla^2 + \mathcal{O},$$

where $\mathcal{O}$ is a differential operator depending only on r. One would need appropriate technical hypotheses on f. Specifically, if we let n denote the minimum of 2 and the order of the differential operator $\mathcal{O}$, then f and all its partial derivatives up to the nth order would have to be continuous.

Appendix B
Proof of the Correspondence between Irreducible Linear Representations of $SU(2)$ and Irreducible Projective Representations of $SO(3)$

In this appendix we prove Proposition 10.6 from Section 10.4, which states that the irreducible projective unitary Lie group representations of $SO(3)$ are in one-to-one correspondence with the irreducible (linear) unitary Lie group representations of $SU(2)$. The proof requires some techniques from topology and differential geometry.

Let us start by stating the definitions and theorems we use from topology. We will use the notion of local homomorphisms.

Definition B.1 *Suppose that M and N are topological spaces, and suppose that $f : M \to N$ is a continuous function. Suppose $m \in M$. Then f is a* local homeomorphism at m *if there is a neighborhood $\tilde{M}$ containing m such that $f|_{\tilde{M}}$ is invertible and its inverse is continuous. If f is a local homeomorphism at each $m \in M$, then f is a* local homeomorphism.

We need a theorem about covering spaces.

Theorem B.1 *Suppose X, Y and Z are topological spaces. Suppose $\pi : Y \to X$ is a finite-to-one local homeomorphism.[1] Suppose Z is connected and simply connected. Suppose $f : Z \to X$ is continuous. Then there is a continuous function $\tilde{f} : Z \to Y$ such that $f = \pi \circ \tilde{f}$.*

---

[1] A function such as π is known as a *covering function*, while a space such as Y is called a *covering space* for X.

For a proof, see [Hat, Proposition 1.30] or [Mas, Theorem 5.1].

Next we introduce the relevant concepts and theorems from differential geometry. First we define local diffeomorphisms.

Definition B.2 *Suppose that M and N are differentiable manifolds[2] of the same dimension, and suppose that $f : M \rightarrow N$ is a differentiable function. Suppose $m \in M$. Then f is a* local diffeomorphism at m *if there is a neighborhood $\tilde{M}$ containing m such that $f|_{\tilde{M}}$ is invertible and its inverse is differentiable. If f is a local diffeomorphism at each $m \in M$, then f is a* local diffeomorphism.

We will appeal to the Inverse Function Theorem.

Theorem B.2 (Inverse Function Theorem) *Suppose that M and N are manifolds of the same dimension, and suppose that $f : M \rightarrow N$ is a differentiable function. Suppose $m \in M$. Suppose the linear transformation $df(m) : T_m M \rightarrow T_{f(m)} N$ is invertible. Then f is a local diffeomorphism at m.*

See Boothby [Bo, II.6] or Bamberg and Sternberg [BaS, p. 237] for a proof of the inverse function theorem on $\mathbb{R}^n$. The corresponding theorem for manifolds follows by restricting to coordinate neighborhoods of m and $f(m)$. We will use the following theorem about group actions on differentiable manifolds.

Theorem B.3 *Suppose M is a differentiable manifold, G is a compact Lie group and (G, M, σ) is a group action. Suppose further that*

   1. *The action is* free, *i.e., if $g \in G$, $m \in M$ and $(\sigma(g))(m) = m$, then $g = I$.*

   2. *The action is* smooth, *i.e., for each $g \in G$, the function*

$$\sigma(g) : M \rightarrow M$$

   is an infinitely differentiable function.

Then the quotient space M/G (defined in Exercise 4.43) is a differentiable manifold, and the natural projection $\pi : M \rightarrow M/G$ is a differentiable function.

A proof of this theorem can be found in [AM, Proposition 4.1.23].

Next, recall the Lie group homomorphism $\Phi : SU(2) \rightarrow SO(3)$ defined in Section 4.3.

---

[2] Also known as C^∞ *manifolds* or *smooth manifolds*.

Proposition B.1 *The function Φ is a local diffeomorphism. In other words, for any $g \in SU(2)$, there is a neighborhood N containing g such that the restriction $\Phi|_N$ is invertible and its inverse is differentiable.*

Proof. First we show that Φ is a local diffeomorphism at $I \in SU(2)$. We use Equation 4.2 to calculate the derivative of Φ at I: for x, y and z near 0 we have, up to first order in x, y and z,

$$\Phi\left(I + \begin{pmatrix} ix & y+iz \\ -y+iz & ix \end{pmatrix}\right) = \Phi\begin{pmatrix} 1+ix & y+iz \\ -y+iz & 1-ix \end{pmatrix}$$

$$= \begin{pmatrix} 1 & 2y & -2z \\ -2y & 1 & 2x \\ 2z & -2x & 1 \end{pmatrix} = I + \begin{pmatrix} 0 & 2y & -2z \\ -2y & 0 & 2x \\ 2z & -2x & 0 \end{pmatrix}.$$

Hence

$$d\Phi(I)\begin{pmatrix} ix & y+iz \\ -y+iz & -ix \end{pmatrix} = \begin{pmatrix} 0 & 2y & -2z \\ -2y & 0 & 2x \\ 2z & -2x & 0 \end{pmatrix}.$$

The kernel of the linear transformation $d\Phi(I)$ from the three-dimensional vector space $T_I SU(2)$ to the three-dimensional vector space $T_I SO(3)$ is trivial, so $d\Phi(I)$ is invertible. Hence by the Inverse Function Theorem (Theorem B.2), Φ is a local diffeomorphism at I.

Next we consider an arbitrary $g_0 \in SU(2)$ and show that Φ is a local diffeomorphism at g_0. Now let N denote a neighborhood of $I \in SU(2)$ on which the restriction $\Phi|_N$ has a differentiable inverse. Since left multiplication by g_0^{-1} is a continuous function on $SU(2)$, the set

$$g_0 N := \{g_0 g : g \in N\}$$

is a neighborhood of g_0. For any $g_0 n \in g_0 N$ we have $\Phi(g_0 n) = \Phi(g)\Phi(n)$. Hence

$$\Phi\Big|_{g_0 N} = L_{\Phi(g_0)} \circ \Phi\Big|_N,$$

where $L_{g_0} : SO(3) \to SO(3)$ denotes left multiplication by $\Phi(g_0)$. Hence the inverse function is

$$\left(\Phi\Big|_{g_0 N}\right)^{-1} = \left(\Phi\Big|_N\right)^{-1} \circ L_{\Phi(g_0^{-1})}.$$

Since $SU(2)$ is a Lie group, the function $L_{\Phi(g_0^{-1})}$ is differentiable; by our choice of N, the function $(\Phi|_N)^{-1}$ is differentiable. Hence $\Phi|_{g_0N}$ has a differentiable inverse, i.e., Φ is a local diffeomorphism at g_0. But g_0 was arbitrary; hence Φ is a local diffeomorphism.   □

Because finite quotients are easier to handle than infinite quotients, it is useful to think of $\mathbb{P}\mathcal{U}(V)$ as a finite quotient of the group $SU(V)$, the set of unitary transformations from V to itself with determinant 1 (Definition 4.2).

Proposition B.2 *Suppose V is a complex scalar product space of finite dimension $n \in \mathbb{N}$. Consider the equivalence relation on the group $SU(V)$ defined by $A \sim B$ if and only if there is a complex number λ such that $\lambda^n = 1$ and $A = \lambda B$. Then $SU(V)/\sim$ is a group and there is a Lie group isomorphism*

$$\mathbb{P}\mathcal{U}(V) \cong SU(V)/\sim .$$

Proof. First we must show that the group operation on $SU(V)$ survives the equivalence. Because we are accustomed to using $[A]$ to denote an element of $\mathbb{P}\mathcal{U}(V)$, we will write elements of $SU(V)/\sim$ as $\{A\}$, where $A \in SU(V)$. Note that if $A_1 = \lambda_A A_2$ and $B_1 = \lambda_B B_2$, with $\lambda_A^n = \lambda_B^n = 1$, then $A_1 B_1 = (\lambda_A \lambda_B)A_2 B_2$, where $(\lambda_A \lambda_B)^n = 1$. So group multiplication survives the equivalence. It follows easily that $SU(V)/\sim$ is a group.

Next we define a function $\Psi \colon (SU(V)/\sim) \to \mathbb{P}\mathcal{U}(V)$ and show it is a group isomorphism. For any $\{A\} \in SU(V)/\sim$, we define

$$\Psi(\{A\}) := [A].$$

Note that $\Psi(\{A\})$ is well defined since any two equivalent elements of $SU(V)$ yield the same element of $\mathbb{P}\mathcal{U}(V)$. The function Ψ is a group homomorphism because, for any $A, B \in SU(V)$ we have

$$\Psi(\{A\}\{B\}) = \Psi(\{AB\}) = [AB] = [A][B] = \Psi(\{A\})\Psi(\{B\}).$$

To see that Ψ is injective, consider $\Psi^{-1}[I]$. If $A \in SU(V)$ and $\Psi(\{A\}) = [I]$, then there must be a complex number λ such that $A = \lambda I$. Notice that

$$\lambda^n = \det(\lambda I) = \det(A) = 1,$$

because $A \in SU(V)$. Hence $A \sim I$, i.e., $\{A\} = \{I\}$. So Ψ is injective. To see that Ψ is surjective, consider $[B]$ for any $B \in \mathcal{U}(V)$. Set $c := \det(B) \neq 0$. Then $\det(c^{-n}B) = 1$, so $c^{-n}B \in SU(V)$. We have

$$\Psi(\{c^{-n}B\}) = [c^{-n}B] = [B].$$

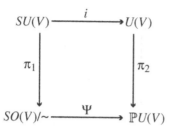

Figure B.1. A commutative diagram for the proof of Proposition B.2. The functions π_1 and π_2 are the natural projection functions. The function i is the inclusion function: any element of $SU(V)$ is automatically an element of $U(V)$.

Hence Ψ is surjective. Since Ψ is an injective and surjective group homomorphism, it is a group isomorphism.

Next we show differentiability. Consider Figure B.1. By construction, the function π_1 is surjective. So given an arbitrary element $c \in SU(V)/\sim$, there is an element $A \in SU(V)$ such that $\pi_1(A) = c$. By Theorem B.3, we know that π_1 is a local diffeomorphism. Hence there is a neighborhood N of A such that $\pi_1|_N$ has a differentiable inverse. The inclusion function is automatically differentiable. Finally, from Theorem B.3 we know that π_2 is a differentiable function. Hence the function

$$\Psi\Big|_N = \pi_2 \circ i \circ \left(\pi_1\Big|_N\right)^{-1}$$

is differentiable. So Ψ is differentiable at c. But c was arbitrary, so Ψ is differentiable.   □

Here is the proof of Proposition 10.6.

Proof. (of Proposition 10.6) First we suppose that $(SU(2), V, \rho)$ is a linear irreducible unitary Lie group representation. By Proposition 6.14 we know that ρ is isomorphic to the representation R_n for some n. By Proposition 10.5 we know that R_n can be pushed forward to an irreducible projective representation of $SO(3)$. Hence $[\rho]$ can be pushed forward to an irreducible projective Lie group representation of $SO(3)$.

Conversely, suppose that $(SO(3), \mathbb{P}(V), \sigma)$ is a finite-dimensional projective unitary representation. We want to show that σ is the pushforward of the projectivization of a linear unitary representation ρ of $SU(2)$. In other words, we must show that there is a Lie group representation ρ that makes the diagram in Figure B.2 commutative, and that this ρ is a Lie group representation.

Consider the function $\sigma \circ \Phi \colon SU(2) \to SU(\mathcal{P}^n)/\sim$. This function is continuous, and its domain $SU(2)$ is simply connected, by Exercise 4.27. Let us show that $SU(2)$ is also connected. Since S^3 is path-connected (any two

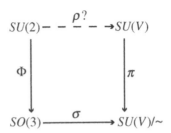

Figure B.2. Commutative diagram for proof that every projective unitary representation of $SO(3)$ comes from a linear representation of $SU(2)$.

points in S^3 lie on a plane through the origin that intersects S^3 in a circle) and $SU(2)$ is topologically equivalent to S^3, we know that $SU(2)$ is connected.

Since the function π is a finite-to-one covering, we can apply Theorem B.1 to conclude that there is a continuous function $\rho: SU(2) \rightarrow SU(\mathcal{P}^n)$ that makes the diagram in Figure B.2 commutative. Note that $\rho(I) = e^{ik/2\pi(n+1)}$ for some integer k. Without loss of generality, we can assume that $k = 0$: if not, replace ρ by $e^{-ik/2\pi(n+1)}\rho$.

Next we will show that ρ is a group homomorphism. Since Φ and σ are group homomorphisms, we know that $\pi \circ \rho$ is a group homomorphism. Hence, for any $g_1, g_2 \in SU(2)$ we have

$$\rho(g_1 g_2) = e^{\frac{ik}{2\pi n}} \rho(g_1)\rho(g_2),$$

for some $k \in \mathbb{Z}$. In fact, we can use this equation to define k as a function of the pair (g_1, g_2). In other words, consider the function $K : SU(2) \times SU(2) \rightarrow SU(V)$ defined by

$$K(g_1, g_2) := \rho(g_1 g_2)\rho(g_2)^{-1}\rho(g_1)^{-1}.$$

Since the function ρ is continuous, so is the function K. Since the domain $SU(2) \times SU(2)$ of K is connected, the range of K must be connected. But the range is a subset of the integer multiples of I, and we know that I lies in the range of K because $K(I, I) = I$. So the range must be equal to $\{I\}$. Hence, for any $(g_1, g_2) \in SU(2) \times SU(2)$ we have

$$\rho(g_1 g_2) = \rho(g_1)\rho(g_2).$$

Hence ρ is a group homomorphism.

Let us show that ρ is differentiable. Consider Figure B.2. Consider arbitrary $g \in SU(2)$. Because π is a local diffeomorphism, there is a neighborhood N of $\rho(g) \in SU(V)$ such that $\pi|_N$ has a differentiable inverse. By

Proposition B.2, the set $\pi[N]$ must be a neighborhood of the point $\pi \circ \rho(g) = \sigma \circ \Phi(g)$. Let $\tilde{N}$ denote the preimage in $SU(2)$ of the set $\pi[N]$ under the function $\sigma \circ \Phi$. Then $\tilde{N}$ is neighborhood of g. On the neighborhood $\tilde{N}$ we have

$$\rho\Big|_{\tilde{N}} = \left(\pi\Big|_{N}\right)^{-1} \circ \sigma \circ \Phi,$$

where all three functions on the right-hand side are differentiable. Hence $\rho|_{\tilde{N}}$ is differentiable, which implies that ρ is differentiable at g. But g was arbitrary; hence ρ is differentiable on all of $SU(2)$.

We have shown that the projective representation σ is the pushforward of the representation ρ, completing the proof.   □

Appendix C
Suggested Paper Topics

- Selection rules and Clebsch–Gordan coefficients.

- Fourier transforms and momentum space.

- Classification of representations of the symmetric group S^n.

- Representations of the Poincaré group and their relation to mass and spin.

- The Peter–Weyl theorem.

- Maximal tori and conjugacy classes of compact groups.

- The Ω^- particle.

- Spin-orbit coupling.

- The hyperfine splitting in hydrogen.

- The crystallographic groups.

- Quarks and representations of $SU(3)$.

- Hilbert spaces (in the mathematical sense).

- The history of the use of hydrogen in modern physics (see Rigden [Ri]).

- Any topic from *Lie Groups and Physics* [St] or *Variations on a Theme by Kepler* [GS].

Bibliography

[AM]   Abraham, R. and J. Marsden, *Foundations of Mechanics, Second Edition*; Addison-Wesley, Reading, Massachusetts, 1978.

[Ar]   Artin, M., *Algebra*; Prentice Hall, Upper Saddle River, New Jersey, 1991.

[Au]   Austen, J., *Pride and Prejudice: An Authoritative Text, Backgrounds and Sources, Criticism (Second Edition)*, Donald Gray, ed.; W.W. Norton & Company, New York, 1993.

[BaS]   Bamberg, P. and S. Sternberg, *A Course in Mathematics for Students of Physics, Volume I*; Cambridge University Press, Cambridge, 1988.

[Bare] Barenco, A., Quantum Computation: An Introduction, in *Introduction to Quantum Computation and Information*, H. Lo, S. Popescu and T. Spiller, eds.; World Scientific, Singapore, 1998.

[Bart] Bartle, R.G., *The Elements of Real Analysis (Second Edition)*; Wiley, New York, 1976.

[BeS]   Bethe, H.A. and E.E. Salpeter, *Quantum Mechanics of One- and Two-Electron Atoms*; Plenum Publishing, New York, 1977.

[Bo]   Boothby, W.M., *An Introduction to Differentiable Manifolds and Riemannian Geometry, Second Edition*; Academic Press, Inc., Orlando, 1986.

[BBE]   Born, H., M. Born and A. Einstein, *Einstein und Born Briefwechsel*; Nymphenburger Verlagshandlung, Regensburg, 1969.

[BBE']   Born, H., M. Born and A. Einstein, *The Born–Einstein Letters*, transl. I. Born; Walker and Company, New York, 1971.

[BtD]   Bröcker, T. and T. tom Dieck, *Representations of Compact Lie Groups*; Springer Verlag, New York, 1985.

[Cal]   Calvino, I., *Cosmicomics*, transl. William Weaver; Harcourt Brace Jovanovich, San Diego, 1968.

[Car]   Carroll, L., *The Annotated Alice: Alice's Adventures in Wonderland and Through the Looking Glass*, introduced and annotated by M. Gardner; Clarkson N. Potter, Inc., New York, 1960.

[Ch]   Chaucer, Geoffrey, *The Canterbury Tales*; http://www.towson.edu/~duncan/chaucer/duallang8.htm.

[Co]   Counterman, C., *MIT 3.091 Atomic and Molecular Orbitals*; http://web.mit.edu/3.091/www/orbs/, 2004.

[Da]   Davis, H.F., *Fourier Series and Orthogonal Functions*; Dover, New York, 1989. (Unabridged republication of the edition published by Allyn and Bacon, Boston, 1963.)

[DeM]   Debnath, L. and P. Mikusinski, *Introduction to Hilbert Spaces with Applications, Second Edition*; Academic Press, San Diego, 1999.

[Di]   Dirac, P.A.M., *The Principles of Quantum Mechanics, Second Edition*; Clarendon Press, Oxford, 1935.

[DyM]   Dym, H. and H. McKean, *Fourier Series and Integrals* (Probability and Mathematical Statistics, Vol. 14); Academic Press, San Diego, 1972.

[ER]   Eisberg, R. and R. Resnick, *Quantum Physics of Atoms, Molecules, Solids, Nuclei and Particles, Second Edition*; John Wiley & Sons, New York, 1985.

[FLS]   Feynman, R.P., R.B. Leighton and M. Sands, *The Feynman Lectures on Physics*; Addison-Wesley, Reading, MA, 1964.

[F]   Fock, V., Zur Theorie des Wasserstoffatoms, *Z. Phys.* **98** (1935), pp. 145–54.

[Fo]   Folland, G.B., *Introduction to Partial Differential Equations, Second Edition*; Princeton University Press, Princeton, 1995.

[FH]   Fulton, W. and J. Harris, *Representation Theory: A First Course*; Springer-Verlag, New York, 1991.

[Go]   Goldstein, H., *Classical Mechanics*; Addison-Wesley, Reading, MA, 1950.

[GS]   Guillemin, V. and S. Sternberg, *Variations on a Theme by Kepler*, AMS Colloquium Publications, Vol. 42, AMS, Providence, 1990.

[Hal58]   Halmos, P.R, *Finite-Dimensional Vector Spaces, Second Edition*; Van Nostrand Co., Inc., Princeton, 1958.

[Hal50]   Halmos, P.R., *Measure Theory*; Van Nostrand Co., Inc., Princeton, 1950.

[Ham]   Hammerstein, Oscar. All Er Nuthin' lyrics, from http://stlyrics.com/lyrics/oklahoma/allernuthinnothin.htm .

[Han]   Hannabuss, K., *An Introduction to Quantum Theory*; Clarendon Press, Oxford, 1997.

[Hat]   Hatcher, A., *Algebraic Topology*; Cambridge University Press, Cambridge, 2002. Also available online at http://www.math.cornell.edu/~hatcher/AT/ATpage.html.

[Hei]   Heilman, C., The Pictorial Periodic Table; http://chemlab.pc.maricopa.edu/periodic/styles.html .

[Her]   Herzberg, G., *Atomic Spectra and Atomic Structure*, transl. Spinks; Dover Publications, New York, 1944.

[Ho]   Hochstrasser, R.M., *Behavior of Electrons in Atoms: Structure, Spectra, and Photochemistry of Atoms*; W.A. Benjamin, Inc., New York, 1964.

[Hu]   Humphreys, J.E., *Introduction to Lie Algebras and Representation Theory*; Springer-Verlag, New York, 1972.

[I]   Isham, C.J., *Modern Differential Geometry for Physicists, Second Edition*; World Scientific, Singapore, 1999.

[Jos]   Joshi, A.W., *Matrices and Tensors in Physics, Third Edition*; John Wiley & Sons, New York, 1995.

[Joy]   Joyce, J., *Ulysses*; Vintage International, New York, 1990.

[Ju]   Judson, H.F., *The Eighth Day of Creation: The Makers of the Revolution in Biology*; Simon and Schuster, New York, 1979.

[La]   Lax, P., *Linear Algebra*; John Wiley & Sons, Inc., New York, 1997.

[L'E]   L'Engle, M., *A Wrinkle in Time*; Farrar, Straus and Giroux, New York, 1963.

[Le]   Levi, P., *The Periodic Table*, transl. Raymond Rosenthal; Schocken Books, New York, 1984.

[Mas]   Massey, W.S., *Algebraic Topology: An Introduction*; Springer Verlag, New York, 1967.

[Mat]   Mather, Marshall III, a.k.a. Eminem, *The Eminem Show*, Aftermath Records, USA, 2002.

[MTW]   Marsden, J.E., A.J. Tromba and A. Weinstein, *Basic Multivariable Calculus*; Springer Verlag, New York, 1993.

[Mi]   Milnor, J., On the Geometry of the Kepler Problem, *American Math. Monthly* **90** (1983) pp. 353–65.

[Mu]   Munkres, J.R., *Elements of Algebraic Topology*; Addison-Wesley, Redwood City, 1984.

[N]   Needham, T., *Visual Complex Analysis*; Clarenden Press, Oxford,1997.

[P]   Pauli, W., Über das Wasserstoffspektrum vom Standpunkt der neuen Quantenmechanik, *Z. Phys* **36** (1926), 336–63.

[RS]   Reed, M. and B. Simon, *Methods of Modern Mathematical Physics I: Functional Analysis, Revised and Enlarged Edition*; Academic Press, New York, 1980.

[Re]   Reid, B.P., *Spherical Harmonics*; http://www.bpreid.com/applets/poasDemo.html, 2004.

[Ri]   Rigden, J., *Hydrogen: The Essential Element*; Harvard University Press, Cambridge, 2002.

[Roe]   Roelofs, L., personal communication.

[Rot]   Rotman, B., *Signifying Nothing: The Semiotics of Zero*; Stanford University Press, Stanford, California, 1987.

[Row]   Rowling, J.K., *Harry Potter and the Sorcerer's Stone*; Scholastic, Inc., New York, 1997.

[Ru76]   Rudin, W., *Principles of Mathematical Analysis, Third Edition*; McGraw Hill, New York, 1976.

[Ru74]   Rudin, W., *Real and Complex Analysis, Second Edition*; McGraw Hill, New York, 1974.

[SS]   Saff, E.B. and A.D. Snider, *Fundamentals of Complex Analysis for Mathematics, Science and Engineering, Second Edition*; Prentice Hall, Upper Saddle River, New Jersey, 1993.

[SA]   Shifrin, T. and M. Adams, *Linear Algebra: A Geometric Approach*; W.H. Freeman and Co., New York, 2002.

[Sim]   Simmons, G.F., *Differential Equations with Applications and Historical Notes*; McGraw Hill, New York, 1972.

[Si]   Singer, S.F., *Symmetry in Mechanics: A Gentle, Modern Introduction*; Birkhäuser, Boston, 2001.

[So]   Sommerfeld, A., *Partial Differential Equations in Physics*, transl. E.G. Straus; Academic Press, New York, 1949.

[Sp]   Spivak, M., *A Comprehensive Introduction to Differential Geometry, Third Edition*; Publish or Perish, Houston, 1999.

[St]   Sternberg, S., *Group Theory and Physics*; Cambridge University Press, Cambridge, 1994.

[Sw]   Swift, J., *Spherical Harmonics*, http://odin.math.nau.edu/~jws/dpgraph/Yellm.html, 2004.

[To]   Townsend, J.S., *A Modern Approach to Quantum Mechanics*; McGraw Hill, New York, 1992.

[Tw]   Tweed, M., *Essential Elements: Atoms, Quarks, and the Periodic Table*; Walker & Company, New York, 2003.

[Wa]   Warner, F.W., *Foundations of Differentiable Manifolds and Lie Groups*; Springer Verlag, New York, 1983.

[We]   *Webster's Encyclopedic Unabridged Dictionary of the English Language*; Portland House, New York, 1989.

[WW]   Whittaker, E.T. and G.N. Watson, *A Course of Modern Analysis*; The Macmillan Co., New York, 1944.

[Wh]   White, H.E., Pictorial Representations of the Electron Cloud for Hydrogen-like Atoms, *Physical Review* **37** (1931).

[Wi]   Wigner, E.P., *Group Theory and its Application to the Quantum Mechanics of Atomic Spectra*, transl. J.J. Griffin; Academic Press, New York, 1959.

Glossary of Symbols and Notation

$:=$	a defining equality, 26
$\hat{K}$	complement of K in $\{1, \ldots, n\}$, 348
$\Im$	the imaginary part of a complex number, 21
$\Re$	the real part of a complex number, 21
$f \circ g$	composition of the functions f and g, 19
$f|_S$	the restriction of the function f to the set S, 19
$\partial_y f$	the partial derivative of the function f with respect to the variable y, 20
τ	natural isomorphism from a complex scalar product space to its dual, 107, 165
τ	complex conjugation on $\mathbb{C}^n$, 325
$\text{sgn}(\sigma)$	sign of the permutation σ, 75
$[a : b]$	element of the projective space $\mathbb{P}(\mathbb{C}^2)$, 300
$[c_0 : \cdots : c_n]$	element of the projective space $\mathbb{P}(\mathbb{C}^{n+1})$, 303
$\hat{f}$	Fourier transform of f, 26
∇^2	the Laplacian operator, 21
Å	angstrom, i.e., 10^{-10} meters , 9
$\hbar$	Planck's constant, 9

H   the Schrödinger operator , 11

E_n the n-th energy eigenvalue of the Schrödinger operator for the electron in the hydrogen atom , 12

V_E eigenspace of the Schrödinger operator corresponding to energy level E , 267

e   charge of the electron, 12

m   mass of the electron, 12

Z   constant factor in Schrödinger operator, 16

$|0\rangle$, $|1\rangle$ basis of kets of a qubit (a.k.a. spin-1/2 particle), 305

$|+\mathbf{z}\rangle$, $|-\mathbf{z}\rangle$ basis of kets for the state space of a spin-1/2 particle, 305

Π_+, Π_- spin up and spin down projection operators, 49

$|+\mathbf{z}\rangle \langle+\mathbf{z}|$ spin up projection operator, 49

ℓ azimuthal quantum number, 11

m magnetic quantum number, 11

n principal quantum number, 10

s spin quantum number, 11

s, p, d, f labels for states of the electron, 11

S^2 the unit two-sphere in $\mathbb{R}^3$, 23

S^3 the unit three-sphere in $\mathbb{R}^4$, 25

C_2 complex scalar product space of continuous square-integrable functions on $\mathbb{R}^3$ whose first and second partial derivatives are all continuous, 365

$\mathcal{Y}_4$ complex scalar product space of spherical harmonics on the three-sphere S^3, 285

$\mathcal{Y}_4^n$ complex scalar product space of spherical harmonics of degree n on the three-sphere S^3, 284

$W^\infty(\mathbb{R}^3)$ complex scalar product space of infinitely differentiable functions with all derivatives in $L^2(\mathbb{R}^3)$, 243

$\mathcal{I}$ complex scalar product space of rotation-invariant functions in $L^2(\mathbb{R}^3)$, 158

$C[-1, 1]$ complex scalar product space of continuous complex-valued functions on $[-1, 1]$, 45

$L^2(\mathbb{R}^3)$ complex scalar product space of square-integrable functions on $\mathbb{R}^3$, 80

$L^2(\mathbb{R}^{\geq 0})$ complex scalar product space of square-integrable functions on the nonnegative real axis, 158

$L^2(S^2)$ complex scalar product space of square-integrable functions on the two-sphere, 84

$L^2(S)$ complex scalar product space of square-integrable functions on a set S, 84

$\mathbb{H}$ complex scalar product space of complex-valued harmonic polynomials in three real variables, 52

$\mathbb{H}^\ell$ vector space of homogeneous harmonic polynomials of degree ℓ in three variables, 53

$\mathcal{P}^n$ complex scalar product space of homogeneous polynomials of degree n in two real variables, 47

$\mathbb{H}^n_4$ complex scalar product space of homogeneous harmonic polynomials of degree n in four variables, 284

$\mathcal{P}^\ell_3$ complex scalar product space of homogeneous polynomials of degree ℓ in three real variables, 47

$\mathcal{Y}^\ell$ complex scalar product space of restrictions of harmonic polynomials of degree ℓ on $\mathbb{R}^3$ to the two-sphere S^2, 53

$\mathcal{Y}$ complex scalar product space of restrictions of harmonic polynomials on $\mathbb{R}^3$ to the two-sphere S^2, 54

$\mathbf{Q}$ the algebra of quaternions, 25

$\{1, \mathbf{i}, \mathbf{j}, \mathbf{k}\}$ a basis for the quaternions, 25

$\mathbf{P}_{\ell,m}$ Legendre function, 29

R_n representation of $SU(2)$ on homogeneous polynomials of degree n, 137

Q_n representation of $SO(3)$ on homogeneous polynomials of even degree n in two variables, pushforward of R_n, 202

$\Psi_{n\ell m}$ spherical harmonic function on S^3, 290

$Y_{\ell,m}$ spherical harmonic function on S^2, 30

$\langle \cdot, \cdot \rangle$ complex scalar product, 82

$\|\cdot\|$ norm, 94

$\mathbb{T}$ circle group, 112

$SO(2)$ group of rotations of the plane, 112

$SO(3)$ group of rotations in three-dimensional Euclidean space, 117

$\rho \cong \tilde{\rho}$ the representations ρ and $\tilde{\rho}$ are isomorphic, 132

$\ker T$ kernel of the linear transformation T, 52

$W^\perp$ the subspace complementary to W inside another vector space, 86

$\Lambda^n V$ alternate tensor product of n copies of the vector space V, 75

$\mathrm{Sym}^n V$ symmetric tensor product of n copies of V, 75

$\mathbb{P}(V)$ projective space over V, 300

$\Pi_{[W]}$ orthogonal projection onto the subspace $[W]$ of projective space, 344

$[W]$ linear subspace of a projective space $\mathbb{P}(V)$, where W is a subspace of V, 303

$[T]$ projectivization of the linear operator T, 304

$\mathbb{P}\mathcal{U}(V)$ projective unitary group of the vector space V, 318

$S/\sim$ the set of equivalence classes in S modulo the equivalence relation $\sim$, 33

$\langle \cdot, \cdot \rangle_*$ complex scalar product on the dual of a complex scalar product space, 107, 165

V^* dual vector space to V, 72, 164

ρ^* dual to the representation ρ, 166

$\mathrm{Hom}_G(V, W)$ fixed points of the natural representation on $\mathrm{Hom}(V, W)$, 169

$V \oplus W$ Cartesian sum of vector spaces V and W, 62

$V \otimes W$ tensor product of vector spaces V and W, 67

$\rho \oplus \tilde{\rho}$ Cartesian sum of representations ρ and $\tilde{\rho}$, 159

$\rho \otimes \tilde{\rho}$ tensor product of the representations ρ and $\tilde{\rho}$, 160

Π_k projection onto the k-th summand of a Cartesian sum, 63

$\mathrm{Hom}(V, W)$ complex scalar product space of linear transformations from V to W, 73, 169

unirrep   unitary irreducible representation, 195

Index

Undergraduate Texts in Mathematics

(continued from page ii)